THE DALAI LAMA

Foreword

The natural environment sustains the life of all beings in the world. However, nearly everywhere these days, it is undergoing extensive damage. Therefore, it is more important than ever that we make an effort to ensure the protection, restoration and replenishment of our environment and its inhabitants. A pure and unspoilt environment is clearly beneficial for everyone.

When the environment becomes damaged and polluted there are many negative consequences. Oceans and lakes lose their cool and soothing qualities so the creatures depending on them are disturbed. The decline of vegetation and forest cover causes the earth's bounty to decline. Rain no longer falls when required, the soil dries and erodes, forest fires rage and unprecedented storms arise. We all suffer the consequences.

Resolving the present environmental, crisis is not just a question of ethics but a question of our own survival. The natural environment is very important not only for those of us alive now but also for future generations. If we exploit it in extreme ways, even though we may make a short term profit from it now, in the long run we ourselves and future generations will suffer.   When the environment changes, climatic conditions also change. When they change dramatically, the economy and many other things change as well. Even our physical health can be greatly affected. Such abuse arises out of ignorance of the environment's importance and the effect on it of climatic change. It is essential to help people to understand this. We need to teach that the state of the environment has a direct bearing on our own benefit.   Through understanding the interdependence of the world and all the beings who live in it, people will be able to adapt their behaviour in such ways that the potential of our bounteous natural realm be nurtured and preserved.

Now, in order to succeed in the protection and conservation of the natural environment, it is important first of all to bring about an internal balance within human beings themselves. These days we are very much involved in the external world, while we neglect the internal world. We do need scientific development and material development in order to survive and to increase the general benefit and prosperity, but equally we need mental peace. From my own limited experience I have found that the greatest degree of inner tranquillity comes from the development of love and compassion. Only a sense of universal altruism can remove the self-centred motives that cause people to disregard each others' needs.

The reflections of scientists and other concerned persons on climate change and our relations to the environment contained in this book under the title Five Years After Kyoto are very valuable. We need such observations to encourage a fresh perspective, especially one that leads to practical action. These days, not only individuals but governments as well are seeking a new ecological order. But unless we all work together, no solution will be found.

March 9, 2004

Preface

Climate change emerged as one of the most debatable environmental issues of the last century. Are we experiencing climate change already? Is it a new trend or has Earth historically experienced change in climate leading to different glacial ages. The Intergovernmental Panel on Climate Change (IPCC) was established in 1988 to assess the scientific, technical and socioeconomic information relevant to understanding the risk of human-induced climate change. Climate change received further international attention during the historic Earth Summit held in Brazil in 1992, at which more than 150 nations were signatories to the United Nations Framework Convention on Climate Change (UNFCCC). The goal of the UNFCCC to promote sustainable development and to reduce the level of greenhouse gas emissions (GHG) to 1990 levels.

UNFCCC defines climate change as "a change of climate which can be attributed directly or indirectly to human activity that alters the composition of the global atmosphere and which is in addition to natural climate variability observed over comparable time periods".

Although certain GHGs are naturally occurring, nonetheless concentration of some GHGs has increased due to human activities such as those that use energy, e.g. global deforestation and agricultural activities among others. The scientific consensus reached by IPCC indicates that incremental GHG emissions caused by human activity since the Industrial Revolution are having a discernible impact on the climate. Though human activity contributes only about 5% of global GHGs (the rest is contributed by natural process), this suffices to disturb the delicate balance of GHGs in the atmosphere and, by extension, the climate. The result is continual warming of the atmosphere and resultant changes in its composition.

This book documents the scientific facts regarding climate change and gives a brief overview of the key developments in the climate change regime, discussing the Kyoto Protocol the Kyoto Protocol - five years after Kyoto and beyond. Contribution discuss the politics of the North and South, and one has used a numerical state parllal equilibrium model that integrates energy markets with an international market for emissions trading to compare the impacts of the Kyoto Protocol both with and without the participation of the United States,

After establishing the scientific base, presenting agreements and policies for climate change in general, and the Kyoto Protocol in particular, the Instruments and Institutions for the Kyoto Protocol reviewed. One contributor points out that an institutional framework at National levels to dual with climate change and implementation of the Kyoto Protocol is missing and should have been included at the very outset.

Another interesting issue discussed in the book is that of human rights of indigenous communities - the severe impacts of global warming already being felt by many indigenous Arctic communities that could give rise to a human rights claim in the Inter-American system. Next, it considers some of the procedural elements of such a claim, with particular emphasis on issues of jurisdiction.

The rapid and large climatic changes can be expected to have far-reaching and, in many instances, unpredictable consequences not only for human societies, but for all forms of life on Earth. For example, a rise in global sea level can threaten coastal cities and settlements throughout the world. Several authors discuss the impact of climate change (associated with environmental, socioeconomic impacts) and what implementation of the Kyoto Protocol would mean to different countries and continents.

While the book is mainly devoted to the science and philosophy behind Climate Change and the Kyoto Protocol - five years after Kyoto, some chapters look beyond Kyoto, making it more inclusive in terms of ecological and sustainable principles in addition to just the economic aspects.

It should be pointed out that scientific uncertainty still exists about Climate Change—whether the natural cycle of global change is accelerated by anthropogenic factors or not. Moreover, much needs to be done to iron out creases in the Kyoto Protocol before it can be wholly implemented.

Velma I. Grover

Contents

Section V
BEYOND KYOTO

Contributors

Byrne, John
Center for Energy and Environmental Policy
University of Delaware
Newark, Delaware, 19716
USA

Alwis, Ajith de
University of Moratuwa
Department Chemical and Process Engineering
Sri Lanka

Glover, Leigh
Center for Energy and Environmental Policy
University of Delaware
Newark, Delaware, 19716
USA

Goldberg, Donald M
Center for International Environmental Law
1367 Connecticut Avenue, NW Suite # 300
Washington, DC 20036-1860, USA
e-mail: dgoldberg@ciel.org

Grover, Velma I.
Natural Resource Consultant
916, 981 Main st. West
Himilton, On, L85, IA8, Canada
e-mail: vgrover@sprint.ca

Gupta, Joyeeta
Institute for Environmental Studies
Vrije Universiteit
Amsterdam
The Netherlands

Hagem, Cathrine
University of Oslo
Dept. Economics
P.O. Box 1095 Blindern, 0317
Oslo, Norway

Hare, Bill
Visiting Scientist
Potsdam Institute for Climate Impact Research (PIK),
Telegrafenberg A31, P.O. Box 601203,
14412 Potsdam, Germany

Hogland, William
Kalmar University
Dept. Technology
Kalmar, Sweden
e-mail: william.hogland@hik.se

Holtsmark, Bjart
Statistics Norway
P.O. Box 8131 Dep, 0033
Oslo, Norway

Huq, Saleemul
Climate Change Programme
IIED,
3 Endsleigh Street,
London
WC1H 0DD, UK

Inniss, Vernese
Center for Energy and Environmental Policy
University of Delaware,
Newark, Delaware,
USA 19716.

Johnke, Brent
Federal Environmental Agency of Germany
Bismarck plate l
Postfaer 330022
D-14191, Berlin
Germany
e-mail: bent.johnke@uba.de

Johnston, Paul
Greenpeace Research Laboratories
Department of Biological Sciences
University of Exeter,
Exeter EX4 4PS, UK.

Kulkarni, Jyoti
Center for Energy and Environmental Policy
University of Delaware
Newark, Delaware, USA 19716.

Kumar, Santosh
Department of Mathematics and Statistics
University of Melbourne
Victoria 3010
Australia

Lobsinger, Alison
Vrije University
Institute for Environmental Studies
Amsterdam
The Netherlands

Marques, Marcia
Rio de Janerio State University, UERJ
Dept. Sanitary and Environmental Energy
PESAGRO-RIO
Rio de Janeiro, Brazil
e-mail: marcia@marques.prs.br.

Mirovitskay, Natalia
Terry Sanford Institute of Public Policy
Box 90239
Durham
NC 27708-0239 USA
e-mail: nataliam@duke.edn

Moinuddin, Khondar
Bangladesh Centre for Advanced Studies
Dhaka, Bangladesh

Müller, Benito
Senior Research Fellow
Oxford Institute for Energy Studies
57, Woodstock Rd.,
Oxford OX2 6FA, UK
e-mail: benito.mueller@philosophy.oxford.ac.uk

Mun, Yu-Mi
Center for Energy and Environmental Policy
University of Delaware
Newark, Delaware, USA 19716

Najam, Adil
Fletcher School of Law and Diplomacy
160, Packard Avenue,
Medford, MA 02215, USA
e-mail: adil.najam@tufts.edn

Oladimeji, A.A.
Federal University of Technology
Department of Biological Sciences
PMB 65
Minna
Nigeria

Rosales, Jon
University of Minnesota
180 McNeal Hall
1985 Buford Avenue
St. Paul, MN 55108
USA
e-mail:rosaφφ31@umn.edn

Rojas, Ana V.
CEDARENA
Apto. 134-2050
San Pedro
Costa Rica
e-mail: anarojas@cedarena.org/ana-v-rojas@hotmail.com

Santillo, David
Greenpeace Research Laboratories
Deptartment of Biological Sciences
University of Exeter,
Exeter EX4 4PS, UK

Smith, Heather A.
Associate Professor and Chair
International Studies Program
University of Northern B.C.
 3333 University Way
Prince George, BC, V2N 4Z9
Canada
e-mail: smith@unbc.ca

Surjadi, Harry
Pesona Depok Blok C No 9
Jl. Margonda Raya
Depok 16432
Indonesia

Toly, Noah
Center for Energy and Environmental Policy
University of Delaware
Newark, Delaware, USA 19716

Umoh, Umoh T
University of Botswana
Deptartment of Environmental Science
Private Bag 0022
Gaborone,
Botswana

Wagner, Martin
Director of International Programs for Earthjustice
Earthjustice,
426, 17th Street, 6th Floor,
Ookland, CA 94612-282
e-mail: mwagner@earthjustice.org

Wang, Young-Doo
Center for Energy and Environmental Policy
University of Delaware
Newark, Delaware, USA 19716

Yasumoto, Akinobu
Executive Director
Global Industrial and Social Progress Research Institute (GISPRI)
3rd Floor, Mitsai-Osk Line Bldg.
Toranomon 2-1-1, Minato-Ku
Tokyo 105-0001, Japan

Section I

INTRODUCTION TO SCIENCE, POLICIES AND POLITICS

CHAPTER 1

Introduction

Velma I. Grover
Natural Resource Consultant; # 916, 981 Main St West; Hamilton,
On, L8S 1A8, Canada. e-mail: vgrover@sprint.ca

Most of us have heard or read about climate change. However, the messages we receive are often conflicting, raising more questions than answers. Is climate change good or bad? Has climate change already started or is it part of our future? Are we doing anything about it? Should we be concerned?

The fact is that we're all part of the problem, and we can all be part of the solution. As individuals, families, communities and stakeholders, we need to understand what is happening and, most importantly, what can be done about it.

This chapter seeks to clear up some of this confusion by providing basic scientific information on the climate change issue. It is also a roadmap for the rest of the book.

It is essential that the climate change issues be properly understood, in order to make wise decisions and take responsible actions not only for our own future, but also our future generations, in response to challenges and opportunities posed by climate change.

Climate change is the most significant environmental problem the world has ever faced. For this reason the global community is working together to share information and ideas to meet this challenge. Five years after Kyoto, the future of the Kyoto Protocol looks bleak at times, especially since the US has refused to ratify it. Thus this book looks at not only the Protocol itself, but related topics: the science behind climate change, climate change treaty and politics, the instruments to be used for implementing the Protocol, human rights issue, and impact of climate change and ratification of the Kyoto Protocol on different countries.

This chapter presents a general introduction to the earth's atmosphere and its different layers (climate change and global warming are related to these atmospheric layers) followed by the science of climate change and some of the causes for it. All the institutions associated with the issue of climate change prior to the Kyoto Protocol are then chronicled and some of the instruments that could be used in implementation of the Protocol targets. Lastly, a few words are said about the impact of climate change and Kyoto Protocol elaborated in country case studies from different continents in Part IV.

The Atmosphere

The small blue and green planet we call home is a very special and unique place. We live on the only planet in our solar system and possibly in the galaxy where life is known to exist. All life exists within a thin film of air, water, and soil about 15 km deep. This spherical shell of life is known as the biosphere. The biosphere can be divided into three layers: the atmosphere (air), the hydrosphere (water), and the lithosphere (rock and soil). It is the unique attributes of the Earth's atmosphere that allows it to be a habitable place for humans, animals, and plants.

A blanket of air, which we call the atmosphere, surrounds the Earth. It reaches over 560 km. (348 miles) from the surface of the Earth. The atmosphere is a mixture of gases and particles that surround our planet. When seen from space, the atmosphere appears as a thin seam of dark blue light on a curved horizon.

The atmosphere is made up of layers (Fig. 1.1) that surround the Earth like rings. 99% of its total mass lies in two regions within the first 50 km above the Earth's surface: the troposphere and the stratosphere. These two regions are of particular importance in understanding the climate system.

The atmosphere serves several purposes: it provides us with the air we breathe, its gases retain the heat that warms the Earth, and its protective layer of ozone shields us from harmful UV rays emitted by the sun. The atmosphere also acts as a reservoir or storehouse for natural substances as well as emissions derived from human activities. Within the storehouse, physical and chemical actions and reactions take place. Many of these can affect our climate and weather systems. Four distinct layers of the atmosphere have been identified using thermal characteristics (temperature changes), chemical composition, movement, and density: the troposphere, stratosphere, mesosphere and thermosphere. Beyond the atmospheric layers lies the exosphere.

Troposphere: The troposphere is closest to the Earth. It extends to about 6 to 17 km above the earth's surface and is thickest at the equator. Temperatures in the troposphere generally decrease as altitude increases. They are warmer nearest the earth, in part because gases in the troposphere are warmed by heat radiated from the earth.

Stratosphere: The stratosphere extends beyond the troposphere, to about 50 km above the earth. Gases in the stratosphere are heated mainly by incoming radiation from the sun; temperature in the stratosphere gradually increases as altitude increases. As a consequence of temperature differences between the troposphere and stratosphere, and the resultant circulation patterns, exchange of air between the two layers is slow.

The stratosphere is also known as the **ozone layer.** The distribution of ozone is closely linked to the vertical structure of the atmosphere. Approximately 90% of all ozone molecules are found in a broad band within the stratosphere. This layer of ozone-rich air acts as an invisible filter to protect all life forms from overexposure to the sun's harmful ultraviolet (UV) rays. Long-term ozone depletion is the result of human activity.

Mesosphere: The mesosphere starts just above the stratosphere and extends 85 km (53 miles) high. In this region, temperatures again fall as low as –93 C.

Thermosphere: The thermosphere starts just above the mesosphere and extends to 600 km (372 miles) high. It comprises two parts – ionosphere (the inner part) and exosphere (the outer part), which gradually merges into space.

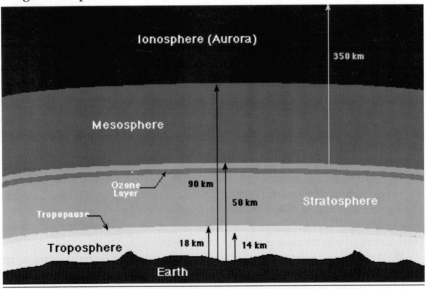

Fig. 1.1: Different layers of the atmosphere depicted diagrammatically.
Source: http://csep10.phys.utk.edu/lect/earth/atmosphere.html

Composition of the Atmosphere

Relatively speaking Earth's atmosphere is about as thin as the skin of an apple. As seen in discussion above about 99% of the Earth's atmosphere occurs within 50 km above the Earth's surface. The remaining 1% extends outward for several hundred kilometers, fading gradually into interplanetary space.

Eighty-one percent of the Earth's atmosphere occurs in the troposphere. Dry air near the Earth's surface consists mainly of nitrogen (78.1% by volume) and oxygen (20.9%) with a small amount of argon (about 0.9%) and carbon dioxide (about 0.035%). Air in the troposphere also contains water vapor and small amounts of "trace gases" such as methane, nitrous oxide, hydrogen, and ozone. Even though many of these gases are present in minute amounts, they cause the atmosphere to act like an insulating blanket around the planet. Without this atmospheric blanket of insulation the Earth's surface would be too cold to sustain life.

The naturally occurring atmospheric gases found in the atmosphere play a key role in determining the climate we experience (and increase in their concentration, mainly attributed to natural as well as anthropogenic causes, is resulting in more dramatic climate changes). In addition to these naturally occurring gases, there are others such as halocarbons (result of industrialization) present in the troposphere now. The chemical composition of the atmosphere is very important and influences climate; addition of gases changes the composition of the atmosphere, which complicates the climate change process.

Science of Global Warming

The terms 'global warming' and 'climate change' are often interchanged. However, these terms are not identical. Climate change includes both warming and cooling conditions. Global warming pertains only to those climate changes related to temperature increase and it is this aspect of climate change that is the focus of this Chapter. In order to understand the science of global warming we must first perceive it in the context of global climate change.

Climate change is a change in "average weather" that a given region experiences. Average weather includes all the features associated with the weather such as temperature, wind patterns, and precipitation. Climate change on a global scale refers to changes in the climate of the Earth as a whole. The rate and magnitude of global climate change over the long term have many implications for natural ecosystems.

In describing climate change, it is important to distinguish long-term changes from short-term ones. Climate processes are influenced by a complex array of interacting elements. These include: the sun, the atmosphere, the oceans and even volcanic activity and changes in topography. The interconnectedness of these elements results in a system that can be hard to predict. At any location, temperatures, rainfall, and other climatic elements can naturally vary a great deal from one year to next and still be considered within the bounds of "normal" climate

Weather

Weather is the state of the atmosphere at a particular time and place. The elements of weather include temperature, humidity, cloudiness, precipitation, wind, and pressure. These elements are organized into various weather systems, such as monsoons, areas of high and low pressure, thunderstorms, and tornadoes. All weather systems have well-defined cycles and structural features, and are governed by the laws of heat and motion. These conditions are studied in meteorology, the science of weather and weather forecasting. Weather differs from climate, which is the weather a particular region experiences over a long period of time. Climate includes the averages and variations of all weather elements. *Source*: http://www.the-cohens.com/marc/HMS-Crew/PDFs/Meteorology.pdf

Climate

Climate is the state of the atmosphere averaged over several years and the expected weather for a particular time of year.

Climate Change

Climate includes all the elements of weather, such as temperature, precipitation, and wind patterns. Climate change refers to changes in the climate as a whole, not just one single element of the weather. Global climate change, therefore, refers to changes in all the interconnected weather elements of the Earth.

Enhanced Greenhouse Effect

If concentrations of greenhouse gases in the atmosphere increase, they will intensify the natural greenhouse effect. This will cause what is referred to as the enhanced greenhouse effect. Scientists agree that if humans continue to add greenhouse gases to the atmosphere through their burning of fossil fuels and clearing of forests, the greenhouse effect will be enhanced. This will cause global temperatures to rise at an unprecedented rate.

Global Warming

Strictly speaking global warming and global cooling refer to the natural warming and cooling trends that the Earth has experienced throughout its history. However, the term global warming has become popularized as the term that encompasses all aspects of human influence on the climate system.

Greenhouse Gases

Most of the greenhouse gases occur naturally in the atmosphere; although they exist in relatively small amounts, their heat trapping properties allow the Earth's surface to be maintained at a temperature that is habitable for humans, animals, and plants as we know them. The main greenhouse gases are water vapor (H_2O), carbon dioxide (CO_2), methane (CH_4), nitrous oxide (N_2O), ozone (O_3), and holocarbons (CFCs, HFCs, etc.)

Greenhouse Effect

The greenhouse effect is the popular term for the warming function the atmosphere plays in the global ecosystem. Just as the glass in a greenhouse holds the sun's warmth inside, so the atmosphere traps the sun's heat near the Earth's surface and keeps the Earth warm. We call this the natural greenhouse effect. It makes the Earth a unique planet for living and growing organisms.

variability. Therefore, one abnormally cool summer or winter occurring after a series of warm summers or winters does not necessarily indicate reversal of a trend. Similarly, one unusually hot season does not by itself prove global warming is taking place. It is often difficult, though critical, to distinguish an important emerging long-term trend from an insignificant short-term irregularity in climate pattern.

Global climate change does occur naturally. The Earth's natural climate system has always been, and still is, constantly changing. Scientists have looked at information recorded over the ages in ancient rocks and ice sheets. It has shown that the Earth has experienced numerous warming and cooling periods over the past one million years.

Aspects of rocks, ice sheets and tree rings provide information regarding past climates. They are referred to as "paleoclimatic indicators". From these paleoclimatic indicators, scientists have been able to infer that global temperatures over the past million years have followed a cycle of fairly regular, long-term variations. Extreme minimum temperatures, corresponding to many major global ice ages, appear to have occurred at roughly 100,000-year intervals for the past 800,000 years. Each of these ice ages, or "glacial periods", was followed by a dramatic 4-6°C warming to an 'interglacial" (i.e., between glacial) period lasting roughly 10,000 to 20,000 years. In addition, within this 100,000-year cycle, smaller, more frequent temperature fluctuations occurred at approximately 20,000- and 40,000-year intervals.

Figure 1.3 shows that CO_2 content has also varied just within a range of 100 parts per million by volume (ppmv) over interglacial periods (which goes to prove that a drastic change in temperature of gases is not needed to cause global warming or cooling).

Many theories have been advanced to explain the cause of these global temperature variations. One hypothesis that is widely accepted suggests that the fluctuations between glacial and interglacial conditions were triggered by changes in the Earth's orbit around the sun. These changes in the Earth's orbit are believed to have occurred at time scales that correlate well with the 20,000- to 40,000-year temperature fluctuations.

The coldest period of the last ice age occurred just over 10,000 years ago, when average annual global temperatures were 4-5°C cooler than they are today. Since the last glaciation, we have entered into a warm period that has been interrupted from time to time with cooler temperatures. The current interglacial period has brought with it relatively stable global surface temperatures. In fact, since the current interglacial period began, the earth's average surface temperature has varied only about one degree from values of today (Fig. 1.4).

One feature of past climate change is the close correlation between temperature and atmospheric concentrations of heat-trapping gases, in

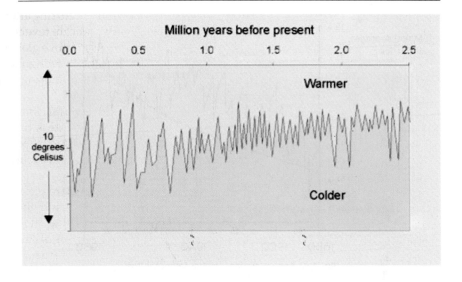

Fig. 1.2: Variations in global temperatures over the past 2.5 million years.

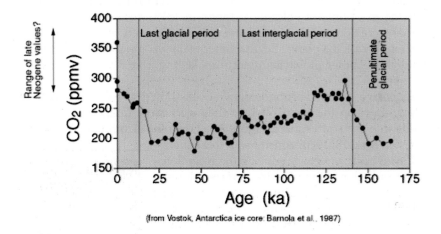

(from Vostok, Antarctica ice core: Barnola et al., 1987)

Fig. 1.3: Atmospheric carbon dioxide contents over the last interglacial-glacial cycle.

particular carbon dioxide (CO_2) and methane (CH_4). Gas bubbles trapped in ancient ice sheets when they formed allow scientists to check how gases in the atmosphere varied with past climate conditions. Gas bubbles in ice cores taken from the Antarctic ice sheet were analyzed and showed that

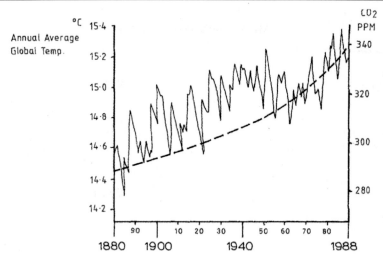

Fig. 1.4: Annual average global temperatures.

when local Antarctic temperatures were at their highest (warmest), the amount of CO_2 and CH_4 in the atmosphere was also at its highest.

For many years, climatologists have also noticed a connection between large explosive volcanic eruptions and short-term climatic change. For example, one of the coldest years in the last two centuries occurred the year following the Tambora volcanic eruption in 1815. Accounts of very cold weather were documented in the year following this eruption in a number of regions across the planet. Several other major volcanic events also show a pattern of cooler global temperatures lasting 1 to 3 years after their eruption.

Explosive volcanic eruptions have been shown to have a short-term cooling effect on the atmosphere if they eject large quantities of sulfur dioxide into the stratosphere. Figure 1.5 shows the eruption of Mount St. Helens on May 18, 1980, which had a local effect on climate because of ash reducing the reception of solar radiation on the Earth's surface. Mount St. Helens had very minimal global effect on climate because the eruption occurred at an oblique angle, ejecting little sulfur dioxide into the stratosphere (*Source*: U.S. Geological Survey).

At first scientists thought that the dust emitted into the atmosphere from large volcanic eruptions was responsible for the cooling by partially blocking the transmission of solar radiation to the Earth's surface. However, measurements indicate that most of the dust

Fig. 1.5: Mount St. Helens eruption May 18, 1980.

thrown into the atmosphere returned to the Earth's surface within six months. Recent stratospheric data suggest that large explosive volcanic eruptions also eject large quantities of sulfur dioxide gas which remains in the atmosphere for as long as three years. Atmospheric chemists have determined that the ejected sulfur dioxide gas reacts with water vapor commonly found in the stratosphere to form a dense optically bright haze layer that reduces the atmospheric transmission of some of the sun's incoming radiation. Some scientists have also attributed this as one of the causes for climate change.

In the last century, two significant climate-modifying eruptions have occurred. El Chichon in Mexico erupted in April, 1982, and Mount Pinatubo in the Philippines during June, 1991. Of these two volcanic events, Mount Pinatubo had a greater effect on the Earth's climate, ejecting about 20 million tons of sulfur dioxide into the stratosphere. Researchers believe that the Pinatubo eruption was primarily responsible for the 0.8°C drop in global average air temperature in 1992. The global climatic effects of the eruption of Mount Pinatubo are believed to have peaked in late 1993. Satellite data confirmed the connection between the Mount Pinatubo eruption and the global temperature decrease in 1992 and 1993. It indicated that the sulfur dioxide plume from the eruption caused a several percent increase in the amount of sunlight reflected by the Earth's atmosphere back to space, causing the surface of the planet to cool.

The preceding discussion indicates that climate change is a natural part of the Earth's climate history. Does human activity have any effect on climate change? Why is it currently being addressed as a critical environmental problem of modern times? If climate varies naturally why is there a concern about climate change? Before answering these questions, let us view the growth in population and then discuss the probable effect of human activities on climate change.

During the period since the last glaciation, the human population has flourished to an unprecedented degree. From the emergence of modern humans 200,000 years ago until about 2,000 years ago, fewer than 250 million people walked the face of the Earth. When Christopher Columbus set sail in search of India about 500 years ago, approximately 500 million peopled the Earth. In another 250 years the number doubled again to 1 billion. The population doubled again to just over 2 billion people in another 125 years. Thus the population has been doubling in lesser and lesser time.

Since our relationship to the Earth has changed so dramatically, obviously the impact of humans on the environment has altered concomitantly. Some of the impacts are: global warming, ozone depletion, loss of living species, and deforestation—all because of the changed relationship between mankind and the Earth's natural balance.

Human activities can disrupt the balance of the natural system that regulates the temperature of the Earth. This natural regulating system is known as the "greenhouse effect". With burning of cheap fossil fuels in the developing world and their adaptation to increasingly sophisticated and mechanized life styles in the developed world, the amounts of heat-trapping CO_2 and CH_4 in the atmosphere have increased. By increasing the amount of these heat-trapping gases, humankind has enhanced the warming capability of the natural greenhouse. This human influence on the natural warming function of the greenhouse effect has become known as the "enhanced greenhouse effect". It is the human-induced enhanced greenhouse effect that causes environmental concern. It has the potential to warm the planet at a rate never before experienced in human history.

Another factor causing significant change in our atmosphere is the rising level of carbon dioxide, due to increased burning of coal and oil, which traps more heat in the atmosphere and thus warms the earth slowly but

The Earth is covered by a blanket of gases which allows light energy from the sun to reach the Earth's surface, where it is converted to heat energy. Most of the heat is reradiated toward space, but some is trapped by greenhouse gases in the atmosphere. This is a natural effect which keeps the Earth's temperature at a level necessary to support life

Human activity—particularly burning fossil fuels (coal, oil and natural gas) and land clearing—is generating more greenhouse gases. Scientists are convinced that this will trap more heat and raise the Earth's surface temperature.

Fig. 1.6: Greenhouse effect
(Source: http://www.greenhouse.gov.au/science/faq/page5.html)

steadily. This phenomenon began with the advent of the Industrial Revolution in the 1800s and has been increasing ever since. Work done by hand and animal power before the Industrial Revolution was taken over by fossil-fuel burning machines during the Revolution, resulting in generation and release of higher amounts of CO_2 into the environment. It is difficult to predict the precise effects of doubling the concentration of carbon dioxide in the atmosphere, but there is no doubt that doubling CO_2 would indeed raise global temperatures.

Besides doubling of CO_2 emissions, concomitant increments in CH_4 emissions have been recorded as a result of expanded rice production and animal husbandry. The concentrations of CO_2 and CH_4 have increased faster since the Industrial Revolution than at any other time in recorded history.

The expected continued temperature rise due to climate change is linked to the observed build-up of greenhouse gases in the atmosphere. This will cause further warming of the Earth's surface over the long term. Whether or not global warming has already begun and is a direct result of the enhanced greenhouse effect still remains within the limits of scientific uncertainty. However, even when expected natural short-term variability, such as unusually cold winters, are taken into account, global temperature trends show that the Earth has, in fact, warmed over the long term.

Most of the natural and human causes of climate change have been depicted in figure 1.7.

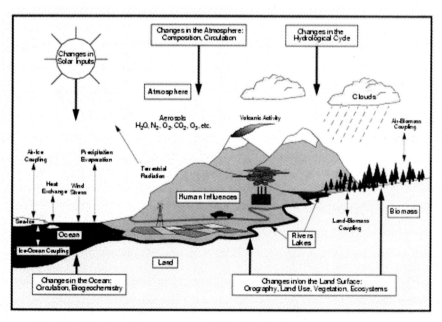

Fig. 1.7: Natural and human causes of climate change

To summarize the ongoing scientific discussion, some of the facts according to the Intergovernmental Panel on Climate Change, January, 2001 are:

- There is new and stronger evidence that most of the warming observed over the last 50 years is attributable to human activities.
- Human influences will continue to change atmospheric composition throughout the 21st century.
- The globally averaged surface temperature is projected to increase by 1.4 to 5.8° C by 2100.
- The projected warming is very likely to be without precedent during at least the last 10,000 years.

"A collective picture of a warming world" includes:

- Over the 20th century:
  - global-average surface temperature increased (0.6° C)
  - temperature in lowest 8 km increased in past 4 decades
  - global-average sea level has risen (0.1- 0.2 m)
  - snow and ice extent have decreased; widespread retreat of glaciers
  - precipitation has increased (up to 1% per decade)
- Since 1750 there has been an increase in:
  - CO_2 (31%), CH_4 (151%), and N_2O (17%)
  - present CO_2 likely not exceeded in the past 20,000 years

Thus, climate change is one of the biggest environmental challenges facing the world today and has/is drawing both local and international attention.

Chronology of Institutional Framework

Some of the institutions involved with the issue of climate change and the treaties and agreements arising therefrom are discussed below:

1. IMO/WMO

Major marine nations used ships to collect weather data about a century and a half ago. The data collected were used to develop a weather forecasting information system to enable ships to steer clear of rough weather and/or storms. This informal arrangement was eventually formalized as the International Meteorological Organization (IMO) in 1873.

The World Meteorological Convention, by which the World Meteorological Organization (WMO) was created, was adopted at the Twelfth Conference of Directors of the IMO in Washington, DC, in 1947. Although the Convention itself came into force in 1950, WMO commenced operations as the successor to IMO in 1951 and, later that year, was established as a specialized agency of the United Nations by agreement between the UN and WMO.

The purposes of WMO are to facilitate international cooperation in the establishment of networks of stations for making meteorological, hydrological and other observations, and to promote rapid exchange of meteorological information, standardization of meteorological observations, and uniform publication of observations and statistics. It also furthers the application of meteorology to aviation, shipping, water problems, agriculture and other human activities, promotes operational hydrology, and encourages research and training in meteorology.

WMO and its predecessor IMO provided the focus for international cooperation in meteorology. This cooperation has ensured free and expeditious exchange of meteorological data and information among the Member countries, which forms the basis of all meteorological activities. From weather prediction to air pollution research, climate change related activities, ozone layer depletion studies and tropical storm forecasting, the World Meteorological Organization coordinates global scientific activity to allow increasingly prompt and accurate weather information and other services for public, private and commercial use, including international airline and shipping industries. WMO's activities contribute to the safety of life and property, socioeconomic development of nations and the protection of the environment.

WMO is constantly endeavoring to improve its programs and services to its member states. Just to quote some examples: The World Weather Watch (WWW) comprises the Global Observing System, the Global Data-processing System, the Global Telecommunication System, Data Management and System Support Activities. Also grouped under the WWW "umbrella" are WMO's satellite and emergency response activities, the latter relating to the coordination and implementation of procedures and response mechanisms for the provision and exchange of observational data and specialized products in the case of nuclear accidents, as well as the Instruments and Methods of Observation Program and the Tropical Cyclone Program (TCP). The World Climate Program (WCP) was introduced to deal with climate and climate change issues which were of major global concern in the 1990s. The collection and preservation of climate data help governments prepare national development plans and determine policies in response to the changing situations.

The WCP was established in 1979. Some of its objectives are: use of existing climate information to improve economic and social planning; improve understanding of climate processes through research, so as to determine the predictability of climate and the extent of man's influence on it; and to detect and warn governments of impending climate variations or changes, either natural or man-made, which may significantly affect critical human activities.

2. Montreal Protocol

The Montreal Protocol was the first treaty to protect the atmosphere from human impacts. The agreement and the manner of its development are quite unique. For instance, research findings were a vital part of the decision-making process and scientific assessments are stipulated in the Protocol every four years as a basis for further decisions on ozone-depleting substances.

Ozone depletion

"Ozone depletion" is the term commonly applied to the thinning of the Earth's protective stratospheric ozone layer. Ozone depletion occurs when naturally occurring stratospheric ozone is being destroyed at a rate faster than it is being created, although natural phenomena can cause temporary ozone losses, chlorine and bromine released from synthetic (human-made) compounds are believed to be the main cause of long-term stratospheric ozone depletion. A number of chlorine and bromine compounds fall into the family of synthetic chemicals known as halocarbons. Chlorine halocarbons (CFCs) have been commonly used in car air conditioners, refrigerators, foams, solvents, and other products.

Research on the ozone layer started as early as the 1930s. Some of the major findings are described here. In the 1970s, concerns arose that stratospheric transport aircraft might damage the ozone layer. In 1974, some scientists, foremost Dr. Sherwood Rowland of the University of California, discovered a dramatic change in the chemical composition of our atmosphere, e.g., dramatic increase in concentration of chlorine throughout the world. This increase in chlorine is attributed to widespread use of chlorofluorocarbons (CFCs). Concomitantly the theory of the role of CFCs in the depletion of the ozone layer was proposed. At the time, CFCs were used in refrigeration, aerosol cans, and some industrial processes. Initially greeted with a great deal of skepticism, further research and monitoring began to convince the scientific community of the validity of the CFC hypothesis. In the late 1970s and early 1980s, some national governments imposed bans on CFCs as aerosol and other propellants in non-essential uses for antiperspirants, hairsprays, and deodorants. In 1977 the United Nations Environment Program (UNEP) established the Coordinating Committee on the Ozone Layer.

In 1981, UNEP acted on a proposal submitted by a group of legal experts and decided to develop a global convention. In 1985, the Vienna Convention on the Protection of the Ozone Layer was signed. The period between the Vienna Convention (March, 1985), and the Montreal Protocol (September, 1987) was characterized by incredible progress. The global scientific community reached a consensus on outstanding matters, while meetings were held in Rome to clarify and quantify the current global emissions of ozone-depleting substances and future trends and new mechanisms for control were discussed. By September, 1987, disagreements and lack of understanding had given way to trust. In turn, the trust offered the prospect of consensus on control measures. Thus on September 16, 1987 the *Montreal Protocol on Substances that Deplete the Ozone Layer* was signed by 24 countries.

On January 1, 1989, the Protocol came into effect. All Parties agreed to meet near-term targets for freezing consumption of key CFCs and halogens at 1986 levels, and reducing consumption by 50% within 10 years. While the Protocol is complex, its most important feature was the dynamic process for controlling ozone-depleting substances in addition to those initially identified in the Protocol. One of the major steps was the amendment to the Montreal Protocol in Copenhagen, in 1992, which resulted in a further acceleration of the phase-out of several ozone-depleting substances. In addition, hydrochlorofluorocarbons, and methyl bromide were added to the list of substances subject to control.

3. Intergovernmental Panel on Climate Change (IPCC)

In order to assess available information on the science, impacts and crosscutting economic and other issues related to climate change, in particular a possible global warming induced by human activities, WMO and UNEP established IPCC in 1988.

The role of the IPCC is to assess the scientific, technical and socioeconomic information relevant for understanding the risk of human-induced climate change. It does not carry out new research nor does it monitor climate-related data. It bases its assessment mainly on published and peer reviewed scientific technical literature.

The IPCC has three Working Groups and a Task Force:

- Working Group I assesses the scientific aspects of the climate system and climate change.
- Working Group II addresses the vulnerability of socioeconomic and natural systems to climate change, negative and positive consequences of climate change, and options for adapting to it.
- Working Group III assesses options for limiting greenhouse gas emissions and otherwise mitigating climate change.
- The Task Force on National Greenhouse Gas Inventories oversees the National Greenhouse Gas Inventories Program.

The Panel meets in plenary sessions about once a year. It accepts/approves IPCC reports, decides on the mandates and work plans of the working groups, the structure and outlines of reports, the IPCC Principles and Procedures, and the budget. It also elects the IPCC Chairman and the Bureau.

The IPCC completed its First Assessment Report in 1990. It played an important role in establishing the Intergovernmental Negotiating Committee for a UN Framework Convention on Climate Change (UNFCCC) by the UN General Assembly.

4. Intergovernmental Negotiating Committee (INC)

The INC, established by the UN General Assembly, conducted negotiations for formation of the Framework Convention on Climate Change. The INC met five times under the chairmanship of Jean Ripert (France), including a resumed fifth session in May 1992 at which the Convention was adopted.

During the fifth session of the INC, the government delegates achieved some success in reaching agreement on some of the issues under negotiation. It was agreed that a number of contentious issues must be resolved before the Earth Summit. With little more than one week of negotiating sessions scheduled before the June conference, there was fear that failure of the INC to produce a significant agreement would greatly diminish the importance of the documents to be signed by the world's leaders in Rio de Janeiro.

Working Group I is responsible for negotiating the preamble, principles, objectives, commitments for stabilization and reduction of emissions, and commitments on financial resources and technology transfer. To date, it has made little progress. Among the most important matters still to be resolved are the establishment of subsidiary bodies, reporting mechanisms, the role of non-governmental organizations, and use of the Global Environment Facility (GEF) or an alternative funding mechanism for implementation of the convention.

Working Group II is responsible for institutional issues.

The following issues must be resolved to successfully negotiate a substantive Climate Change Convention: An intersessional meeting must be convened between those countries that have been asked to make specific CO_2 reduction commitments (primarily the OECD countries). The current mode of decision-making must shift from a consensus-based to a majority-based process to prevent minority views from impeding progress. Developing countries need to continue to push the industrialized nations to make firm, meaningful commitments. All participants must shift their thinking from a national to a global perspective.

The INC met a further six times after the adoption of the Convention to prepare for the first session of the Conference of the Parties (COP 1), this time under the chairmanship of Raúl Estrada-Oyuela (Argentina). Over this period, the INC developed recommendations for decisions on the implementation of the Convention to be taken at COP 1, including the financial mechanism, reporting obligations and the adequacy of commitments. The INC met for the last time in February, 1995, before handing over the results of its work to COP 1 in March/April of that year. So far 9 meetings have taken place and the last meeting, COP-9, was held in Milan, Italy in Dec., 2003.

5. United Nations Conference on Environment and Development (UNCED)

The very first international environmental conference to voice concerns over growing pollution was held in June 1972 at Stockholm—the United Nations Conference on the Human Environment. It set a goal of establishing a new and equitable global partnership through creation of new levels of cooperation among States, key Sectors of societies and people and to work toward international agreements which respect the interest of all and protect the global environmental and development system. Building on the same sentiments another conference—UNCED—was held in June, 1992 at Rio de Janeiro. The UNFCCC was adopted in 1992 and became effective in 1994. It provides the overall policy framework for addressing the climate change issue.

The IPCC has continued to provide scientific, technical and socioeconomic advice to the world community, and in particular to the 170-plus Parties to the UNFCCC through its periodic assessment reports on the state of knowledge of causes of climate change, its potential impacts and options for response strategies. Its Second Assessment Report, Climate Change 1995, provided key input to the negotiations, which led to adoption of the Kyoto Protocol by the UNFCCC in 1997. The IPCC also prepares Special Reports and Technical Papers on topics for which independent scientific information and advice is deemed necessary and it supports the UNFCCC through its work on methodologies for National Greenhouse Gas Inventories. The Third Assessment Report "Climate Change 2001" provides a comprehensive and up-to-date assessment of policy-relevant scientific, technical, and socioeconomic dimensions of climate change. It concentrates on new findings since 1995 and pays greater attention to the regional (in addition to the global) scale, and non-English literature.

The aforesaid report assisted governments in making important policy decisions in the negotiations and eventual implementation of the UN Framework Convention on Climate Change, which was signed by 166 countries at the UN Conference on Environment and Development (Rio de Janeiro, 1992). The Convention was ratified in December, 1993 and entered into force on March 21, 1994.

The Atmospheric Research and Environment Programme coordinates and fosters research on the structure and composition of the atmosphere, the physics and chemistry of clouds and weather modification research, tropical meteorology research and weather forecasting. This major Programme aims to help Members to implement research projects and to disseminate relevant scientific information, drawing the attention of Members to outstanding research problems of major importance such as atmospheric composition and climate change.

6. Kyoto Protocol

When the governments adopted the UNFCC, it was quite obvious that its commitments would not be sufficient to seriously tackle climate change. At COP 1 (Berlin, March/April, 1995), in a decision known as the Berlin Mandate, Parties therefore launched a new round of talks to decide on stronger and more detailed commitments for industrialized countries. After two-and-a-half years of intense negotiations, the Kyoto Protocol was adopted at COP 3 in Kyoto, Japan, on December 11, 1997. The Protocol sketched the basic features of its "mechanisms" and compliance system but did not flesh out the all-important rules of how they would operate. The 1997 Kyoto Protocol shares the Convention's objective, principles and institutions, but significantly strengthens the Convention by committing Annex I Parties to individual, legally binding targets to limit or reduce their greenhouse gas emissions. The targets cover emissions of the six main greenhouse gases, namely:

- Carbon dioxide (CO_2);
- Methane (CH_4);
- Nitrous oxide (N_2O);
- Hydrofluorocarbons (HFCs);
- Perfluorocarbons (PFCs); and
- Sulphur hexafluoride (SF_6)

The maximum amount of emissions (measured as the equivalent in carbon dioxide) that a Party may emit over the commitment period in order to comply with its emissions target is known as a Party's *assigned amount*. To achieve their targets, Annex I Parties must put in place *domestic policies and measures*. The Protocol provides an indicative list of policies and measures that might help mitigate climate change and promote sustainable development. Parties may offset their emissions by increasing the amount of greenhouse gases removed from the atmosphere by so-called carbon "sinks" in land use, land-use change, and forestry (LULUCF) sector. However, only certain activities in this sector are eligible. These are *afforestation, reforestation,* and *deforestation* (defined as eligible by the Kyoto Protocol) and *forest management, cropland management, grazing land management,* and *revegetation*.

The Protocol also establishes three innovative "mechanisms" known as *joint implementation, clean development mechanism,* and *emissions trading*. These are designed to help Annex I Parties cut the cost of meeting their emissions targets by taking advantage of opportunities to reduce emissions, or increase greenhouse gas removals, that cost less in other countries than at home. (http://unfccc.int/resource/process/guideprocess-p.pdf)

Thus far in this chapter we have discussed atmosphere, the science behind climate change, and the institutions associated with climate (change) from a historical perspective. The following chapters in this book give in-depth analysis of some of these topics.

Following this introductory chapter, Müller in Chapter 2, **Global Climate Change Regime: Taking Stock and Looking Ahead**, argues that some widespread Northern misconceptions notwithstanding, the FCCC regime is unlikely to succeed unless the key Southern (equity) concern of (sharing) human impact burdens is firmly incorporated in its agenda for the coming years.

Hagem and Holtsmark in **Kyoto Protocol: Insignificant Impact on Global Emissions** present some likely impacts of the Kyoto Protocol related to global greenhouse gas emissions, fossil fuels markets, and the emerging market for emission permits. To evaluate the impacts of implementing the Kyoto Protocol, authors have taken their point of departure in a numerical static partial-equilibrium model that integrates energy markets with an international market for emissions trading. This model is then used to compare the impacts of the Kyoto Protocol both with and without the participation of the United States.

This is followed by **Climate Negotiations from Rio to Marrakech: An Assessment** by Gupta and Lobsinger who provide a brief overview of the key developments in the climate change regime, discussing the Kyoto Protocol and beyond.

After establishing the base of science, agreements and policies for climate change in general, and the Kyoto Protocol in particular, Part II looks at the Instruments and Institutions for the Kyoto Protocol. In **Kyoto Protocol and Desirable Institutions for Mitigation of Global Warming** Yasumoto discusses what is required in future and proposes an active framework that can fully attain the spirit of the Kyoto Protocol and lead to sustainable development. Global warming caused by greenhouse gas is the issue that involves the entire Earth, where every country "has a right to, and should, promote sustainable development". Because of this, institutions to mitigate global warming should be "cost-effective so as to ensure global benefits at the lowest possible cost". The author feels the Kyoto Protocol should have mentioned the framework of domestic institutions right from the beginning.

In the next chapter, **Johnston** et al outline technological proposals to accelerate the sequestration of the main greenhouse gas, carbon dioxide, into oceanic reservoirs, either through direct disposal of fossil fuel-derived CO_2 in the water column or on the seabed or through attempts to enhance biological uptake through large-scale ocean fertilization programs. These proposals are evaluated against a number of criteria necessary for acceptance of ocean disposal/sequestration as a sustainable contribution to the goals and commitments of the Kyoto Protocol to UNFCCC, as well as consistency with other international legal instruments. The chapter concludes that the ultimate objective of the UNFCCC, namely stabilization of greenhouse gas concentrations at levels that would prevent dangerous interference with the climate system, cannot be met through such. Although theoretically capable of reducing atmospheric CO_2 levels in the coming century to a similar level as emission reductions at the source strategy, CO_2 disposal in the ocean will result in higher CO_2 levels, and hence larger climate changes and sea-level rise in future centuries. Given the threat these technologies present to effective implementation of the Kyoto Protocol and to sustainable energy and waste management more broadly, efforts should be made to ensure that ocean disposal/sequestration of CO_2 not be used to offset commitments to emissions reductions for greenhouse gases.

Rosales in **The Idea of Social Progress and Emissions Trading** argues that emissions trading has a history that predates and is deeper and more ubiquitous than the economic efficiency rationale initially posed by Coase (1960) and his followers. Emissions-trading has a structured epistemology and ontology that becomes evident with deeper analysis. This longer and broader consideration of emissions trading gives a better understanding of the emissions trading discourse—how it is formed and maintained as negotiated in the UNFCCC. When the philosophical underpinnings of the institutions surrounding emissions trading are considered, one better understands why certain policy options or mechanisms are offered, supported, and implemented. This analysis can also be used to identify dissention, as these structures are negotiated and temporal. The systems of thought ushered in with the Enlightenment, stemming from paganism, were once subordinate to the dominant or hegemonic Catholic Church in Europe. Similarly, there are systems of thought today that remain subordinate or invisible but that may gain support as the UNFCCC climate change system negotiates itself into new forms.

Johnke in **Status Report on CO_2 Emissions Saving through Improved Energy Use in Municipal Solid Waste Incineration Plants** examines the question of whether and to what extent the use of available energy potential in MSW incineration plants is possible and can be improved in terms of plant technology, in order to thereby reduce climate-relevant CO_2 emissions. This analysis is geared to the current state of energy production technology in existing installations. The results of this report are also intended to show what contribution the use of existing energy potential in MSW incineration plants can make to a reduction of CO_2 emissions within the framework of the targets specified in the national climate protection program (Germany). The national climate protection program sets out measures and targets for CO_2 emission reduction until 2010. An emission reduction of the order of 18 – 20% (about 180,000 – 200,000 Gg of CO_2) is expected by 2005 based on the CO_2 emission reduction measures taken since 1990. According to BMU Umivelt 11/2000, from 1998 onwards an additional CO_2 reduction potential of 50,000 – 70,000 Gg needs be tapped in order to achieve the target of a 25% reduction by 2005.

Part III is devoted to issues relating to human rights. Goldberg and Wagner in **Petitioning for Adverse Impacts of Global Warming in the Inter-American Human Rights System** examine the options for bringing a human rights complaint against the United States. Because international human rights law gives states primary responsibility for ensuring the protection of human rights, this analysis has been confined to a claim against the United States. However, because some human rights institutions have in some instances recognized the responsibility of corporations for human rights violations, global warming may be a basis for claims against U.S. corporations as well. The authors first discuss the severe impacts of global warming already being felt by many indigenous Arctic communities that could give rise to a human rights claim in the Inter-American system. Some of the procedural elements of such a claim are then considered, with particular emphasis on issues of jurisdiction. Some of the rights protected by the Inter-American human rights system that might be implicated in a human rights claim are then examined and the remedies possibly available to claimants discussed.

Impacts

Rapid and large climatic changes can be expected to have far-reaching and, in many instances, unpredictable consequences not only for human societies, but also for all forms of life on Earth. There could be many direct effects of large and rapid climate change. A rise in global sea level, which would threaten coastal cities and settlements throughout the world, is possible. Also, an increase in intensity of severe weather conditions as well as frequency of heat waves and droughts is possible. In addition, the resultant shift in climate zones could be so rapid that many plant and animal species would not be able to adopt quickly enough to ensure their survival. Change in precipitation in some areas is depicted in Figure 1.8.

Climate changes are causing havoc throughout the world—either floods in Bangladesh and Mozambique or droughts elsewhere; El Nino is causing problems in various places, e.g. Indonesia.

Another disturbing consequence of human-induced climate change is that it may permanently affect the Earth's climate system. The elements that influence the climate such as oceans, forests, and clouds interact in complex ways. Once interrupted by the initial effects of global warming, they may not be readily able to provide the same temperature regulating functions to which life on Earth has become accustomed.

Part IV of the book presents case studies of some countries in which the impact of climate change and the Kyoto Protocol has already been evidenced.

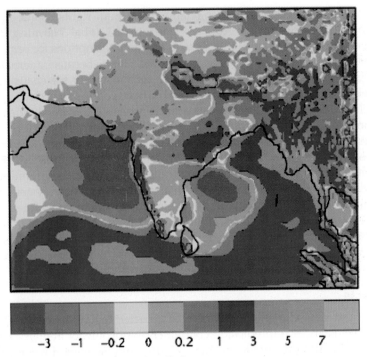

-3   -1   -0.2   0   0.2   1   3   5   7

Fig. 1.8: Precipitation change (mm/day)

In the first case study, **Climate Change and Water Resources Management in Semi arid Southern Africa,** Umoh et al. examine water resources management in semi arid Southern Africa under the increasing effects of climate change. The water resources in the sub-region, climatic variability, and current trend of drought and desertification with reference to recent climatic changes are examined and management strategies adopted for controlling the situation described. Lastly, some suggestions are proffered for effective water resource management.

In **Adaptation to Effects of Climate Change in Southern Africa,** Umoh et al. indicate that timely adaptation strategies in agriculture, wildlife resources, rangeland and livestock production, human health, hydrology and water resources, and Drought and Desertification could contain the effects of the impacts of climate change.

Huq and Moinuddin in **Climate Change, Vulnerability and Adaptation in Bangladesh** discuss a country singularly vulnerable to climate change. Bangladesh's long coastline, vast low-lying landmass, high population density and nature-dependent traditional agricultural practices make it highly susceptible to any climate change. It is likely that the disadvantaged and poor community would suffer worse than the better-off strata of society. The chapter further discusses the impact of climate change on various sectors such as agriculture, fisheries etc. It is clear that climate change is very bad news for Bangladesh in the long term. Nevertheless in the short (and medium) term there are also some opportunities for the country to seize if one is clever and far-sighted enough to realize them and take appropriate actions. Bangladesh has been taking a somewhat passive posture in the international negotiations on climate change. As a member of the Group of Seventy-Seven (G77) it takes a common position with the rest of the G77 group but seldom takes a leadership role on any issue. So, under these circumstances what can Bangladesh actually do either to prevent the worst from happening or coping with it when it does? The answer is indeed much if strategies were developed and implemented right now.

Smith in **Seeking the Middle Ground between More and Less: A Canadian Perspective** begins with a brief analysis of some competing perspectives on Canada and multilateralism. She explains why Canada engages in multilateralism and investigates its practice and consequences. She then examines US influence on Canada's orientation, by first giving an analysis of Canada's activities in the early stages of the climate issue, followed by the period between the first Conference of Parties in Berlin and the US withdrawal from the Kyoto Protocol. Canada's behavior after the US withdrawal from the Kyoto Protocol in 2001, including Canadian participation at the Sixth Conference of Parties resumed, Canadian ratification and bilateral activities between Canada and the US are then discussed. A historical overview allows the author to trace Canadian policy development and Canada's position vis-à-vis the United States over a fourteen-year period and thereby assess how the tensions noted above played out in practice. Finally, the conclusion reflects on the lessons of the case study in light of the American withdrawal. While Kyoto may provide the "architecture" for the continuation of future negotiations, one should not expect the US to return to the fold in the immediate future. Rather, the United States will strike bilateral and trilateral deals with states such as Canada, Mexico and Australia, intervene in future international discussions when it feels its interests are undermined, and rely on its allies who have ratified the Kyoto Protocol to further ensure that future negotiations do not threaten American interests. Canada will be one such ally.

Chapter 14, **Costa Rica and Its Climate Change Policies: Five Years after Kyoto**, by Rojas briefly explains the Costa Rican policy on climate change. In order to better understand the negotiating positions presented in international forums by this country, it is necessary to make reference to the Berlin Mandate and the Pilot Phase of Activities Implemented Jointly or AIJ Projects, because it is this experience that helped Costa Rica distinguish the relevant negotiating issues during and after COP 3 in 1997.

After the Kyoto Protocol was signed, Costa Rica focused its national policy on the development and marketing of AIJ projects and possible Clean Development Mechanism or CDM Projects. An important number of the AIJ projects already developed and/or approved in the country were related to avoiding deforestation. Therefore, the political agreement has had an important effect on the country's portfolio and possible CDM projects. Another impact of the latest agreement is the importance other issues related to climate change have gained. For example, now there's more openness to discuss the country's vulnerability and to find ways to include public participation in these discussions. The chapter concludes with the outcome for Costa Ricans after COP7. It presents the impacts the Executive Board and the approved rules of the Protocol will have on the climate change policies to be implemented in Costa Rica.

Gupta in **India and the Climate Convention: The Challenge of Sustainable Development** outlines India's role in the evolving climate change negotiations. The author feels that although the negotiation dynamics have changed considerably since the start of the negotiations ten years ago, India's policy remained constant and this stagnation reflects to what extent how out of touch Indian negotiators are with the domestic developments in the field of energy and forestry and with the international mood in relation to the South's demands for assistance and equity. The chapter highlights the key strengths, weaknesses, opportunities and threats India faces in relation to the environmental treaty and argues that as long as India sees the climate change negotiations as a platform for finger pointing exercises, it will fail to capitalize on the developments in international environmental law on the one hand, and domestic developments on the other.

Indonesia: The suffering nation by Surjadi describes the impact of climate change on Indonesia and the significance of the Kyoto Protocol and implementation of certain mechanisms, e.g. the CDM.

The next chapter **"Trump Card in Kyoto Pact: Russia's Interests and positions on the Global Climate Change Regime"**, by Mirovitskaya maps the evolution of Russian government stand on Kyoto Protocol. She gives a comprehensive overview of the Russian government stand, domestic policies and negotiation process, especially in light of the confusion if Russia is still part of Kyoto Protocol or not. She further looks at the benefits and liabilities on Russia on its participation in the global climate regime.

In **Sri Lanka: Its Industry and Challenges in the Face of Climate Change**, Alwis explains the background to industrial development in the country and looks at the energy issue, which is currently crippling it. The energy equation, which is linked to the climate issue, is thus examined in detail considering past patterns and future expectations. Current industrial performance is reviewed and steps to be taken to mitigate adverse environmental discharges to realize desired national goals are discussed. The estimated contribution to the Greenhouse Gas Inventory from industrial activities in Sri Lanka is compared with the overall national estimate.

Marques and Hogland in **Environmental and Socioeconomic Impacts Associated with Climate Changes in Brazil** argue that climate change, so far, has not received priority in the environmental policy agenda in developing countries, where basic sanitation is still the main environmental concern. However, indications show that populations of developing countries are the most vulnerable to climate changes. Severe environmental as well as socioeconomic impacts are foreseen as a consequence of synergy between climate change and regional vulnerability due to inappropriate land use and occupation, littoralization/urbanization process, highly dense population in cities and poverty. Severe environmental and socioeconomic impacts due to climate changes in Brazil are described (e.g. precipitation decrease in the north, drought in the northeast and flooding in south Brazil during *El Niño*, with reverse effect in the same regions during *La Niña*). Increasing deforestation of Amazon forests is expected to contribute significantly to the regional changes in the hydrological cycle. The authors recommend urgent inclusion of the topic *climate change* in the governmental agenda and improvement of the strategic impact assessment associated with it.

Najam in **The Future of Global Climate Change Policy: Developing Country Priorities Beyond Kyoto** start with a review of the concerns that the developing countries of the South have about the direction that the global climate change regime is taking. They then identify a set of key Southern interests for future negotiations on the subject and conclude by arguing that although the WSSD is unlikely to make substantial headway in this direction, it could launch a process that would revitalize the global climate change debate and make the regime more inclusive as well as more efficacious.

In **Reclaiming the Atmospheric Commons: Beyond Kyoto**, Byrne et al. say that due analysis of the situation led them to conclude that the Kyoto Protocol-Marrakech Accord is unlikely to improve climate justice or sustainability. The alliances underlying the COP-negotiated agreements appear to be largely economic in character, not ecological. As a result, its outcomes are better predicted as elements of neoliberal globalization strategy than as commitments to ecological principles, values or goals. If international negotiations are to avoid being coopted as a venue merely for deciding economic advantage, the authors argue that two principles — equity and sustainability — must guide deliberations. More broadly, the authors feel that the paradigm of *capitalization,* which guided nature-society relations in the industrial era, needs to be replaced with one that reclaims our climate and atmosphere as elements of a global *commons.*

Thus, the book has tried to address various issues related to climate change—some scientific facts and the Kyoto Protocol but it is important to mention that scientific uncertainty still exists (it is unclear whether the natural cycle of global change is accelerated by anthropogenic factors). Also, much needs to be done to iron out creases in the Kyoto Protocol before it can be implemented. Also, steps need to be taken to control atmospheric emissions to slow the global warming effect; we owe this to our future generation. For this a multidisciplinary approach and understanding needs to be developed for a broader perspective and clearer picture (specially for policy-making).

CHAPTER 2

Global Climate Change Regime: Taking Stock and Looking Ahead[*]

Benito Müller
Senior Research Fellow, Oxford Institute for Energy Studies,
57 Woodstock Rd, Oxford Ox2 6FA UK,
www.oxfordclimatepolicy.org/www.oxfordenergy.org
e-mail : benito.mueller@philosophy.oxford, ac.uk.

Introduction[1]

Climate change may well be the biggest and most complex environment-related problem for international co-operation this century and beyond. In the last ten years, the issue has been the focus of intense and, given its complexity, remarkably successful global negotiations under the United Nations Framework Convention on Climate Change (FCCC). The focus of these negotiations has been firmly on establishing a multilateral emission mitigation regime. This 'mitigation agenda' found its culmination to date in the Kyoto Protocol and the recently finalised Marrakech Accords, which will enter into force after Russian ratification. This article argues that — not withstanding some widespread Northern misconceptions — the FCCC regime is unlikely to succeed unless the key Southern (equity) concern of (sharing) human impact burdens is put firmly on its agenda for the coming years.

The Phenomenon

Global climatic changes are nothing new. The last 500 millennia, for example, have seen regular cycles in the Earth's climate, alternating between ice ages and interglacial periods (Fig. 2.1). Indeed, everything else being equal, evidence suggests that we are at the peak of one of these main interglacial periods, which accounts for the worry in the late 1970s about the onset of another ice age.[2] Yet these worries were not particularly acute. After all,

---

[*] Originally published in O.S. Stokke and O.B.Thommerren (eds.), Yearbook of International Co-operation on Environment and Development 2002/2003 (YICED 02/03). London: Earthscan, 2002, pp. 27-38.

the main cycle—with a temperature variation of 12ºC—had a cooling period of over 80,000 years. *Après nous le déluge* becomes less problematic at these timescales, both as statement and as attitude.

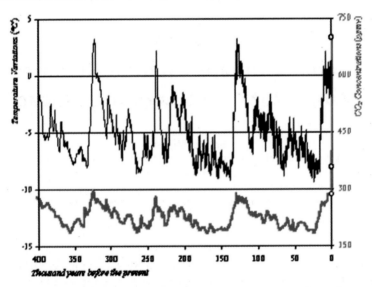

Fig. 2.1: CO_2 concentrations and temperature variations (from Present/ from Pre-historic Temperature and CO_2 Concentrations: Petit *et al.,* 1999). CO_2 Concentrations: Pre-industrial (= 280ppm), Current (1998 = 365ppm), 2100 Projections (= 540 - 970ppm, IS92a = 710ppm): IPCC TAR1.

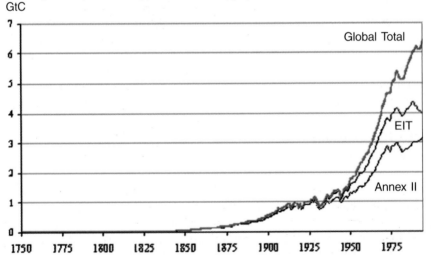

Fig. 2.2: CO_2 emissions, fossil-fuel burning, cement manufacture, and gas flaring. 1751-1998 (from G. Marland, T.A. Boden, and R.J. Andres (2001), 'Global, Regional, and National Annual CO_2 Emissions from Fossil-Fuel Burning, Cement Production, and Gas Flaring: 1751-1998 (revised July 2001)', <http://cdiac.ornl.gov/ftp/ndp030/region98.ems>)

This situation, however, has since changed dramatically, as witnessed in the recent *Third Assessment Report* of the Intergovernmental Panel on Climate Change (IPCC).[3] The global average surface temperature –having increased by about 0.6°C over the 20th century—is projected to increase between 1.4 to 5.8°C over this century, at a rate "very likely [≥90%] to be without precedent during at least the last 10,000 years." The threat of an impending ice age has given way to concerns about much more immediate climatic changes in the "opposite direction." The reason is that in the course of the last century, mankind has unintentionally become a force to be reckoned with in influencing the Earth's climatic system. It graduated –or blundered– from "climate-taker" to "climate-maker".

Fundamental Distinctions

The most general distinction between the causes of the current climatic changes is thus between "natural" on the one hand, and "anthropogenic" ("human-induced", "man-made"), on the other. A paradigm of natural climate variations are the ice-age cycles of geological time scales, some of which prove to be closely correlated with anomalies in the terrestrial orbit.[4] Yet there are other natural causes which can lead to changes in regional and global climates.

Take the phenomenon of "volcanic winters". The sulfur dioxide emissions of the volcanic eruption on the Aegean island of Thera (Santorini) in 1628 B.C.[5], for example, have been used to explain the average global cooling of 1.5°C over the following one hundred years,[6] which, in turn, has been suggested as one of the key factors in the downfall of the Minoan civilization during the first half of the 16th century B.C.[7] Other natural climate change events have been identified as having had equal, if not worse social impacts –the 3 to 5°C cooling following the Toba (Indonesia) eruption of about 73,000 years ago apparently almost spelled the end of humankind.[8]

Anthropogenic causes, in turn, are largely based in human energy use and agricultural practices relating to the emission of greenhouse gases. Rice cultivated under flooded conditions generates methane emissions into the atmosphere due to the decomposition of organic matter. Deforestation reduces the absorption of carbon dioxide (CO_2) through vegetation growth. However, the biggest anthropogenic cause of climate change by a long way is not these agricultural practices, but the use of fossil carbon –coal, oil and gas– as combustion fuels in all economic sectors: transport, domestic heating, industrial production, electricity generation, and so on.

There will obviously be differences in the relative shares of CO_2 emissions for these sectors within a country, but arguably the most significant differences are not within but between countries. In 1998, for example, the CO_2 emissions per head of population ranged from 20,000 kg for the United States at one end of the spectrum, to least-developed countries such as Sierra Leone with 110 kg, at the other.[9] Given the importance of energy in economic growth and the historic worldwide reliance on fossil energy sources, it is not surprising to find that over the last century (Fig. 2.2), industrialized countries (the "North" = OECD and the economies in transition of the former Soviet Union and Eastern Europe) have collectively emitted five times the emissions of the developing world (the "South"),[10] a fact which gives some idea about the regional distribution of causal responsibilities for (potentially inevitable) anthropogenic climate change impacts.[11]

The reason for drawing the distinction between anthropogenic and natural causes lies in the possibility of attacking a root cause of the problem: while it is well within our ability to reduce greenhouse gas emissions, it is unlikely that our "geo-engineering" skills will ever be able to control volcanic activity, let alone the terrestrial orbit around the Sun. However, people must not only be singled out as causes but also as recipients of climate change impacts. The fact of the matter is, climate change is only a problem because of adverse impacts on life systems. And this is true regardless of whether the impacts are anthropogenic or not.

As it happens, climate change impacts are divided not only with respect to their cause ("natural" versus "anthropogenic"), but also relative to who or what they affect, namely "social" or "human impacts" on human systems ("society"), on the one hand, and "ecological ones" or natural ecosystems ("nature") on the other. One and the same cause can obviously give rise to a variety of impacts, both on different social systems –social groups, countries or regions—and different natural ecosystems, such as tropical rain forests or coral reefs. Giving rise to both types of impacts is common to many pollution problems. What distinguishes climate change is the nature and potential seriousness of its human impacts. These impacts transform the issue from a purely environmental into an environment- and development-related problem. Moreover, its anthropogenic components additionally introduce issues of interpersonal justice between those who cause the impacts and those who suffer from them.

Story To Date: An Environmental(ist) Pollution Agenda

International Response: IPCC, FCCC, and Kyoto Protocol[12]
Knowledge of "greenhouse gases" and a "greenhouse effect" is far from new. 'As early as 1827, the French scientist Fourier[13] suggested that the earth's atmosphere warms the surface by letting through high-energy solar radiation but trapping part of the longer wave heat radiation coming back from the surface. ... At the end of the nineteenth century the Swedish scientist Arrhenius[14] postulated that the growing volume of carbon dioxide emitted

by the factories of the Industrial Revolution was changing the composition of the atmosphere, increasing the proportion of greenhouse gases, and that this would cause the Earth's surface temperature to rise."[15]

However, it took the international community until the late 1970s to take an interest in the phenomenon, with the first World Climate Conference taking place in 1979 under the aegis of the World Meteorological Organization (WMO). Driven by further rise in public concern in developed countries about *industrial pollution*—smog, acid rain, toxic rivers and lakes, etc.—a series of international meetings led in 1988 to the formation of the Intergovernmental Panel on Climate Change (IPCC) "to assess the scientific, technical and socioeconomic information relevant for the understanding of the risk of human-induced climate change."[16] To date, the IPCC has published three Assessment Reports –the latest in 2001– which have been extremely influential in shaping the global climate-change agenda. After considerable debate about the findings of the 1990 *First Assessment Report*, the ministerial segment of the Second World Climate Conference (1990) called for initiation of a UN-negotiated climate-change regime.

The initial phase of regime formation very speedily culminated at the 1992 Rio Earth Summit in the UN Framework Convention on Climate Change (FCCC). In light of the excellent and detailed exposition of this Convention and its related legal instruments in the reference section of the *Yearbook*, there is no need to introduce it here in detail except for three key "architectural elements".

- **Article 2** defines: 'The ultimate objective of this Convention ... is to achieve, ... stabilization of greenhouse gas concentrations in the atmosphere at a level that would prevent dangerous anthropogenic interference with the climate system. Such a level should be achieved within a timeframe sufficient to allow ecosystems to adapt naturally to climate change, to ensure that food production is not threatened, and to enable economic development to proceed in a sustainable manner."

- **Article 3** (on Principles) stipulates in the first paragraph: "The Parties should protect the climate system for the benefit of present and future generations of humankind, on the basis of equity and in accordance with their common but differentiated responsibilities and respective capabilities. Accordingly, the developed country Parties should take the lead in combating climate change and the adverse effects thereof."[17]

The equity-related differentiation principles regarding *responsibilities* and *capabilities* of Article 3 found their way into the architecture of the Convention primarily through the introduction of two lists of countries: *Annex I*, containing the industrialized countries with their significant historical emission records (Fig. 2.2), and *Annex II* with the affluent industrialized countries. For example, the Parties included in Annex I commit themselves in conformity with the degree of their responsibility in

- **Article 4.2** to adopt policies and measures "with the aim of returning individually or jointly to their 1990 levels those anthropogenic emissions of carbon dioxide and other greenhouse gases not controlled by the Montreal Protocol" by the end of the 1990s, thus demonstrating "that developed countries are taking the lead in modifying longer term trends in anthropogenic emissions consistent with the objective of the Convention."
- **Article 4.4**, in turn, demands of Annex II Parties to "assist the developing country Parties that are particularly vulnerable to the adverse effects of climate change in meeting costs of adaptation to those adverse effects."

To be clear, the target of returning to 1990 levels by 2000 was stated as an aspiration without legally binding status. On October 15, 1992, the United States of America—preceded only by three small island states—was the first major country, North or South, to ratify the Convention, which itself came into force on March 21, 1994.

In April 1995, the first session of the Conference of the Parties (COP1) in Berlin adopted what became known as the "Berlin Mandate".[18] In it, the Parties concluded that the Annex I commitments in Art. 4 of the Convention were not adequate and agreed to begin a process "to take appropriate action for the period beyond 2000." This process was, *inter alia*, meant "to set quantified limitation and reduction objectives within specified timeframes, such as 2005, 2010 and 2020" for Annex I Parties, and "not introduce any new commitments for Parties not included in Annex I," thus reaffirming the need for Annex I leadership in conformity with the demands on equity by the existing differences in causal responsibility.

The ensuing negotiations—carried out under the aegis of the Ad Hoc Group on the Berlin Mandate (AGBM)—found their culmination in the morning of December 11, 1997 at the third session of the COP in Kyoto, when the chairman of the negotiations, Ambassador Estrada-Oyuela, declared the Kyoto Protocol to be unanimously agreed.[19] The Protocol's key response to the Berlin Mandate was set down in two Annexes—listing greenhouse gases (Annex A) and *legally binding* percentage reduction figures (Annex B)—and in **Article 3**:

3.1 The Parties included in Annex I shall, individually or jointly, ensure that their aggregate anthropogenic carbon dioxide equivalent emissions of the greenhouse gases listed in Annex A do not exceed their assigned amounts, calculated pursuant to their quantified emission limitation and reduction commitments inscribed in Annex B and in accordance with the provisions of this Article, with a view to reducing their overall emissions of such gases by at least 5 percent below 1990 levels in the commitment period 2008 to 2012.

3.2 Each Party included in Annex I shall, by 2005, have made demonstrable progress in achieving its commitments under this Protocol.

In keeping with the Berlin Mandate, the Protocol did not introduce emission targets—or QELRCs (Quantified Emission Limitation and Reduction Commitments)—for developing countries. And while it fell short of providing Annex I targets for the year 2020 mentioned in the Mandate, it did provide for additional, post-2012 commitment periods:[20]

3.9 Commitments for subsequent periods for Parties included in Annex I shall be established in amendments to Annex B to this Protocol, ... The Conference of the Parties ... shall initiate the consideration of such commitments at least seven years before the end of the first commitment period.

The *Yearbook* FCCC reference is witness to the fact that there is much more to the architecture of the emission mitigation regime introduced by the Protocol than just these targets and timetables enshrined in Article 3. And while it is not possible to characterize these features in more detail in the present context, some of the most recent achievements cannot be left completely unmentioned. Having achieved a political breakthrough at the extraordinary COP6-bis session in Bonn in July 2001, the negotiators reconvened for the seventh regular COP session at Marrakech in November where they succeeded in specifying the operational details of the Protocol sufficiently for it to become technically ratifiable.

The negotiations at Bonn and Marrakech were dominated by four distinct yet related problem areas, three of which concerning the "flexibilities" were built into the Kyoto mitigation regime. Their success became manifest, for example, in the adoption of eligibility criteria for the three "Kyoto mechanisms"—emissions trading, joint implementation, and the Clean Development Mechanism (CDM), see *Yearbook* FCCC reference— and in the election of a CDM Executive Board, to facilitate a prompt start of CDM transactions. A second flexibility issue dominating the debate was the nature and volume of permissible greenhouse gas "sinks" through land-use (change) and forestry activities. A compromise on how much of the carbon absorbed from the atmosphere could be counted against the Kyoto emission targets was reached with the intention of enabling ratification of some key countries such as Japan and Russia. The third mitigation issue which exercised people's minds during the negotiations was the Kyoto target compliance. The compliance regime proposed under the Kyoto Protocol is one of the strongest of any multilateral treaty and its institutional structure was sufficiently clarified for the language to become ratifiable, not-withstanding a postponement of a decision on its legal nature to after entry-into-force.

The fourth key issue area raised during the negotiations at Bonn and Marrakech was capacity building, technology transfer and adverse climate change effects on developing countries as described, in particular, in Articles 4.8 and 4.9 of the FCCC. The COP decided to establish a Climate-Change Fund and a Least-Developed Country Fund under the Convention to complement the Adaptation Fund established in Bonn under the Kyoto Protocol.

There are already quite a number of detailed readily available studies of what has become known as the Marrakech Accord and its socioeconomic and environmental implications.[21] For the purposes of this paper, the overall conclusion to be drawn is that the task of finalising the operational details of the Protocol has been completed, which puts the question of "adequacy of commitments" again at center stage, particularly in the run-up to the "second commitment period" negotiations scheduled in Art 3.9 to start not later than 2005.

National Implementation

The issue of adequacy of commitments is not new, and while negotiators felt themselves bound by the remit of the Berlin Mandate, other stakeholders did not. In July 1997—five months before the Kyoto Conference—the US Senate, for example, passed the Byrd-Hagel Resolution (S.R. 98) stipulating the United States should not be a signatory to any FCCC protocol which would "mandate new commitments to limit or reduce greenhouse gas emissions for the Annex I Parties, unless the protocol ... also mandates new specific scheduled commitments to limit or reduce greenhouse gas emissions for Developing Country Parties *within the same compliance period.*"[22]

On March 13, 2001, US President Bush withdrew from the Kyoto process for precisely such adequacy of commitment reasons. Indeed, his specific opposition to the Protocol was "because it exempts 80% of the world, including major population centers such as China and India, from compliance, and would cause serious harm to the U.S. economy. ... there is a clear consensus that the Kyoto Protocol is an unfair and ineffective means of addressing global climate change concerns."[23]

What is the current state of Parties' emissions relative to the objectives set in the Convention and the Kyoto Protocol? As it happens, collectively, developed countries have already met their (implied) Kyoto target of a 5% reduction in 1990 greenhouse gas emissions by 2008-2012.[24] This may seem curious, in particular since the United States—the world's single largest greenhouse gas emitter—has made no particular headway in complying with the objective stated in Article 4.2 of the Framework Convention (let alone with its Kyoto commitment): Far from having returned to their 1990 target level, US emissions at the end of the last decade overshot this level by around 12%,[25] and the predicted trend (Fig. 2.3) will hardly satisfy the Convention's stipulation that industrialized country "policies and measures will demonstrate that developed countries are taking the lead in modifying longer term trends in anthropogenic emissions."[26]

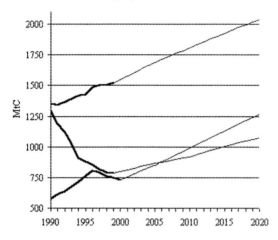

Fig. 2.3: Carbon dioxide emissions. 1990-2020 (from Benito Müller (2001) 'Fatally Flawed Inequity')

The reason why collectively Annex I still manages to be below the implied Kyoto mitigation requirement is the (unintentional) overachievement of the so-called "economies in transition" (EITs), i.e., the countries of Eastern Europe and the former Soviet Union. In industrialised countries, economic collapse is more often than not correlated with a reduction in greenhouse gas emissions. Between 1990 and 1999, the Russian Federation for example experienced a drop in real GDP of 45%[27] with a concomitant reduction of CO_2 emissions of 36%. The EITs collectively reduced their emissions over the same period by 39% from 1,300 MtC to 790 MtC (Fig. 2.3),[28] at a "cost" to the economies of $420 bn—or $823 tC^{-1}— as it were.[29]

In light of the overwhelming majority of studies predicting a traded carbon permit price of less than $ 100 tC^{-1} [30], this has been a costly way to abate. By contrast China, having turned around its emissions in 1996, thereafter mirrored the EIT reductions (Fig. 2.3) while continuing to enjoy an annual economic growth of between 7 and 9%[31]. More precisely, unlike most Annex I countries, China managed (under no obligation) to reverse its emissions, leaving them by the end of the decade 9% lower than their 1996 peak and 27% up on the 1990 benchmark—less than half the previously predicted 67% increase[32]—without prejudice to its remarkable economic growth.

The economic collapse of the EITs was obviously not due to a climate change policy. And yet is worth mentioning this "carbon cost," if only to highlight that the resultant surplus permits—often referred to as "hot air"

—have not been some free windfall to the countries involved. Or, put differently, that this "hot air" is not necessarily the sort of ill-gotten gain it is sometimes portrayed as in arguments defending the environmental integrity of the regime.

Nonetheless, the collective return of Annex I emissions to 1990 levels can hardly be claimed to be the result of policies and measures, demonstrating that developed countries are taking the lead in modifying longer term trends in anthropogenic emissions, as demanded in Article 4.2 of the Convention. The conclusion thus has to be that the Convention's aspirational ("voluntary") target setting has not been a success.

As for the Kyoto Protocol, it is obviously too early to judge compliance with its legally binding targets. Moreover—due to its international flexibility mechanisms—the issue could not be discussed in terms of these simple domestic emissions. And yet since countries are not generally inclined to sign, let alone ratify an international treaty, without some confidence of being able to comply with legally binding provisions, it seems that most of the Annex I Parties bar the United States consider compliance possible, given their declared intention to ratify by WSSD.

Equity: Northern Perspective

Concerning equity, one issue has dominated the debate to date, namely quantified developing country emission targets – the issue of "meaningful developing country participation," as it has somewhat euphemistically become known in the US context. Developing countries have had some success in demanding on grounds of differentiated responsibilities that industrialized countries must take the lead in adopting legally binding emission targets. However, this has by no means been universally accepted. As a matter of fact, a rejection on grounds of (i) unfair cost distribution and (ii) environmental ineffectiveness has, as mentioned above, led to arguably the greatest setback to the global climate change effort to date: the US administration's withdrawal from the Kyoto regime.

Ad (i). The (perceived) "enormity" of any cost is inevitably in the eye of the beholder. In the case of the United States, a study supported by the American Petroleum Institute which had considerable impact on American perceptions of the Kyoto Protocol[33] predicted what has become accepted as a "worst case" estimate for US mitigation costs under the Kyoto Protocol, namely a 2% reduction of gross domestic product from "Business as Usual" (BaU). Whether such a change in the way of life is bearable or not is one thing, but it is and remains a matter of life *style*. And it is difficult to see how even this sort of maximum life-style impact could turn the absence of developing country targets into an unfair competitive disadvantage, given the projected increase in the North-South welfare gap for the period (Fig. 2.4).

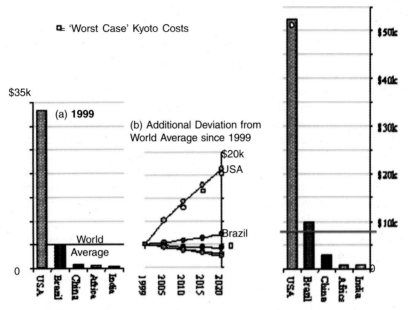

Fig. 2.4: Per capita GDP projections. BaU and Kyoto costs. 1999-2020 ('000 US 1997 $) (from Benito Müller, Axel Michaelowa, and Christiaan Vrolijk, 2001, Rejecting Kyoto, p. 4.)

Ad (ii). It is thus not surprising that the Bush administration's rejection of the Kyoto Protocol on grounds of imposing these "unfair costs" has not found a great deal of empathy in the rest of the world. The environmental integrity point, however, has had more of a following. Indeed, the fact that the Kyoto Protocol is unable to deliver the objective of the Convention is universally accepted. Yet, most people involved in the debate also realize that it was never meant to be more than a first, albeit important, step in this direction.

Outside the Bush administration, concern about the environmental integrity of the multilateral regime has led to a focus on designing the mitigation regime of the envisaged second commitment period (2013–17) and beyond. In particular—seemingly unaware of recent developments in China[34] and the possibility that with sufficient momentum the decarbonization of industrialized country economies is likely to spillover to the rest of the world (whether they want it or not, with or without targets)[35]—many environmentally concerned protagonists have exercised their minds about including *developing* countries in a second commitment period target system in order to ensure the environmental integrity of the regime.

A large number of proposals have been made as to how this might be achieved, many of which explicitly deal with the issue of distributive justice (often forcefully raised by developing country stakeholders). Some of them are based on *ex ante* allocations of country quotas ('assigned amounts'), such as the "grandfathering"[36] and "per capita"[37] proposals, and their mixtures of both diachronic (e.g. "contraction and convergence"[38]) and synchronic (e.g. "preference score"[39]) varieties. Others involve more "flexible" targets based for example on "emission intensities",[40] or "price caps".[41] Studies and publications on the merits and shortcomings of these and many other proposals for introducing developing country mitigation targets are too numerous to be introduced, let alone properly discussed, within the confines of this article. However, information is readily available,[42] which is why we shall now turn to an issue which appears to be less appreciated but arguably as important, namely the question of why emission *mitigation* has managed to dominate the multilateral climate change debate to date.

Environmental Protection and the Concept of "Sustainable Development"

The dominance of emission mitigation in the international climate change debate is reflected in the proportion of text afforded to this issue in the language of the international treaties. While there are some articles both in the Convention and the Protocol which are concerned with other matters, the majority deals with mitigation issues such as international transfers of emission quotas ("flexibility mechanisms"), land use and land-use change ("sinks"), regime for complying with the quantified emission targets, compilation of national emission data ("National Communications") —to name just some of the issues which have exercised many a mind in the past couple of years.

The agenda to date has been about the emission mitigation burdens for a variety of reasons, some more pragmatic, others more philosophical. At the pragmatic end of the spectrum is the fact that greenhouse gas emissions can readily be fitted into an existing paradigm in the industrialized North: air and atmospheric pollution. This has been recognized as a problem (smog, acid rain, etc.) by governments in industrialized countries for many decades and most of them have introduced elaborate institutional structures (environment protection agencies etc.) to deal with it. While it is not altogether clear whether the problem of mitigating greenhouse gas emissions is best served by a subsumption under the air pollution paradigm, the fact remains that in most countries, particularly in the North, climate change has been handed over to institutions dealing primarily with protection of the natural environment.

Another pragmatic reason –with "philosophical undertones," as it were– lies in the possibility to attack the problem (anthropogenic climate change) at its root cause. While there have been voices suggesting that it might be better to spend the effort and money on improving adaptive capacities rather than reducing emissions,[43] the majority view by far is that since it is possible to attack the root cause of the problem, it is better to do so than to deal solely with the effects, especially if the cause is people who, after all, can be held responsible for their actions.

Yet, arguably the most fundamental reason for the mitigation focus in the current regime is to be found at the philosophical end of the spectrum: the perceptions of the very nature of climate change, the views of "what it is really all about". More specifically, the focus on mitigation is the result of a dominant Northern perception of things. To understand the nature of this perception and the way it arose, it may be useful to turn briefly to the nature and history of the closely related concepts of "sustainability" and "sustainable development".

It is rare that the creation of a concept is precisely dated. "Sustainable development," according to Ashok Khosla,[44] was launched on March 5, 1980 in the *World Conservation Strategy* prepared jointly by the World Conservation Union (IUCN), WWF (formerly the World Wildlife Fund) and the UN Environment Program (UNEP). Two things are worth highlighting in the present context. For one there is the notion's impeccable ecological parentage, exemplified in the IUCN's declared mission: "to influence, encourage and assist societies throughout the world to conserve the integrity and diversity of nature and to ensure that any use of natural resources is equitable and ecologically sustainable."[45]

And then there is the date itself: predating the Rio Earth Summit (1992) as well as the Brundtland Commission's report on *Our Common Future* (1987), it marked the end of a decade of intense public concern about industrial pollution. Almost by definition, the public concerned was that of the industrialized North. Significantly, the decade began with the Conference on Human Environment (UNCHE, Stockholm, 1972), the first UN forum concerned with global environment and development needs. Although UNCHE "indicated that 'industrialized' environmental problems, such as habitat degradation, toxicity and acid rain, were not necessarily relevant issues for all countries ... it was the pending environmental problems that dominated the meeting and led to wider public environmental awareness."[46] It thus seems safe to say that in the 1980s 'sustainable development' was about environmental or ecological sustainability. It was about living (consuming) "within ones ecological means." Or, to use a health metaphor, it was appropriate to obesity clinics, but not to famine relief.

Returning to the climate change concerns of the developed North, it stands to reason that their emergence in the late 1970s and early '80s at the height of popular concerns about industrial pollution of the local environment is responsible for the "ecological view" of the problem. A typical example of this (still prevailing) view is the most recent edition of *Social Trends,* a flagship survey of the UK Office for National Statistics.

Table 2.1: Environmental Concerns. England and Wales. 2001

	(Percentage of 'personally very worried')
Disposal of hazardous waste	66%
Effects of livestock methods	59%
Pollution in rivers and seas	55%
Pollution in bathing waters and on beaches	52%
Traffic exhaust fumes and urban smog	52%
Loss of plants and animals in the UK	50%
Ozone layer depletion	49%
Tropical forest destruction	48%
Climate change/global warming	46%

Source: Table 11.1 in National Statistics (2002), *Social Trends*, p.180.

Climate change is given some prominence, namely under the *Air and Atmospheric Pollution* (*sic!*) section of Chapter 11 on *The Environment*. Table 2.1 shows the degree of populations' "worriedness" about the issue, but more importantly in the present context, it clearly demonstrates with its juxtapositions what sort of problem climate change is perceived to be. Indeed, according to this official survey, 'climate change is recognised as one of the greatest threats to our environment".[47]

The most recent *Annual Report* of the US Council of Economic Advisers—just to give a non-Euro centric example—characterizes climate change as a "potential problem [which] spans both generations and countries, implicating simultaneously the environment, on the one hand, and the world's fundamental economic reliance on fossil fuels ... on the other".[48]

Climate change in the industrial world is thus mainly perceived as a problem of polluting the environment, of degrading ecosystems. As such, its essence is seen to be that of a wrongful act against "Nature." Accordingly, environmental effectiveness—the capacity to "make good" the human-inflicted harm on Nature—becomes a key criterion in assessments of climate change measures. The chief victim from this perspective is nature, mankind's role is primarily that of culprit. And while climate impacts on human welfare are regarded as potentially life-style-threatening, they are taken to be self-inflicted and hence largely "deserved". Environmental integrity ("to do justice to Nature") is the overriding moral objective.

To be sure, these views are by no means inappropriate,—in the *Northern* context. Industrialized countries still have to learn how to live sustainably, in the original environmental meaning of the term. And this lesson must include a drastic reduction of greenhouse gas emissions as the uppermost objective. Yet, this real need for emission mitigation in the industrial context should not blind one to the possibility that for others, the "climate change reality" may be fundamentally different.

Looking Ahead: A Human(ist) Impact Agenda?

Equity: A North-South Divide

While there has been some technological progress since the Minoan late bronze age –with a concomitant increase in adaptive capacity– the fact that a mere 1.5°C change may have been sufficient to precipitate the collapse of one of the most advanced civilizations of the time might give food for thought, given the range of 1.4 to 5.8°C projected for this century. The Summary for Policymakers of the IPCC's recent *Synthesis Report* reinforces such unease, not only about impacts but also about their distribution.

The reality of climate change for the South (Box: *Southern Realities*) is quite different from the one experienced in the North (see above). For many, if not most developing countries, the phenomenon of climate change, e.g. volcanic eruptions, floods and earthquakes, is *not* really a problem of sustainable development (in the technical sense of learning "to live within one's environmental means"); it is primarily a matter of natural disaster management. The only difference between climate change impacts and other natural disasters is the possibility of anthropogenic attribution, the issue of human causal responsibility. As such, the phenomenon, unlike, say, earthquakes, comes arguably within the remit of corrective interpersonal justice regarding damages and restitution.

Given its governmental approval,[49] it is significant that the *Synthesis Report* Summary for Policymakers does mention disproportionate impacts on developing countries (Box: *A Question of Equity*). However, it is equally telling of the Summary to stop short of referring to the problem in terms of "responsibilities", and instead focus on the fact that additional mitigation may reduce the severity of impacts (Box: *A Question of Responsibility?*). There can be no doubt that the need to adapt must be minimized –at the very least for those parties who are largely innocent—and that the effort required to do so must be carried by those who are, if not guilty, then at least largely causally responsible. And yet, as we are beyond the point of being able to prevent impacts altogether, one question can no longer be avoided: who is going to bear the burden of the residual, unavoided impacts?

Given the expected distribution of these impact burdens and its discrepancy with causal responsibilities, it should not be surprising that a recent study[50] found this to be the one key equity concern of developing country governments. In contrast to the perception in the North, climate change in the South has come to be seen primarily as a problem of harm to human beings, harm which is largely *other*-inflicted, and not life-*style*, but *life*-threatening in character.

What may be more of a surprise is the finding that in the Northern hemisphere—where discussions on equity have been spearheaded largely by nongovernment stakeholders (academic, NGO)– the main equity

The Third Assessment Report: Synthesis Summary for Policy-makers

Southern Realities

Recent regional changes in climate, particularly increases in temperature, have already affected hydrological systems and terrestrial and marine ecosystems in many parts of the world. ... Preliminary indications suggest that some social and economic systems have been affected by recent increases in floods and droughts, with increases in economic losses for catastrophic weather events.[Question 2]

Reductions of greenhouse gas emissions, even stabilization of their concentrations in the atmosphere at a low level, will neither altogether prevent climate change or sea-level rise nor altogether prevent their impacts.[Question 6]

When considered by region, adverse effects are projected to predominate for much of the world, particularly in the tropics and subtropics.[Question 3]

A Question of Equity

The impacts of climate change will fall disproportionately upon developing countries and the poor persons within all countries, and thereby exacerbate inequities in health status and access to adequate food, clean water, and other resources. Populations in developing countries are generally exposed to relatively high risks of adverse impacts from climate change. In addition, poverty and other factors create conditions of low adaptive capacity in most developing countries. [Question 3]

The impact of climate change is projected to have different effects within and between countries. The challenge of addressing climate change raises an important issue of equity.[Question 6]

A Question of Responsibility?

Mitigation and adaptation actions can, if appropriately designed, advance sustainable development and equity both within and across countries and between generations. Reducing the projected increase in climate extremes is expected to benefit all countries, particularly developing countries, which are considered to be more vulnerable to climate change than developed countries. Mitigating climate change would also lessen the risks to future generations from the actions of the present generation.[Question 6]

[The development of planned adaptation strategies to address risks and utilize opportunities can complement mitigation actions to lessen climate change impacts. However, adaptation would entail costs and cannot prevent all damages. The costs of adaptation can be lessened by mitigation actions that will reduce and slow the climate changes to which systems would otherwise be exposed.[Question 6]

Source: Intergovernmental Panel on Climate Change (IPCC) (2002), *Climate Change 2001: Synthesis Report -- Summary for Policy Makers*, Question 2, <http://www.ipcc.ch/pub/tar/syr/index.htm>.

problem is considered to be the issue of allocating emission mitigation targets. Moreover, this is often taken to be a problem mainly because it is seen to be a *sine qua non* for an expansion of the mitigation regime to developing countries, itself seen as necessary to guarantee the environmental integrity of the regime.

For the South, the issue of sharing their impact burdens equitably is much closer to home than injuries to coral reefs or other nonhuman life systems: it is an issue of interpersonal justice, an issue of human perpetrators and human victims. The Southern view has been succinctly summarized by Sokona, Najam and Huq:

The third assessment report of the Intergovernmental Panel on Climate Change has made it abundantly clear that even if the Kyoto Protocol is implemented in full, the impacts of global climate change will start being felt within the next few decades and that the most vulnerable communities and countries are those which are already the poorest and least able to adapt to these changes. The threat is especially pressing for the least developed countries and the small island developing countries, where any economic development they may be able to achieve in the next few decades is in real danger of literally being swept away due to human-induced climate change. In the past, climatic disasters such as floods, cyclones and droughts may have been attributable to nature alone; in future they will definitely have a component that is human induced. More importantly, it is also clear that the contribution of these countries to the climate change problem is minuscule. The result is that those who have been least responsible for creating the crisis are most at risk from its ravages.[51]

If the Northern protagonists are prepared to "do justice to Nature," then they should also be prepared to do the same for their fellow human beings in the South. In other words, the environmentalist agenda which has so far dominated the international climate change regime has to be complemented by a humanist[52] agenda, addressing the very real concerns of climate change impacts on human beings. What we need is not just a regime with environmental-, but also *human integrity*.

An Impacts and Adaptation Instrument?[53]

Even though the dominant Northern environmentalist agenda has left its mark at the very heart of the multilateral framework,[54] there are some articles of the Framework Convention which would seem to permit redressing the balance:

> FCCC Art.3 (Principles).2. The specific needs and special circumstances of developing country Parties, especially those that are particularly vulnerable to the adverse effects of climate change ... should be given full consideration.
>
> FCCC Article 4 (Commitments).4. The developed country Parties and other developed Parties included in Annex II shall also assist the developing country Parties that are particularly vulnerable to the adverse effects of climate change in meeting costs of adaptation to those adverse effects.[55]

But to be perfectly clear: this need to redress the balance regarding human impacts does not supplant the need for further emission reductions in the second commitment period and beyond! And while the bulk of these will have to remain in the industrialised world, the view that developing country emissions need to be addressed cannot be ignored.

However, there are other ways of addressing these emissions in the first decades of this century than a simplistic transferral of the Northern model by asking developing countries to take on quantified emission limitation and reduction commitments.

For one, as mentioned earlier, sufficiently strong Annex I commitments could have technology spill-over effects which could deal with the issue of DC emissions without the need for quantified constraints. Even if industrialised countries should feel worried about their capacity to generate such spill-overs, there are ways of introducing quantified developing country targets which do not impose disproportionate obligations on them. For example, the North could accept a quantity of 'Certified Emission Reduction Obligations' (CEROs) -to be undertaken in developing countries under the existing Clean Development Mechanism, one of the Kyoto flexibilities - as part of *Annex I commitments*. For the sake of economic efficiency and North-North equity, these CEROs could be tradable and grandfathered. Indeed, to avoid South-South inequities, a number of tradable 'CER permits' (CERPs) -permits to generate CERs— greater or equal to the total of CEROs for the commitment period could be distributed among developing countries on a per capita basis. While it is not certain whether such a scheme would be acceptable, the fact remains that there are ways of addressing developing country emissions without imposing obligations disproportionate with their responsibilities.

During the high-level segment at COP7 in, Marrakech, Thiru T.R. Baalu, India's Minister for Environment and Forests, left no doubt about his government's view on these matters:

> 'The efforts so far have been focussed on mitigation. In the coming decades, adaptation needs to be given much greater attention. The next decade, Mr. President, therefore should see concrete implementation of existing mitigation commitments and active consideration and action on adaptation to the adverse impacts of climate change.'

It thus did not come as a complete surprise that at COP5 in New Delhi, 'the issue of adaptation was given major significance by Indian Prime Minister Atal Behari Vajpayee in his inaugural address as well as in the final Delhi Declaration. As a result of all this attention in Delhi, COP8 was being referred to (unofficially) as the "Adaptation COP'".[56] Contrary to disappointments expressed by certain Northern and NGO stakeholders,[57] COP8 may indeed prove to be a turning point in the climate change negotiations towards more of a balance between the key concerns of North (mitigation) and South (impacts/adaptation) - although it would be somewhat premature to assume with Joke Waller-Hunter, executive secretary of the FCCC, that at COP8 'adaptation has been brought on a equal footing with mitigation.'[58]

Although COP8 did not result in a 'New Delhi Mandate' to negotiate an Impacts and Adaptation instrument under the FCCC - which some had previously argued for[59] — the Indian government did put forward the idea of a 'Protocol on Adaptation.'[60] And while this first attempt failed, it stands to reason that substantive progress on such an instrument would facilitate the Kyoto successor negotiations. Whether India will wish to pursue its idea and take such a lead, and whether the rest of the world would be willing to follow remains to be seen. The fact remains that the international climate change regime under the Framework Convention can only hope to achieve its objective if it addresses these humanist concerns by being as much about innocent humans as it is about healthy eco-systems.

Notes and References

[*] Originally published as 'The global climate change regime: Taking stock and looking ahead' in O. S. Stokke and O. B. Thommessen (eds.), 2002, Yearbook of International Co-operation on Environment and Development 2002/2003. London, Earthscan, pp. 27-38.

[1] The author is grateful to Brian Buck, Joanna Depledge, Sebastian Oberthür, and Olav Stokke for their help and critical review.

[2] See, e.g.: Fred Hoyle, 1981. Ice. Hutchinson, London.

[3] See, e.g.: Joanna Depledge 2002. The Third Assessment Report of the IPCC, Royal Affairs Briefing Paper. Royal Institute of International Affairs, London.

[4] See, e.g.: John Imbrie and Katherine Palmer Imbrie, 1997. Ice Ages, Solving the Mystery, (Harvard University Press, Boston, MA; or Richard B. Alley, 2000. The Two-Mile Time Machine: Ice Cores, Abrupt Climate Change and Our Future, Princeton University Press , Princeton, NJ.

[5] Sturt W. Manning, 1999. A Test of Time: The Volcano of Thera and the Chronology and History of the Aegean and East Mediterranean in the Mid Second Millennium BC. Oxbow Books, Oxford, UK. <www.rdg.ac.uk/~lasmanng/testoftime.html>.

[6] 1647 B.C: +0.65°C, 1559 B.C: –0.9°C, relative to present. J.R. Petit, J. Jouzel, D. Raynaud, N.I. Barkov, J.-M. Barnola, I. Basile, M. Bender, J. Chappellaz, M. Davis, G. Delayque, M. Delmotte, V.M. Kotlyakov, M. Legrand, V.Y. Lipenkov, C. Lorius, L. Pepin, C. Ritz, E. Saltzman, and M. Stievenard 1999. 'Climate and Atmospheric History of the Past 420,000 Years from the Vostok Ice Core, Antarctica', Nature, 3 (99): 429-436. Data Source: Historical Isotopic Temperature Record from the Vostok Ice Core. <http://cdiac.esd.ornl.gov/ftp/ trends/temp/ vostok/vostok.1999.temp.dat>.

[7] "... the eruption on Thera could have lowered annual average temperatures by 1 to 2 degrees across Europe, Asia and North America. ... the summer temperatures would have dropped more - suggesting years of cold, wet summers and ruined harvests". Jessica Cecil 2001. 'Ancient Apocalypse: The Fall of the Minoan Civilisation. <http://www.bbc.co.uk/history/ancient/apocalypse_minoan1.shtml>. For more details on the eruption see Chapter 5 of Floyd W. McCoy and Grant Heiken 2000. 'Volcanic Hazards and Disasters in Human Antiquity', Special Paper 345 Geol. Soc. of Amer., Boulder, CO.

[8] Michael R. Rampino and Stanley H. Ambrose 2000, 'Volcanic Winter in the Garden of Eden: The Toba Supereruption and the Late Pleistocene Human Population Crash'. (In: McCoy and Heiken. 2000. 'Volcanic Hazards and Disasters in Human Antiquity' pp. 71-82.

[9] G. Marland, T. A. Boden, and R. J. Andres 2002, 'Global, Regional, and National Fossil Fuel CO_2 Emissions. Carbon Dioxide Information Analysis Center, Oak Ridge National Laboratory, Oak Ridge, TN. <http://cdiac.esd. ornl.gov/trends/emis/em cont.htm >.

[10] World Resources Institute (WRI). Contributions to Global Warming Map. <http://www.wri.org/ climate/contributions map.html>.

[11] However, one has to be cautious in interpreting such figures. If, for example, like me, one is of the opinion that these responsibilities need to be compared in terms of average yearly per capita emissions, the Northern responsibility increases 15-fold over that of the South.

[12] For more details on the institutional structure and procedural rules see, for example: Joanna Depledge 2002. A Guide to the Climate Change Process. Climate Change Secretariat, Bonn, Germany.

[13] Baron Jean-Baptiste Joseph Fourier (1768-1830), French mathematician, physicist, and Egyptologist.

[14] Svante August Arrhenius (1859-1927), Swedish physical chemist.

[15] Michael Grubb with Christiaan Vrolijk and Duncan Brack 1999, The Kyoto Protocol: A Guide and Assessment. Royal Inst. of Int. Affairs, London, p. 4.

[16] Intergovernmental Panel on Climate Change. About IPCC', <http://www.ipcc.ch/about/about.htm>.

[17] "Adverse effects of climate change" means changes in the physical environment or biota resulting from climate changes which have significant deleterious effects on the composition, resilience or productivity of natural and managed ecosystems or on the operation of socioeconomic systems or on human health and welfare [FCCC, Definitions].

[18] United Nations Framework Convention on Climate Change (UNFCCC). 1995. The Berlin Mandate: Decision 1/CP.1 <unfccc.int/resource/docs/cop1/07a01.htm>.

[19] Michael Grubb. 1999. The Kyoto Protocol, p. 111.

[20] Indeed, according to Art.9 a review of the Kyoto Protocol 'in the light of the best available scientific information and assessments on climate change and its impacts, as well as relevant technical, social and economic information' might have to be undertaken as early as 2003, assuming entry into force by WSSD.

[21] Suraje Dessai 2001. The climate regime from the Hague to Marrakech: Saving or sinking the Kyoto Protocol. Working Paper 12. Tyndall, Norwich UK. <http://www.tyndall.ac.uk/publications/working papers/ working papers.shtml>. Thomas Legge and Christian Egenhofer. 2001. After Marrakech: the regionalisation of the Kyoto Protocol, CEPS Commentary, CEPS, Brussels. <http://www.ceps.be/Commentary/Nov01/ Marrakech.htm>. Asbjørn Torvanger, 2001. An evaluation of business implications of the Kyoto Protocol. Report 2001-05. CICERO, Oslo. <www.cicero.uio.no/publications/detail.asp?1690>. O. Blanchard, P. Criqui, and A. Kitous 2002. 'After the Hague, Bonn and Marrakech : the future international market for emissions permits and the issue of hot air', (IEPE), Grenoble <http://www.upmf-grenoble.fr/iepe/textes/ Cahier27Angl.pdf>. M.G.J. den Elzen and A.P.G. de Moor. 2002. 'The Bonn Agreement and Marrakesh Accords: An updated analysis. RIVM report 728001017/2001. RIVM, Bilthoven NL <http://www.rivm.nl/ieweb/ieweb/Reports/rep728001017 marrakech.pdf>. Donald Goldberg and Katherine Silverthorne. 2002. 'The Marrakech Accords. ABA Newsletter 5(2) January. <http://www.abanet.org/ environ/committees/climatechange/newsletter/jan02/goldberg.html>. Andreas Löschel and ZhongXiang Zhang 2002. 'The Economic and Environmental Implications of the US Repudiation of the Kyoto

Protocol and the Subsequent Deals in Bonn and Marrakech. <http://papers.ssrn.com/ abstract=299463>. Christiaan Vrolijk 2001. 'The Marrakesh Accords'. RIIA Meeting Report, RIIA, London. <http://www.riia.org/Research/eep/ cop7meeting.pdf>. TERI. 2001. Review of COP-7. <http://www.teriin.org/climate/cop7.htm>. Christiaan Vrolijk. 2002. A New Interpretation of Kyoto: The Hague, Bonn and Marrakesh. RIIA Briefing Paper. RIIA, London (in press). Christiaan Vrolijk, 2002. The Marrakesh Accords: A brief point by point descriptions and comments. Annex to Ch. Vrolijk, 2000. A New Interpretation of Kyoto, (in press).

[22] Emphasis added.

[23] <http://www.whitehouse.gov/news/releases/2001/03/20010314.html>.

[24] See, e.g. Sebastian Oberthür and Hermann E. Ott, with Richard G. Tarasofsky. 1999, The Kyoto Protocol. International Climate Policy for the 21st Century, Springer, Berlin, p. 273.

[25] 1990:1,355 MtC; 1999: 1,520 MtC.

[26] FCCC, Art. 4.2 (a).

[27] Measured in local currency units, Source: International Monetary Fund (IMF). 2001. World Economic Outlook 2001, Real Gross Domestic Product, local currency (LCU bn). <http://www.imf.org/external/ pubs/ft/weo/2001/01/data/index.htm>.

[28] MtC = Million ("Mega") ton of Carbon. 1 unit C = 3.67 units CO_2. Data Source: Energy Information Administration (EIA). World Carbon Dioxide Emissions from the Consumption and Flaring of Fossil Fuels, 1980-1999. < http://www.eia.doe.gov/emeu/international/environm.html#IntlCarbon >

[29] 1997 US $. Source: Energy Information Administration (EIA) 2001. International Energy Outlook 2001. <http://www.eia.doe.gov/oiaf/ieo/appendixes.html#appen>. Table A3. World Gross Domestic Product (GDP) by Region, Reference Case, 1990-2020 (Billion 1997 Dollars).

[30] See, e.g. Chapter 16 in Ulrich Bartsch and Benito Müller. 2000. Fossil Fuels in a Changing Climate: Impacts of the Kyoto Protocol and Developing Country Participation. Oxford University Press, Oxford, UK.

[31] IMF 2001. World Economic Outlook 2001. Real Gross Domestic Product, Constant Prices (billions of local currency units).

[32] International Energy Agency. 2000. World Energy Outlook.

[33] Benito Müller 2000. 'Congressional Climate Change Hearings: Comedy or Tragedy?' <http://www.wolfson. ox.ac.uk/~mueller>.

[34] David G. Streets, Kejun Jiang, Xiulian Hu, Jonathan E. Sinton, Xiao-Quan Zhang, Deying Xu, Mark Z. Jacobson, James E. Hansen, 2001. 'Recent Reductions in China's Greenhouse Gas Emissions. Science 294: 1835-7.

[35] Michael Grubb, Chris Hope, and Roger Fouquet. 2002. 'The Climatic Implications of the Kyoto Protocol: The Contribution of International Spillover. Climatic Change (in press).

[36] Allocation in proportion to baseline (benchmark) emission figures.

[37] Allocation in proportion to baseline (benchmark) population figures.

[38] See, e.g.; A. Meyer, 2000. Contraction & Convergence: The Global Solution to Climate Change. Green Books Ltd., Dartington, UK; or A. Agarwal and S. Narain, 2000. 'Addressing the Challenge of Climate Change: Equity, Sustainability and Economic Effectiveness. In: M. Munasinghe and R. Swart (eds.) 2000. Climate Change and Its Linkage with Development, Equity, and Sustainability, IPCC, Geneva.

[39] Benito Müller, 1999. Justice in Global Warming Negotiations: How to Obtain a

Procedurally Fair Compromise. Oxford Institute for Energy Studies, Oxford, UK.

[40] See, e.g.: Kevin A. Baumert, Ruchi Bhandari, and Nancy Kete, 1999. 'What Might a Developing Country Climate Commitment Look Like?. Climate Notes. World Resources Institute, Washington, DC.

[41] See, e.g., William Pizer, 1999. 'Choosing Price or Quantity Controls for Greenhouse Gases', Climate Issues Brief 17 Resources for the Future, Washington, DC.

[42] See, e.g., Intergovernmental Panel on Climate Change (IPCC) 2001. [TAR3] Climate Change 2001: Mitigation, Cambridge University Press, Cambridge. M. Munasinghe and R. Swart eds. 2000. Climate Change and Its Linkage with Development, Equity, and Sustainability, Benito Müller, Axel Michaelowa, and Christiaan Vrolijk 2001. Rejecting Kyoto: A Study of Proposed Alternatives to the Kyoto Protocol (Climate Strategies, London. <http://www.climate-strategies.org/rejectingkyoto2.pdf>. Sijm, Jos, Jaap Jansen and Asbjørn Torvanger, 2001. Differentiation of mitigation commitments: the Multi-Sector Convergence approach, Climate Policy 1(4) 481-497; Lasse, Ringius, Asbjørn Torvanger and Bjart Holtsmark 1998. 'Can multi-criteria rules fairly distribute climate burdens? OECD results from three burden sharing rules. Energy Policy 26(10): 777-793; or Cédric Philibert and Jonathan Pershing 2001. Considering the Options: Climate Targets for All Countries. Climate Policy, 1: 211-227.

[43] Thomas C. Schelling. 1997. 'The Cost of Combating Global Warming,' Foreign Affairs, November/ December, 1997.

[44] Ashok Khosla. 2001. 'The Road from Rio to Johannesburg', Millennium Papers, 4 UNED Forum, London.

[45] World Conservation Union (IUCN) 2000, 'About IUCN', <http://www.iucn.org/2000/about/content/ index.html>.

[46] Rosalie Gardiner. 2001. 'Earth Summit 2002 Explained', Earth Summit 2002 Briefing Paper. UNED Forum, London, 1. <http://www.earthsummit2002.org/Es2002.PDF>.

[47] Ibid. 183. To be sure, not all the concerns listed in Table 1 are just about natural eco-systems, and it would be simplistic to expect them to be, given the complex interactions involved. Nevertheless, the issues listed are overall significantly closer to an ecological paradigm than to a natural disaster one.

[48] Council of Economic Advisers 2002. Annual Report. U.S. Government Printing Office, Washington, DC, p.244. <http://w3.access.gpo.gov/eop/index.html>.

[49] 'This summary, approved in detail at IPCC Plenary XVIII (Wembley, UK, September, 24-29, 2001), represents the formally agreed statement of the IPCC concerning key findings and uncertainties contained in the Working Group contributions to the Third Assessment Report. Ibid.

[50] Benito Müller (in press, August 2002). Equity in Climate Change: The Great Divide Oxford Institute for Energy Studies, Oxford.

[51] Youba Sokona, Adil Najam, and Saleemul Huq 2002, 'Climate Change and Sustainable Development: Views from the South. Int. Instit. for Environment and Development, London, p. 2.

[52] Apart from designating a philosophical and literary movement which originated in northern Italy in the second half of the fourteenth century, "humanism" is also used to designate "any philosophy which recognizes the value or dignity of man". The Encyclopedia of Philosophy, 4. Paul Edwards (ed.) Collier MacMillan Publ., London, 69f.

[53] For more on the issues discussed in this section see: *Benito Muller, Framing Future Commitments: A Pilot Study on the Evolution of the UNFCCC Greenhouse Gas Miti-*

gation Regime: Oxford OIES, 2003 Benito Muller 'An FCCC Impact Response Instrument as part of a Balanced Global Climate Change Regime'; June 2002 www.OxfordClimatePolicv.org): Benito Muller Responding to climate change impacts, 27 June 2003 http://www.scidev.net/dossiers/

[54] For example, while both human and ecological impacts are mentioned in the central passages from the Convention mentioned earlier, there seems to be an (unconscious) ranking putting the latter before the former.

[55] Article 4.4 interestingly, is not a commitment on Annex I (large emitters), but on Annex II (affluent) Parties, which arguably means that it is based on the ability to pay ('solidarity') principle, thus not dependent on the ability to separate anthropogenic from natural impacts.

[56] Saleemul Huq, "The Adaptation COP', Dhaka: *The Daily Star,* 15 Nov. 2002, www.dailystamews. com/200211/15/n2111509.htm#BODY377

[57] 'While the high level ministerial delegation at the Eighth Conference of the Parties (COP-8) of the United Nations Framework Convention on Climate Change (UNFCCC) unanimously endorsed the Delhi Declaration, green activists underscored that the environment is the loser. Even the European Union (EU) expressed disappointment with the end results, saying that the text lacks action and a vision for the future. The EU acknowledged the forward looking technical results, though. [...]

Not everybody is buying everything, though. The green lobby is the most upset of the lot. Said Jennifer Morgan, director of climate change programme at the World Wildlife Fund (WWF), "We are quite disappointed. The document lacks urgency to grapple with long term challenges of climate change. It's not a step backwards, but there is nothing forward looking either." Added her colleague Liam Salter, a climate change officer at WWF, "The Delhi Declaration is pretty weak. Most of it is recycled text. It does not move the process forward. There is no significant improvement. There has been progress on technical issues, though." There seems to be a unanimous consensus amongst green groups. Steve Sawyer, climate policy director at Greenpeace, also added, "The document does not amount to much. They did manage to get the language right to some extent by getting in, Kyoto Protocol, but it's not a forward looking document. It's missed opportunity." The fossil fuel lobby gets most of the flak for it. Kate Hampton, international coordinator of climate change campaign at Friends of the Earth, said:

"It's frightening that fossil fuel interests have hijacked the conference. They are the most powerful vested interest group in the world." Of course, one can blame it on the pressure of getting results with a consensus as well. Said K S Gurmit Singh, executive director of a Malaysian NGO called CETDEM, "There is fatigue. The political pressures have gone off. We just had the World Summit on Sustainable Development. Reaching a consensus is not easy. "[Rajiv Tikoo, 'Climate Meet Boils Down To Lukewarm Response,' The Financial Express, 3 Nov. 2002, www.financialexpress.comi

[58] Reported in: Rajiv Tikoo, 'Climate Meet Boils Down To Lukewarm Response,' Bombay: The Financial Express, 3 Nov. 2002, www.fmancialexpress.com

[59] Benito Muller 'A New Delhi Mandate?,' Climate Policy 2,2002:241-2.

[60] '2) To initiate further action necessary for global, regional and sub-national assessment of adverse effects and steps to facilitate implementation of adaptation measures. Such action should include the adoption of a Protocol on Adaptation; [Draft of the Delhi Declaration tabled within G-77 and China at COPS on 26 October 2003]

CHAPTER 3

Kyoto Protocol: Insignificant Impact on Global Emissions

Cathrine Hagem[a] and Bjart Holtsmark[b],
[a]University of Oslo, Department of Economics, P.O. Box 1095 Blindern,
0317 Oslo, Norway; [b] Statistics Norway, P.O. Box 8131 Dep, 0033 Oslo, Norway.

Introduction

Some likely impacts of the Kyoto Protocol related to global greenhouse gas emissions, fossil fuel markets and the emerging market for emission permits are discussed. The Kyoto Protocol, negotiated in 1997, requires all industrialized countries to limit their emissions of a basket of six greenhouse gases (or groups of gases) for the period 2008–2012[1]. In all, the agreement requires that the average annual emissions of industrialized countries in the period 2008–2012 not exceed 95% of 1990 emissions.

All industrialized countries, including the United States, have signed the agreement. However, the agreement will not enter into force until it has been ratified by at least 55 countries, and these ratifying countries must have contributed at least 55% of the industrialized world's CO_2 emissions in 1990.[2]

The Kyoto Protocol left undecided a number of rules required to make it operational. After a series of negotiation rounds the seventh Conference of the Parties to the Climate Convention (COP-7) took place in Marrakech in October/November 2001, which reached agreement on rules for implementation of the Protocol. However, President George W. Bush has made it clear that he does not intend to send the Kyoto Protocol to the Senate for ratification. And irrespective of Bush's position, the Protocol is unlikely to have garnered the necessary 2/3 majority in the Senate to achieve ratification. Hence, it is unlikely that the United States, despite the signature from the previous administration, wishes to join the effort to bring the Protocol into force.

President Bush's decision has ignited a discussion among the remaining industrialized countries about whether or not to implement the Protocol in the absence of the United States. Because the United States was responsible for "only" 36% of the industrialized countries' CO_2 emissions in 1990, the Protocol can—in principle—enter into force without ratification by the United States. This would, however, require virtually all of the other major industrialized countries – including the Russian Federation, which was responsible for 17% of the industrialized countries' CO_2 emissions—to ratify the Protocol.

To evaluate the impacts of implementing the Kyoto Protocol, we have taken our point of departure from a numerical static partial equilibrium model that integrates energy markets with an international market for emissions trading. The model is used to compare the impacts of the Kyoto Protocol both with and without the participation of the United States. The main results from the model calculations are :

- If the Kyoto Protocol is implemented without the participation of the United States,
  — then the effect on global emissions will be reduced from small to insignificant,
  — and the international permit price will be reduced to one-third what it would have been if the United States had upheld its commitments.
- The outcome of the Bonn and Marrakech negations regarding credits from forest activities weakened the agreement even further.

Description of the numerical model and dataset

We apply a static partial equilibrium model that emphasizes the links between the fossil fuel markets and a market for emission permits under the Kyoto Protocol.[3] The model divides the world into 32 countries and regions. In each country or region, a numeraire good is produced using four inputs: oil, coal, gas, and non-CO_2 climate gases. The three fossil fuels are modeled as substitutes, while the marginal product of non-CO_2 gases is independent of the use of fossil fuels. The assumed production technology yields a linear demand function for all inputs.

There are five markets for fossil fuels: one global oil market, one global coal market, and due to high transport costs, three regional gas markets (North America, Asia, and Europe including the Russian Federation). Furthermore, there is an international market for emission permits across all the industrialized countries with emission caps under the Kyoto Protocol. The coal market and the gas market in North America and Asia are assumed to be competitive, while strategic behavior is built into the models of the European gas market, the oil market and the market, for emission permits. In the oil market it is assumed that OPEC behaves strategically and restricts its oil supply in order to increase the oil price. In the European gas market the Russian Federation is assumed to be the price setter. Concerning the market for emission permits this market is as a starting point assumed to be competitive. Later on analyses are done of Russian behavior to exhibit its market power here.

The model determines equilibrium prices in the fuel markets and the market for emission permits as well as the various country and regional export and import of fossil fuels and emission permits.

The model is calibrated to a scenario of the world economy and world energy markets in the year 2010. The "business as usual" (BaU) scenario was constructed by taking the *Conventional Wisdom* (CW) scenario developed by the European Commission (1996) as the baseline. The total growth in emissions from 1990 to 2010 in the USA, in the other industrialized countries, and in the rest of the world is set at 24, 8, and 50%, respectively.

The abatement cost follows implicitly from formulation of the demand functions and the elasticity of demand for each fuel in each country. There is no consensus in the literature about demand elasticities in fossil fuel markets. Estimates range from –0.15 (Smith *et al.*, 1995) to less than –1.0 (see e.g. Golombek and Bråten, 1994 and Golombek, Hagem and Hoel, 1995). For lack of decisive evidence, we have chosen a middle road by assuming average demand elasticities of –0.5 for all fossil fuels. Demand elasticities for oil and coal have, however, been differentiated across countries in order to reflect the different structures of fuel demand in them. Using detailed information from the IEA (1995), the consumption of oil and coal in each country was divided into two parts, one inelastic and one presumably more elastic.

Projected producer prices in 2010 were taken directly from the European Commission study (op.cit), except in the case of the gas market, where the study reports only one gas price. We have taken the gas price to be the European gas price, while the other gas prices were calculated under the assumption that relative gas prices between the three markets would be as projected by the IEA (1998).

Consumer prices in the BaU scenario were obtained by adding existing fiscal taxes to the producer prices.

As for fuel supply, it is generally recognized that the supply of coal is more elastic than the supply of other fuels. We have followed Golombek and Bråten, (1994), by assuming supply elasticities of 2.0 for coal producers and 0.75 for both gas producers and competitive oil producers.

The model does not incorporate the possibility for carbon sequestration. The implication of this limitation of the model is discussed below.

Calculating impacts of the Kyoto Protocol: different scenarios

After the U.S. withdrawal from the Protocol negotiations continued and the Marrakech Accords specify how the Protocol will be implemented if it enters into force. The rules of the Kyoto Protocol regarding allowable amounts of crediting from forest activities and rules for international emissions trading have important impacts on the likely achieved global emission reductions, the distribution of emission reductions across countries, and the cost of the Kyoto Protocol.

The Marrakech Accords defines four different types of emission permits. First, we have Assigned Amount Units (AAUs) which each gives the right to emit one ton of CO_2. Each industrialized country is eligible to issue a number of AAUs equal to its Assigned Amount specified in Annex B in the Kyoto Protocol. The parties are free to transfer or import as many AAUs as preferred, so long as this complies with a number of requirements. AAUs could also be banked to a future commitment period.

Second, Emission Reduction Units (ERUs), which are issued after application of the Joint Implementation mechanism. The host country, an industrialized country, should issue and transfer to the investor a number of ERUs according to the attained emission reductions while a corresponding number for AAUs in the host country are cancelled. In this study Joint Implementation and ERUs are not discussed further. The reason is that in a simplified model framework, ordinary trade with AAUs makes JI and ERU-trade redundant.

Third, Certified Emission Reductions (CERs), which are accrued credits from emission reduction projects carried out in developing countries through the Clean Development Mechanism (CDM).

Fourth, Removal Units (RMUs), which are issued in relation to carbon sequestration activities (landuse, land-use change, and forestry (LULUCF)). There are no constraints on generating RMU from afforestation, reforestation, and deforestation activities. Hence, in order to minimize the cost of the agreement, the Parties would take advantage of these carbon-sequestration possibilities and carry out activities for which the costs are less than reducing their emissions of greenhouse gases. Since our model does not include the possibilities for carbon sequestration, we shall to some extent overestimate the cost of the Protocol and with that respect the permit price.

Although there are no constraints on generating RMU from afforestation, reforestation, and deforestation activities, a ceiling on eligible forest management activities is specified for each industrialized country. In total the industrialized countries are allowed to issue 256 million RMUs from forest management activities, of which the Russian Federation and Canada may issue 121 and 44 million, respectively.

It is commonly argued that carbon sequestration through forest management, within the limits of the specified ceilings, would occur even without the Kyoto Protocol. If this is the case, RMUs generated from forest management is a so-called "no-regret" option that leads to no real net emission reductions compared to the "Business-as-Usual" scenario.

The parties eligible to take part in the mechanisms are free to transfer or acquire the different emission permits (AAUs, RMUs, CRUs, and CERs). The cost of the Kyoto Protocol and the trade in permits are highly dependent on how the market for the different types of emission permits function. Furthermore, whether the RMU generated from forest management can be perceived as a no-regret option or not, is decisive for the environmental impact of the agreement.

In order to study these effects we computed the impact of the Kyoto Protocol under different assumptions regarding: (i) the potential for acquiring CERs after investments in emission reduction projects in developing countries; (ii) whether Russia exerts market power in the permit market; and (iii) whether RMUs generated from forest management can be perceived as a no-regret option or not. Furthermore, to evaluate the impact of the United States withdrawal from the Protocol, we did a comparison of the likely effects of the Kyoto Protocol with and without participation from the US.

Acquiring emission permits through the Clean Development Mechanism involves higher transaction costs than buying permits on the international permit market. There are three main reasons for this. First, there is a levy on CDM. Two percent of the certified emission reductions issued for a CDM project activity go toward financing the Kyoto Protocol adaptation fund. Second, another fee will be introduced to cover administrative expenses of the CDM. And last but not least, investment in specific projects involves search costs in order to find suitable projects and administration cost for the investor and/or the host during the project period. These transaction costs increase the costs of CDM projects and reduce the number of CERs on the permit market. Especially, if the international permit price is very low, the transaction cost of acquiring CERs may, for most of the potential emission reduction projects, exceed the international permit price. In that case, emission reduction through the CDM would only be implemented to a very limited extent. Due to the transaction cost, probably only implementation of quite large emission reduction projects would be profitable. Hence, CDM can only be used to take advantage of a certain part of the low-cost abatement opportunities in the developing world. It follows from this discussion that it is difficult to predict the use of the CDM mechanism without specific knowledge of the transaction costs for all potential CDM projects. In our numerical model we have not incorporated transaction costs of emission credits acquired through CDM. To get some idea of how CDM affects the outcome of the Kyoto Protocol we considered the two extreme outcomes: (a) the transaction costs of CDM projects are so high that fulfilement of the emission constraint in the Kyoto Protocol does not involve emission reductions from the developing countries through CDM, and (b) transaction costs are negligible for all abatement options.

In our numerical model, the outcome (a) corresponds to a situation wherein the industrialized countries are not allowed to use the CDM as a means to meet their emission reduction obligation, while (b) corresponds to a situation where the developing countries have Assigned Amounts (AAUs) equal to their BaU-emissions and are allowed to trade these on the international market. Situation (a) overestimates the cost of the Kyoto Protocol, while (b) underestimates the cost.

It has often been argued that CDM to some extent will be a loophole which, due to control problems, might place a considerable supply of low-cost CERs on the market that are not really based on emission reductions. In that case, the CDM mechanism leads to less global emission reductions. (See *inter alia* Bohm, 1994, and Hagem, 1996 for a discussion of the possible adverse impact of CDM on global emissions). Here we ignore this possibility and assume that all CERs sold on the market are met by a corresponding emission reduction.

Table 3.1. summarizes the characteristics of the different numerical simulations (scenarios) we carried out in addition to the Business as Usual scenario, described above.

Table 3.1. Characteristics of the different scenarios

Scenario	U.S. participation	Are RMUs from forest management no-regret options?	Does Russia exert market power in the permit market?	Is CDM included?
1	✔			
2				
3		✔		
4		✔	✔	
5		✔		✔
6		✔	✔	✔

Scenario 1 is a scenario as though the US were still a party to the Protocol. It does not take into account CRU-trading or the possibility for acquiring RMUs from no-regret options. The Russian Federation exerts no market power in the permit market in this scenario.

In scenario 2 the U.S. has withdrawn from the Protocol, but in all other respects scenario 2 is similar to scenario 1.

In scenario 3 the mentioned possibilities for issuing of RMUs from no-regret options is included. Otherwise this scenario is similar to scenario 2.

Scenario 4 extends scenario 3 by letting the Russian Federation exert its market power in the permit market.

Scenario 5 extends scenario 3 by including the Clean Development Mechanism and CER-trading into the model. Scenario 5 does not assume that Russia exercises market power in the permit market. Scenario 6 therefore extends scenario 5 by letting the Russian Federation take advantage of its market power in this market.

Consequences of US withdrawal

To evaluate the impact of the US withdrawal we compare scenarios 1 and 2. Hence, no-regret options or carbon sequestration activities for acquiring RMUs are not included in this simulation, neither is the CDM mechanism.

As mentioned in the introduction, the Kyoto Protocol requires industrialized countries to reduce emissions in the period 2008–2012 by an average of 5% in relation to 1990 levels. Given the figures used in this article, this would require industrialized countries to reduce their emissions by an average of 12.8% from current BaU emissions.[4]

The global emissions reduction resulting from the Protocol would, however, be significantly less. This is because the Kyoto Protocol places no restrictions on emissions from developing countries. Emissions from developing countries constituted 45% of the global emissions in 1990, and these emissions are expected to increase by 50% up to the period 2008–2012 if the Kyoto Protocol does not enter into force. Furthermore, because the Kyoto Protocol will result in a drop in the market price (producer price) of fossil energy, developing countries are expected to increase their emissions even more should the Protocol enter into force (carbon leakage). The model calculations estimate that all in all the global emissions reduction during the first commitment period would be about 5.5% (compared to BaU-emissions), if the United States implements the agreement.

The model calculations predict that the Kyoto Protocol without the United States would reduce the global emissions by only 0.9% relative to BaU. The reason why the global effect of the climate agreement would become so small if the United States drops out is related to three conditions. First, the US emissions of greenhouse gases made up a large portion of the industrialized countries' emissions in 1990 (32%). Second, in accordance with the Kyoto Protocol, the United States must reduce its emissions by 7%, which is higher than the average of 5% for the industrialized countries as a whole. Third, the expected increase in BaU emissions in the United States is higher than average for the industrialized countries. If the United States were to uphold its Kyoto commitments, emissions would be 27% lower than in the BaU scenario. The comparable figure for the other industrialized countries is (on average) 3.4%.

Thus, implementing the Kyoto Protocol without the participation of the United States means that the global impact becomes minimal, not only because the United States was responsible for a large share of the industrialized countries' total emissions in 1990, but also because the Kyoto Protocol set particularly stringent abatement requirements for the United States in the period 2008–2012 compared to BaU emissions. The left diagram in Figure 3.1 illustrates the distribution of emission reductions and the flows in the quota market given the participation of the United States. With respect to emission trading, we see that the United States is the main buyer while the Russian Federation and other Eastern European countries are the main sellers. The United States also represents a large portion of the emission reductions, nearly 1,100 million tons[5] of CO_2 equivalents compared to a total global reduction of just below 2,300 Mt CO_2 equivalents in this scenario.

The estimated emission increases that take place in developing countries is so-called carbon leakage related to the drop in the price of fossil fuels, mainly coal. Due to high CO_2 content, the price drop on coal is, on average, 7.1% in this scenario. The producer price of oil in contrast drops only 1.7%. The gas price in the Pacific region is only slightly affected. In total the carbon leakage is 190 Mt CO_2 equivalents in this scenario.

The emission reductions in the Russian Federation and the other Eastern European countries are less than the permit export from these countries. This is a case of selling permits without undertaking corresponding domestic reductions and is related to the generous quota allotments for these countries, which allows them to sell hot air.

In scenario 1 the permit price is calculated to be US$ 16 per ton CO_2.[6]

The left diagram in Figure 3.1 illustrates how important the United States is for the Kyoto Protocol. Without the participation of the United States, a large portion of the emission reductions is eliminated, at the same time as the demand for quotas on the international market drops dramatically.

The right diagram in Fgure 3.1 illustrates the results of the US withdrawal. Global abatement is reduced from 2,300 Mt to less than 400 Mt. Even though the participation of the United States would have meant a US abatement of just over 1,000 Mt, the withdrawal of the United States means that global emissions will be almost 2,000 Mt higher. This is derived clearly from the fact that the quota allocated to the United States is, in this particular scenario, almost 2,000 Mt less than the country's BaU emissions. The permits that the United States would have imported now become available to other countries.

The withdrawal of the United States results in the permit price falling from US$ 16 per ton CO_2 equivalent to only US$ 4.70. The considerable price drop corresponds well with the picture we have already painted of how important the U.S. emissions cap is to the Kyoto Protocol.

Emission trading in this scenario is mainly based on hot air. Only the Russian Federation, Ukraine, and the other Eastern European countries are sellers in this scenario and export in total 871 mill. AAUs. Abatement, on the other hand, carried out by the permit exporters in this scenario constitutes all together only 209 Mt CO_2 equivalents, far less than the extent a permit export should indicate.

Naturally, compared with scenario 1 carbon leakage is less important in scenario 2 due to smaller price drops in the fossil fuel markets. The price drops in the coal and oil markets are 1.0 and 0.4 in this scenario leading to a total leakage of 45 Mt CO_2 equivalents. One-third of this leakage is to the U.S. where consumers now could take advantage of slightly lower fossil fuel prices.

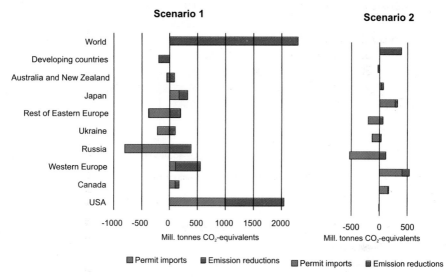

Fig. 3.1: Simulated emission reductions and flows in the quota market in scenario 1 and scenario 2. (In scenario 2 the U.S. has withdrawn from the Protocol.) Black bars on the right-hand side represent permit import, while black bars on the left-hand side represent permit export.

Inclusion of forest management and Russian market power

In the previous section the consequences of US withdrawal from the Protocol was analyzed. In this section we continue to assume that the US has withdrawn from the Protocol, but in addition analyze the consequences of the rules for getting emission credits from forest management activities, which we now assume would have been carried out even in the absence of the Kyoto Protocol (no-regret options). These credits will considerably increase the supply of permits and consequently lead to a price drop in the permit market. The question is then to what extent the Russian Federation, as a large permit seller, could increase the permit price by exercising its market power in the permit market. Before this question is analyzed the inclusion of credits from forest management is studied.

As already mentioned, a Party included in Annex B may have additions to its assigned amount as a result of forest management under Article 3.4 in the Kyoto Protocol. However, these additions must not exceed certain values. In total, the additions may not exceed 256 M, of which the Russian Federation could add 121 M, Canada 44 M, Japan 48 M. The other countries have only minor maximum values for forest management additions.

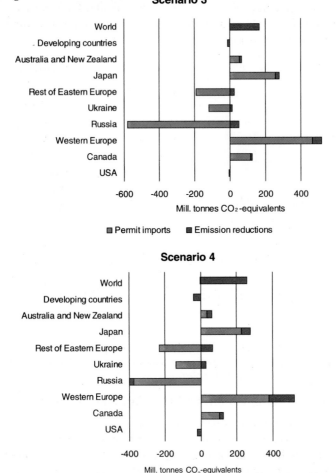

Fig. 3.2: Simulated emission reductions and flows in the permit market in scenarios 3 and 4. The scenario includes extended assigned amounts from forest management. Black bars on the right-hand side represent permit import, while the black bars on the left-hand side represent permit export.

In scenario 2 total emission reductions in countries with caps was 439 Mt CO_2 eqv. This number could serve as an indicator of the real demand for AAUs in this scenario. A total extension of the assigned amounts of 256 mill. CO_2 eqv. is therefore a significant increase in the supply of permits. When these assigned amounts are made available on the market, the permit drops from US $ 4.70 to 2.00 per ton CO_2. If forest management projects have also been carried out in a BaU scenario, the abatements in countries with caps are reduced by 256 Mt CO_2 eqv. Total abatement is then 183 Mt CO_2 eqv., while global emission reductions are slightly smaller (164 Mt) due to some leakage in both the US and the developing countries.

So far we have assumed that the Russian Federation does not exercise its market power in the permit market. In scenario 3, however, the Russian Federation has a market share at 65% in the permit market. Hence, it is likely that the Russian government will regulate its permit export in order to increase the permit price in the first commitment period, not least because AAUs may be banked to a later commitment period.

The right diagram in Figure 3.2 illustrates the flows in the permit market and emission reductions/leakages if the Russian Federation maximizes its net income from permit export. Such a policy would mean a reduction in Russian permit export from 582 to 375 M AAUs. This would increase the permit price from 2.0 to 5.1. In other words, in this situation the Russian Government has a considerable market power that is likely to be exercised.

Inclusion of the Clean Development Mechanism

In the previous two sections CDM was ignored. This means that we have overestimated the cost of implementing the Kyoto Protocol, and hence also overestimated the permit price, because none of the low-cost abatement opportunities in the developing countries have been utilized.

In this study we have modeled CDM as though the developing countries have assigned amounts equal to their BaU-emissions. This would only give a correct picture of the impact of CDM if there were no transaction costs connected to the acquirement of CERs from CDM projects. As argued above, the cost of acquiring CERs is larger than the pure costs of abatement. Hence, in this scenario we overestimate the potential for cost savings through the CDM mechanism.

In scenario 5 (Fig. 3.3) CDM is included in the model while the Russian Federation does not exercise its market power in the permit market. This gives a permit price at US$ 0.90 and a global emission reduction at 175 Mt CO_2 equivalents. Of this emission reduction, 102 Mt is carried out in the developing countries. Hence, in scenario 5 the emission reductions in the industrialized countries is only 81 Mt CO_2 equivalents.

Scenario 5

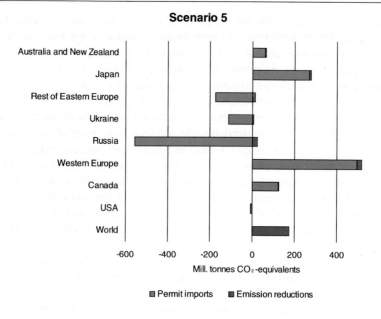

■ Permit imports   ■ Emission reductions

Scenario 6

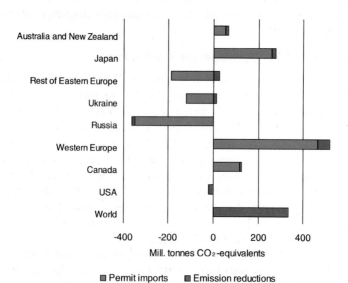

■ Permit imports   ■ Emission reductions

Fig. 3.3: Simulated emission reductions and flows in the permit market in scenarios 5 and 6. The scenarios include extended assigned amounts from forest management and permits from the CDM. Black bars on the right-hand side represent permit import, while black bars on the left-hand side represent permit exports.

A permit price at US$ 0.90 is not a very satisfactory outcome to the Russian Federation. It is therefore likely that this country will do what is possible in order to increase the permit price. However, when CDM is taken into account the Russian market power in the permit market is somewhat reduced because an increased permit price will generate new CDM projects.

In scenario 6 (Fig. 3.3) the Russian Government restricts its permit import in order to maximize its net income from permit export in the first commitment period. This implies that the Russian permit export is reduced from 555 to 350 million permits. However, due to the reduced market power the permit price is only increased to US$ 2.10.

Summary and conclusion

Table 3.2 summarizes the results. When the United States withdraws from the agreement, calculations show that the permit price drops by 2/3. This dramatic price drop is a clear illustration of much of what has been said about the importance of the participation of the United States to the Kyoto Protocol.

Table 3.2: Effects of Kyoto Protocol in different scenarios

Scenario	Permit price (US$/AAU)	Net cross border trade (M. AAUs, CERs, and RMUs)	Emission reductions in countries with caps (Mt CO_2 eqv.)	Global emissions reduction (Mt. CO_2 eqv.)
1	16.0	1377	2482	2293
2	4.7	871	439	395
3	2.0	891	183	164
4	5.1	741	317	259
5	0.9	945	81	175
6	2.1	891	180	393

The third and fourth columns of Table 3.2 show the impact of the Protocol on emission reductions within participating countries and on global emissions. We can state unequivocally that emission reductions would be negligible without the participation of the United States. The global emission reductions for the scenarios without participation from the US (scenarios 2 – 6) are in the magnitude of only 0.4 – 0.9% of Business-as-Usual emissions.

On comparing scenario 3 with scenario 4, and scenario 5 with scenario 6, it can be seen the Russian Federation's exercise of market power in the permit market (scenario 4 and 6) leads to higher global emission reductions than a situation wherein the Russian Federation did not exploit its market power (scenarios 3 and 5). The main reason for this is that the Russian

Federation in order to exert its market power has to keep a share of its AAUs out of the market.

In our calculations we have considered the possibility for acquiring RMUs from forest management projects that would be profitable to implement even without the generation of RMUs. From Table 3.2 we find the impact on global emission reduction of crediting no-regret forest management projects by comparing scenarios 2 and 3. We see that the possibility of acquiring RMUs from no-regret forest management activities decreased global emission reduction by 60%.

It is important to note that in our model simulations we have not taken into consideration the possibilities and possible gain from transferring part of the assigned amount of emission permits to the next target period. The Kyoto Protocol puts no constraints on transferring AAUs to subsequent periods. Obviously, the international permit price in the first target period (2008 – 2012) is not only dependent on the country's abatement costs in that period, but also on the expectations about the permit price in the subsequent target periods. It is not profitable for a country to sell permits in the first target period for a price below the expected (discounted) permit price in the second period. We found in our calculations very low permit prices. Unless countries expect that the target for emission reductions in the second Kyoto period will be quite close to the target for the first Kyoto period, no country will sell its permits at the price found in our calculations. If we ignore the impact of uncertainty regarding future emission targets and abatement costs, and assume that the target for emission reductions increases over time, the present value of the market price for permits will be identical across all Kyoto periods. Hence, in order to give a correct estimate of the permit price in the first Kyoto period, we need to take into account the outcome of the negotiations for emission reductions in future Kyoto periods. This is beyond the scope of this study, however. For the time being it is hard to predict the future for international negotiations for global emission reductions. What we can say, however, is that if the Kyoto Protocol should lead to any significant global emission reductions, the target for abatement must be increased significantly in future, which again would lead to higher abatement costs and hence higher permit prices. Hence, if we observe permit prices at around $ 5 per ton CO_2 equivalents in 2008, we can draw the conclusion that the international community has not yet managed to design an agreement that has a significant impact on combating global warming[7].

Footnotes

1.   The expression "industrialized countries" as used herein means those countries listed in Annex B in the Kyoto Protocol. Annex B lists the countries with emission caps and the size of these caps.
2.   Ratification means that the country's legislative body approves the agreement.

3.   A more detailed description of the model and some possible impacts of the Kyoto Protocol with the participation of the United States is found in Holtsmark (1997) and Holtsmark and Mæstad (2000).

4   Emissions for industrialized countries as a whole are expected to increase by 9 percent from 1990 to the first commitment period (2008–2012) if the Protocol is not implemented, that is, under a "business as usual" scenario (BaU).

5   "Tons" herein refers to 1000 metric kilogram ("metric tons").

6   The literature provides a number of different estimates of quota prices resulting from the Kyoto Protocol. In a special edition of the *Energy Journal* (see Weyant, 1999), various studies are presented of the impacts of the Kyoto Protocol. The introductory chapter by Weyant and Hill (1999) compares eleven different model calculations of the quota price in a nonrestricted market between Annex B countries. The quota prices vary from US\$ 7 to US\$ 61 per ton CO_2. In eight of the model calculations the quota price was under US\$ 20 per ton CO_2.

7   In this argumentation we ignore the possibility for the discovery of new low-cost abatement options.

References

Bohm, P. 1994. On the Feasibility of Joint Implementation of Carbon Emission Reductions, Research Paper in Economics:2. Department of Economics, University of Stockholm, Sweden.

European Commission, 1996. European Energy to 2020 – a Scenario Approach. Directorate General for Energy (DG XVII), Brussels.

Golombek, R and Bråten, J, 1994. Incomplete international climate agreements. Optimal carbon taxes, market failures and welfare effects. Energy Journal 15: 141-165.

Golombek, R, Hagem, C. and Hoel, M, 1995. Efficient Incomplete International Climate Agreements. Resource Energy Econ. 17: 25-46.

Hagem, C. 1996. Joint implementation under asymmetric information and strategic behavior. Environ. Res. Econ. 8: 431–447.

Holtsmark, B. 1997. Climate agreements: Optimal taxation of fossil fuels and the distribution of cost and benefits across countries. Working Paper 1997-05. CICERO

Holtsmark, B. and Mæstad, O. 2000. Emission trading under the Kyoto Protocol — effects on fossil fuel markets under different regimes. Energy Policy 30 (3), 207-218.

IEA, 1995. Energy Prices and Taxes. OECD/IEA. Paris

IEA, 1998. World Energy Outlook. IEA/OECD. Paris

Smith, C. Hall, S. and Mabey, N. 1995. Econometric modeling of international carbon tax regimes. Energy Econ. 17: 133-46.

UNFCCC, 1999. Submission by Germany on Behalf of the European Community, Its Member States, and Croatia, Bulgaria, Czech Republic, Estonia, Hungary, Latvia, Lithuania, Poland, Slovak Republic and Slovenia on Emission Trading (Art 17 KP.). Available at http://www.unfccc.int/resource/docs/1999/sb/misc03a3.pdf.

Weyant, J. (ed.) 1999. "The costs of the Kyoto Protocol: a multi-model evaluation." Energy Journal, Special Issue, International Association for Energy Economics.

Weyant, J and Hill, J. 1999. Introduction and Overview in Weyant J. (ed.) "The costs of the Kyoto Protocol: a multi-model evaluation", Energy Journal, Special Issue, International Association for Energy Economics.©

CHAPTER 4

Climate Negotiations from Rio to Marrakech: An Assessment

Joyeeta Gupta and Alison Lobsinger
Vrije Universiteit, Institute for Environmental Studies, Amsterdam,
The Netherlands.

Introduction

The greenhouse gas problem is probably one of the most serious long-term issues that faces humankind today. This is because of the potentially disastrous effects it can have on the global climate and on sea level, which in turn can affect not only basic necessities such as food and water, but also economic development and social infrastructure.

It is also clear that the problem is seen as fairly serious by national governments. More countries are involved in the negotiations than ever before; 186 of 194 current independent countries worldwide participate in the regime[1]. These countries have been actively engaged since the agreements made at Rio de Janeiro, at the United Nations Conference on Environment and Development in 1992. The seriousness is also evident in the professional approach to dealing with the problem. An Intergovernmental Panel on Climate Change (IPCC) has been established to assess the latest state-of-the-art information on science, impacts, economics and policies of climate change. The five yearly reports of this Panel feed into the negotiating process and provide the justification for taking action. The international community also takes these issues seriously in that two legally binding agreements have already been adopted, and every year countries come together several times to meet in the context of these treaties to discuss the progress made and the necessary steps for the future. Thus there is a lot of "noise", a lot of movement within the context of the climate change regime.

And yet there is increasing fear that all this progress is not necessarily moving the regime toward problem-solving. Instead it would appear that new conflicts are appearing and new definitions and instruments being created to help individual countries avoid responsibilities under the regime. What will become evident in the course of this chapter is that although there is considerable commitment and evidence of action to deal with the climate change issue, there is considerable friction between countries in the North and between the North and the South. The impact of the regime on the problem of climate change and also on the relations between countries is assessed here.

Science of climate change

With industrialization, countries have begun to use large quantities of energy. The bulk of this energy comes from fossil fuels, e.g. coal and oil. In the generation of this energy, large quantities of carbon dioxide are emitted. This gas is a greenhouse gas that traps the heat from the sun in the lower atmosphere. In the process, the earth becomes warmer and this problem is termed global warming.

Carbon dioxide, methane[2] and nitrogen oxide are the most important greenhouse gases, with other greenhouse gases, such as chlorofluorocarbons (CFCs: CFC13, CFC12, etc), perfluorocarbons, sulfur hexachloride (SF_6), water vapor and ozone, also playing a role.[3] These gases have different lifetimes in the atmosphere, ranging from a week for water vapor, all the way to 50 to 200 years for carbon dioxide, the most significant of the greenhouse gases. Consequently, even if greenhouse gas emission levels are reduced, their concentration in the atmosphere is not immediately affected. To stabilize or reduce atmospheric concentrations of these gases can only be achieved by drastically reducing emissions.

Carbon dioxide is also emitted in the course of deforestation; alternatively afforestation leads to an absorbtion of carbon dioxide. This brings us to the issue of sinks. A sink is any process or mechanism which removes a greenhouse gas, aerosol, or a precursor of a greenhouse gas from the atmosphere (Watson et al., 2000).

The problem of global warming is expected to lead to an increase in the average global surface temperature, a rise in sea level leading to inundation and salt water intrusion, extreme weather such as storms and cyclones, and changes in precipitation regimes. Houghton et al. (2001) have reported that the global average surface temperature has increased by 0.6±0.2ºC since 1990. Snow cover has decreased by 10% since the 1960s, average sea level has risen by 0.1 to 0.2 m over the 20th century, precipitation has increased by 0.5 to 1.0% in the 20th century in the mid and high latitudes of the Northern Hemisphere and by 0.2 to 0.3% per decade in tropical land areas. However, it has decreased by approximately 0.3% in the subtropical regions of the Northern Hemisphere. Models predict an increase of 1.4-5.8ºC in the period 1990 – 2100. In the same period the sea level is likely to rise by 0.09-0.88m. Rain patterns are likely to change and ocean currents may change (Houghton et al. 2001).

Thus, simply stated, emissions of greenhouse gases lead to the accumulation of these gases, which further lead to the climate change problem. However, since the climate system is not a linear system, the story of global warming is far more complex. Anthropogenic greenhouse gas emissions account only for 3% of total emissions of greenhouse gases, and sceptics argue that the scientific models cannot reflect the complexity and the negative and positive feedback effects in the real climate system.[4]

The question regarding who is most likely to be affected by the potential impacts of climate change is a difficult one to answer. The scientific models present results in terms of global and regional averages and the local impacts are sometimes very difficult to discern (Watson et al., 1998). However, there are few doubts that those most hard hit will be the poor, and they will thus be the most vulnerable to climate change. About 1.7 billion people (30% of the global population) have inadequate access to water, and this number is expected to rise to 5 billion by 2025. If the temperature rises marginally, developed nations could experience both economic losses and gains, according to the IPCC. However, if temperatures rose more than just marginally, even developed nations would experience losses; therefore if temperatures rise marginally this will enhance the disparities in income between the rich and the poor. In terms of financial losses, the rich have more to lose. If we translate this information in relation to countries, it becomes clear that the 38 small independent island countries are extremely vulnerable to the problem of climate change. The 53 African states are also most likely to be hit hard by water and food security issues, while most of them have negligible emissions barring South Africa. Latin America and Asia are also likely to be affected but to different degrees.

The discussion of impact also raises the question of who is responsible and who should take action? The bulk of anthropogenic greenhouse gases are emitted by developed nations in the course of energy generation, through land use, and in the industrial, cement and transportation sectors. Over the years 1990 to 1999, the top five emitters of carbon from fossil fuels were the United States, Russia, Germany, China and the United Kingdom. The US emitted 30.3% of the total emissions, the European Union (EU) consisting of 15 countries emitted 22.1%, Russia 8.9%, China 7%, Japan 3.7%, Ukraine 2.3% and India 2%. In terms of emissions per capita in 1999, the US emitted 5.6, Russia 2.7, the EU 2.4, China 0.5, Japan 2.4, Ukraine 2.1 and India 0.2 tons carbon (WRI, 2002). The top 20 countries emitted 83.1% and the rest of the world 16.9%.

This, of course, provides only a partial picture of the situation of greenhouse gas emissions. There are several other gases emitted and these gases are also absorbed by sinks. Data regarding these other gases and regarding sinks are inadequate and so a complete comprehensive picture of emissions globally is difficult to provide.

An initial response on the basis of emission data is that the top emitters need to take action first. The question is then - should they be the top emitters in terms of total emissions or on a per capita basis? Clearly from the perspective of trying to deal with the problem of climate change, the top emitters should be targeted. From the perspective of fairness, one may argue instead that the top emitters on a per capita basis are the ones to be focused on (Dasgupta, 1994). In terms of sectors, it appears that the sectors that emit the most emissions are power generation, cement production, energy intensive sectors, transport, and building design. There are a number of options to reduce the emissions of greenhouse gases. There are literally hundreds of technologies and practices that can reduce end-use consumption of energy in countries. About 1000-1100 Mt C eqv y^{-1} can be saved in the building sector including appliances used in this sector. Most of these reductions can be achieved at negative net direct costs. Energy and material efficiency in industry can lead to a saving of 1,300-1,500 Mt C eqv y^{-1} where more than half of the energy efficiency can be achieved at net negative costs. Options in the energy supply and conversion sector can cost up to US $ 100 Mt C eqv y^{-1}, but can lead to a reduction of 350—700 Mt C eqv y^{-1}. In the transport sector there are several options that could lead to a reduction of 300-700 Mt C eqv y^{-1}, but might cost from $25-50 per t C. Agricultural options to reduce non-CO_2 gases could cost anywhere between $0-100 per t C eqv (Metz et al., 2001: 7).

The climate change problem is "global, long-term (up to several centuries), and involves complex interactions between climatic, environmental, economic, political, institutional, social and technological processes…. Developing a response to climate change is characterized by decision-making under uncertainty and risk, including the possibility of non-linear and/or irreversible changes" (Metz et al., 2001: 2)".

The agreements

Simultaneous with the publication of the first IPCC reports on climate change, the United Nations General Assembly established a negotiating committee to negotiate a treaty on climate change. The pre-negotiation documents[5] indicated that the climate change problem was caused by all countries; however, countries that were more developed had a much higher contribution to the problem by virtue of the fact that they used much more fossil fuel based energy in their economies. These developed countries also had more resources and technologies to deal with the problem. At the same time, the climate change problem was likely to affect all countries in the world and the less developed countries would by virtue of their lack of resources have greater difficulties dealing with the climate change problem. It was generally felt that instead of watching the developing countries use available resources to develop, the developed countries should help these countries adopt more modern technologies which would be less polluting (SWCC, 1990).

On the basis of these ideas, the initial negotiations centered on the types of responsibilities countries should adopt and how these responsibilities should be divided between developed and developing countries. This led to the drafting of the United Nations Framework Convention on Climate Change (FCCC) that was adopted in 1992, and came into force in 1994. At present 186 governments and the European Community are parties to the Convention.

The Convention divides the global community into three main groups, Annex I, Annex II and the Non-Annex I countries. Annex 1 countries consist of 41 industrialized nations (including the European Union) of the world including some countries in transition from East and Central Europe. These countries, by virtue of their advanced development status, have high greenhouse gas emissions on a per capita basis. Annex II consists of the 25 richest Annex I countries (including the EU and Turkey[6]). These countries have high emissions and are very wealthy. The Non-Annex I countries consist of the remaining 153 countries—mostly developing countries and some countries from Central Europe.

The "ultimate objective" of the Convention is "stabilization of greenhouse gas concentrations in the atmosphere at a level that would prevent dangerous anthropogenic interference with the climate system. Such a level should be achieved within a time-frame sufficient to allow ecosystems to adapt naturally to climate change, to ensure that food production is not threatened and to enable economic development to proceed in a sustainable manner" (Article 2, FCCC 1992). No details as to level and year were specified because of lack of scientific and political consensus about what the specific level of concentration should be. While there were some that argued that greenhouse gas concentrations should not be allowed to exceed a doubling of pre-industrial levels, others argued that this was not feasible and a more reasonable goal should be adopted, and therefore no levels were specified.

Article 3 of the Convention discusses five guiding principles that are to be the basis of the allocation of responsibilities between countries. These principles state that Parties should adopt policies on the basis of the common but differentiated responsibilities and capacities of different countries (Article 3.1). This principle thus recognizes the different situations and levels of development in different countries and in particular highlights the situation of the most vulnerable countries (Art. 3.2). This refers to the 38 independent small island countries that are likely to be seriously affected if there is a rise in the sea-level or there are extreme weather events. The Convention is also based on the "precautionary" principle. According to this principle, countries should take cost-effective measures "to anticipate, prevent or minimize the causes of climate change and mitigate its adverse effects" (Art. 3.3). This principle is an important idea because it authorizes countries to take action even in the absence of scientific certainty, as long as the risks are likely to be serious and irreversible. At the same time, the principle is abstract in that it is difficult to give a precise meaning to it.[7] The Convention also recognizes that the Parties have a right to sustainable development, that economic development is necessary for preparing policies on climate change, and requests countries to integrate climate policy in national development programs. The Convention also clearly specifies that policies should be consistent with the notions of an open international economic system (Art. 3.5).

The Convention further describes the obligations of countries in Article 4. Countries listed in Annex I were requested to adopt measures to bring their CO_2, CH_4, and N_2O emissions back to 1990 levels by the year 2000. Although this target began as a strongly worded commitment, it was weakened during negotiations until it became an aspirational one. It also calls on Annex II countries to provide technologies and new and additional financial[8] resources to developing countries to help them meet the agreed costs of complying with the preparation and implementation of their national reports.

Under the Convention, countries are encouraged to cooperate in the field of scientific research, education, and public awareness. The Convention established five different bodies to ensure its smooth implementation. These include the Conference of the Parties which consists of all Parties (countries that sign and ratify the Convention) and which meets once a year to take decisions. A secretariat deals with the day-to-day management of the tasks under the Convention. There are two bodies—the Subsidiary Body on Scientific and Technological Advice and the Subsidiary Body on Implementation—which are supposed to focus on matters relating to science and implementation respectively. A financial mechanism was established under the Convention to provide financial assistance to developing countries. This financial mechanism is located within the Global Environment Facility established by the World Bank, the United Nations Development Program and the United Nations Environment Program.

In the event of questions relating to the implementation of the FCCC, the Parties are encouraged to establish a multilateral consultative process (Article 13). When there are disputes between Parties, these are to be resolved through peaceful means as specified in international law. The

Convention is also not set in stone. It can be amended, or there can be Protocols negotiated in relation to it, or Annexes can be added to it.

The bottom line is that while the Convention provides a broad framework for action, it has ambiguous principles and weak targets. This has allowed countries, both those who considered climate change a threat and those who did not perceive any immediate risk, to ratify it. Developing countries in particular were quite enthusiastic because of the promise of technological and financial assistance (Gupta, 1997). However, for all that the principles are ambiguous and the targets and commitments weak, the Convention still allowed hope to all parties. Ratifications followed quickly, with the United States being one of the first nations to ratify, and the Convention came into force in 1994.

Kyoto Protocol

With the entry into force of the Convention, there were two meetings of the Conference of the Parties during which it was decided that the existing commitments did not go far enough and that there was need for a Protocol with binding quantitative commitments. At the First Conference of the Parties, the Parties signed the Berlin Mandate, in which they recognized that the commitments of the Convention were not strong enough. In the two years leading up to the Third Conference of the Parties in Kyoto in December 1997, the Parties met to discuss stronger commitments. At Kyoto, the European Union arrived wanting to negotiate for strong binding targets, while the United States was not willing to move beyond stabilization. Japan, as host to the negotiations, was caught between the need to demonstrate leadership, and the fear that being an energy-efficient economy, taking such a leadership stance would lead to unacceptable costs to their economy (Oberthür and Ott, 1999; Gupta and Grubb, 2000; Grubb et al., 1999). However after much negotiation the Kyoto Protocol to the United Nations Framework Convention on Climate Change was adopted.

The Kyoto Protocol (KP) presents a variety of policies and measures that all countries are invited to consider. These include energy efficiency policies, protections of sinks and reservoirs, sustainable forestry practices, afforestation and reforestation, sustainable agriculture, environmentally sound technologies, encouragement of reforms in relevant sectors, and controlling emissions from the transport sector, to mention just a few.

The center-piece of the Kyoto Protocol consists of the targets adopted. As a result of negotiations it was finally agreed that developed countries would jointly reduce their net emissions (emissions from sources minus removals by sinks) of 6 greenhouse gases by 5.2% in the period 2008 to 2012 in relation to emission levels in 1990. The six gases are carbon dioxide, methane, nitrous oxide, HFC, SF_6 and PFCs. The Protocol does not list separate targets for each individual gas but instead a combined target for the basket of gases, expressed in CO_2 equivalence. The Kyoto targets call on the US to reduce its emissions by 7% in 2008-2012, the EU by 8%, and Japan by 6%. Ukraine and Russia are allowed to stabilize their emissions, while Australia, Iceland, and Norway are allowed to increase their emissions by 8%, 10% and 1% respectively.

By using the term net emissions it allows countries to take into account afforestation, reforestation, and deforestation and other agreed land use, land-use change and forestry activities in order to meet their quantitative commitments. As a result, countries are expected to provide data regarding their carbon stocks and changes in the carbon stocks since 1990.

The Protocol allows countries to achieve their targets via the use of the "flexible mechanisms". These include Joint Implementation, Emission Trading, and the Clean Development Mechanism. Joint Implementation allows developed countries to make investments in Eastern and Central European countries in return for emission reduction credits (if there are emission reductions in the host country, the investing country receives the credits). Emission trading was included to allow the emission targets of countries (Assigned Amounts as they are called in the Protocol) to be made tradable. If a country emits less than its assigned amount it can sell the remainder to another country that has overused its assigned amount. The Clean Development Mechanism allows countries and companies from developed countries to invest in sustainable development projects in the developing world in return for emission credits. In addition there is a mechanism called Activities Implemented Jointly. This mechanism, launched in 1995, is a precursor to Joint Implementation and the Clean Development Mechanism and allows countries to participate on a voluntary basis in such schemes during a pilot phase in which no crediting is allowed. This pilot mechanism also continues to operate (Climate Change Secretariat, 2001).

The parties to the Protocol have to develop national systems for calculating emissions, and were requested to develop cost-effective programs to improve the quality of the national inventories, and to cooperate in technology, science and the development of education and training programs. Developed countries in Annex II were also expected to provide new and additional financial resources to meet the agreed fixed incremental costs of developing countries in meeting their specific obligations.[9]

After Kyoto

While the Protocol has presented commitments that Parties will have to meet, should they choose to ratify it, the negotiations in Kyoto left much unfinished business. The Parties to the Protocol had to outline operational details as well as address the vulnerability of developing nations and develop guidelines for reporting. At the Fourth Conference of the Parties, in Buenos Aires, Argentina, the Parties outlined a schedule for outlining the operational details of the Protocol and for strengthening implementation of the Convention. The deadline in this agreement, called the Buenos Aires Plan of Action, was set for the Sixth Conference of Parties. At the sixth Conference of the Parties in the Hague in 2000, the meeting broke down without agreement (Grubb and Yamin, 2001). At the resumed Sixth Conference of the Parties in Bonn in July 2001, the Parties adopted a Political Declaration which was elaborated upon at the Seventh Conference of the Parties, in Marrakech.

In contrast to the situation after the negotiation of the climate convention, the developed countries appeared to be extremely reluctant to ratify the Kyoto Protocol. This is not in itself surprising because unlike the Convention, the Protocol includes hard, binding quantitative commitments against which the performance of developed countries can be measured. But the key reason for the reluctance can be found in the post-Kyoto politics. Although the ruling democrats in the US were supportive of the Protocol, the US Senate had made clear prior to the Kyoto negotiations that they would not support quantitative commitments because of the potential impacts on the US economy and because these were not going to be effective since the developing countries were exempted from such quantitative targets. In the meanwhile US politics focused on developing appropriate climate policies. During this period 1997-2000, the other developed countries were hesitant to ratify the Protocol because if the US did not do so, this would increase the costs of implementation for them. This is because domestic measures to reduce greenhouse gas emissions would lead to an increase in the price of exported commodities and such commodities would be less competitive in the international market. Nevertheless, when the George W. Bush administration clarified in 2001 that the US was withdrawing from the process, the rest of the world was very incensed and united to try and push the process further. This led to the adoption of a document that explains the rules of implementing the Kyoto Protocol and there were hopes that the developed countries would now ratify the Protocol.

Marrakech Accords

At Marrakech, a set of Accords was drawn up that covered a number of key issues, thus paving the way for ratification of the Kyoto Protocol. The Marrakech Accords tries to meet many of the developed and developing country concerns. In relation to the former, detailed decisions were taken with regard to the flexibility mechanisms in terms of eligibility of country and projects and in terms of accounting and verifying the emission reductions.

In relation to the developing countries, decisions were taken to promote the implementation of capacity building. Capacity building is learning-by-doing, and should be country-driven, continuous, progressive and iterative, and undertaken in an effective, efficient and programmed manner. To promote the implementation of technology transfer, the accords call for special emphasis on assessing country needs, assessing technology information, developing enabling environments, promoting capacity building, and developing mechanisms and institutions for technology transfer and the establishment of an expert group on technology transfer. The Accords include a decision to increase funding in the GEF and to establish three new funds—a climate change fund, a least-developed countries fund, and an adaptation fund, of which the last is to be financed by 2% of the proceeds of CDM projects (except when undertaken in a least-developed country). The issue of defining sinks and which activities can be included under the Protocol was particularly challenging. A Compliance Committee was also established under the Accords.

An assessment: the procession of Echternach[10]—three steps forward, two steps backwards

The climate change regime thus far is assessed here in terms of its achievements and challenges. What is clear is that every year new agreements are made, sometimes legally binding and sometimes supportive decisions, in an effort to keep the coalition of countries together to address this complex and challenging issue. What this section makes clear is that, every time the regime appears to move forward, a few years later it appears to move backward.

Environment

The ultimate objective in the FCCC, while clear, is weak because it does not specify concentration levels and target years within which this objective is to be achieved. While initially people were talking about stabilizing concentrations at double pre-industrial levels at 450 ppmv, now the discussion has moved to 550 ppmv and even 650 ppmv.

The idea that the North would reduce its own emissions can be compared to the benchmark in the original IPCC (Houghton et al., 1990) report which stated that in order to stabilize concentrations and to make room for the developing countries to grow, the developed countries would have to reduce their CO_2 emissions by at least 60%. The first target in the FCCC, prepared with much care in the prenegotiation process, to stabilize emissions of developed countries at 1990 levels in 2000 was weakened so much that it was unclear whether any legally binding target was included in the Convention. The follow-up target in the KP might in relation be seen as weaker because although it aims at a 5.2% reduction, it includes three new gases, flexible base years with respect to these gases, the possibility of including sinks, and the potential of using the flexible mechanisms. In particular, the potential of "hot air" threatens the integrity of the environmental goals in the Protocol. Hot air refers to the fact that Ukraine and Russia have been given commitment levels for 2008-2012 that they are unlikely to be able to achieve since the economies of both countries have collapsed considerably since the fall of the Berlin Wall. If these predictions are right, then both countries will have surplus assigned amounts left, which they could easily sell to the other developed countries. Such a transaction would result in a transfer of financial resources and emissions, but may not necessarily lead to a real and additional reduction of emissions. Hence, recent studies indicate that the implementation of the Kyoto Protocol "will have only a very modest impact on the rate and pace of climate change" (Grubb et al., 1999: 158).

Ratification reluctance

Let us then examine the climate change problem in terms of the transatlantic issues. Soon after ratifying the Convention people began to realize what the consequences of embarking on such a negotiating process could be. Technologies may exist to deal with the climate change problem, but they are expensive, and reordering society and its production and consumption patterns would not be simple.

Nations began to realize that it would be difficult to reduce emissions if they were already energy efficient and that if efficiency were to be increased, it would prove very expensive. In addition, the influential business and industry groups began to resist change. All of this led to growing political reluctance of Northern nations to take on measures to reduce their emissions.

In the lead up to the third Conference of the Parties the Byrd-Hagel Resolution was introduced in the United States. This called on the US not to accept binding quantitative targets until and unless key developing countries also participated meaningfully in the negotiations, especially because of the increased costs associated with taking action in the United States. This led to enhanced pressure on the developing countries to take up voluntary commitments (Clinton, 1997). At the same time, the South insisted that it was the responsibility of the developed countries to take action and that only then would the South follow. Initially the rest of the developed world argued that they would only ratify if the US did so. This threatened deadlock appeared to evaporate the moment the US boldly stated that it was going to withdraw from the regime. The rest of the developed countries felt as if the entire process was not being taken seriously by the US and have now gone ahead with the Marrakech Accords and with the political will to ratify. However, the damage has been done. The Marrakech Accords number some 150 pages, but whether they really help to address the climate change problem remains a key question.

Most of the developed countries are key contributors to the problem of climate change. The challenge is that it is seen as impractical to take far-reaching unilateral measures to deal with the issue. This is because unilateral measures may lead to an increase in the price of energy, and this may lead to an increase in the price of the products generated. Such products will become less competitive in the international market and this might affect the economic welfare of individual countries. This has been the key argument behind the desire that all (developed) countries should simultaneously deal with the problem of climate change, so that a level playing field is created. With the withdrawal of the United States from the regime, the effectiveness of the regime is put at considerable risk. This is not just because of the fact that the US contributes 30% of the global carbon emissions (36.1% according to the Annex to the Kyoto Protocol), but also because US reluctance casts a damper on action to be taken by other countries. In addition, in an effort to minimize the costs to themselves, countries began to define and redefine sinks so as to minimize the extra effort that had to be taken by them.

A close look at the costs of climate change policy to the developed countries could provide useful insights. If we exclude emissions trading, most studies reveal that the costs of implementing the Kyoto Protocol would be about 0.2 – 2% of GDP in 2010. If emissions trading is included, the costs fall to 0.1% to 1.1% of projected GDP. Most of these models do not include CDM, negative cost options, ancillary benefits, or targeted revenue recycling (Metz et al., 2001). All these indicate that the costs of complying with the Protocol are very small, although in absolute terms a 0.5% loss of GDP amounts to about $ 85 billion (Grubb et al., 1999).

North-South issues

Several issues are of relevance from a North-South level. The original North-South bargain as expressed in the idea that the North would lead in reducing its own emissions and would help the South with financial assistance and technology transfer has been considerably weakened. As a result of the guarantees to the right of growth of developing countries, they were initially attracted to the bargaining table, and they have stayed there. For many of the developed nations this promise was made because they felt responsible for contributing to the problem and felt it was in their power to deal with it. It was not entirely altruistic, however, as they were able to move away from "liability" and the "polluter pays principle", to being the "good guy" in international negotiations.

In relation to the FCCC (Gupta, 1997) and the KP (Gupta, 2001), irritation increased regarding the weakening of the commitments of the developed countries and that ratification of the Protocol was linked to voluntary commitments by the developed countries (Gupta, 1998). In relation to the financial assistance to developing countries, the resources available are far from adequate. While the UN has recommended that developed countries make 0.7% of the GNP available as Official Development Assistance (ODA), ODA resources have decreased since 1993. The percentage of ODA is 0.23 % of GNP (IPCC, 2000). While new and additional resources were seen as necessary to help developing countries face the additional burden of environmental issues, estimated at $ 125 billion available every year at the United Nations Conference on Environment and Development (Agenda 21, 1992), less than $10 billion has been made available to the Global Environment Facility in the last ten years. Despite the fact that many United Nations Economic Commission of Europe treaties[11] call for the polluter pays principle to be central to determining who pays how much, in the context of the climate change treaty, this principle does not play a significant role (Agarwal et al., 1999). Instead it is left to the willingness of countries to contribute funds. This reduces the amount of financial resources made available to help developing countries deal with the problem of climate change and its impacts.

While the developing countries wanted an independent financial mechanism under the Convention, the developed countries insisted that financial resources be channeled via the GEF or not at all (Gupta, 1995). This problem has eased to some extent, since there are now three new funds to be operated under the Convention. The Marrakech Accords tried to deal with this problem by creating the three new funds; but the creation of new funds does not in itself imply that funding will be forthcoming. Only the adaptation fund receives 2% of the proceeds of CDM projects and thus is assured a regular source of income. However, since the CDM is the only one of the three flexibility mechanisms that contributes to this fund, it appears as though only North-South cooperation is taxed; this increases the price of North-South cooperation vis-a-vis North-North cooperation while arbitrarily exempting other flexibility mechanisms from such a tax.

During preparations for the first Conference of the Parties, Joint Implementation began to receive considerable attention as a concept that would allow developed nations to undertake emission reduction projects wherever in the world it would be cheapest to do so. Most of the developing nations were opposed to this because they feared that the developed world would come into their countries and buy off the cheapest reduction options, leaving only expensive options over for them, and because this would reduce the incentive to reduce emissions domestically (Maya and Gupta, 1996). However, the increasing pressure and the perception of some developing countries that joint implementation would bring funding and technologies led to the acceptance of Activities Implemented Jointly at the first Conference of the Parties. Despite the fact that there was no complete evaluation of this pilot phase and that the developing countries had wanted a Clean Development Fund to be established, financed by the fines from countries in noncompliance, the Clean Development Fund was renamed as the Clean Development Mechanism and metamorphosed into a new form of Joint Implementation that did not resemble the original fine proposal except in name.

In terms of technology transfer not much progress has been made thus far. In an effort to set this right, the Marrakech Accords established an expert group on technology to give clear guidance on this issue. While clearly there is an injunction on countries to take action, there are few incentives in place. There are, however, some disincentives. Countries have to report on action taken in relation to these activities and lack of action, when reported, can be an embarrassment.

On a North-South level, the original grievances of the South have not disappeared; if possible they have become more contentious and mired in a wealth of decisions and policies that do not seem to lead anywhere concrete. On the other hand, the Marrakech Accords, although delightfully vague in terms of the concrete options available for developing countries, do take an important first step in the direction of providing some content to the issues of financial assistance and technology transfer.

Part of the North-South bitterness also arises from the fact that the classification of countries into the two groups is not based on scientific criteria. The South is, in general, represented by the Group of 77(G-77) established within the UN in 1964 to represent the interests of the South. The membership of this body has now increased to 130 countries and this in itself calls for a huge amount of coordination and cooperation. At the same time, the non-Annex I group consists of 153 countries, i.e., an additional 23 countries.

A closer examination of this group indicates that it is far from homogeneous. It is increasingly becoming clear that some of these countries fall into the category of very rich countries. Singapore, for example, is among the richest countries in the world. The Bahamas, Cyprus, Brunei, Israel, Kuwait and the United Arab Emirates too fall into the category of rich countries with per capita incomes above $10,000 per year. At the same time, if we look at the group of developed countries in Annex I, we see that Bulgaria, Croatia, Hungary, Latvia, Lithuania, Romania, Slovakia, Ukraine, the Czech Republic, Estonia, Poland and the Russian Federation fall into the category of middle income countries. Thus while the task of dividing responsibilities between the rich and the poor countries of the world would appear simple, the rich among the poor, and the poor among the rich complicate the negotiating process (Gupta et al., 2001). This makes the developed countries annoyed that the newly industrializing countries, and countries that appear to be very rich (e.g, Argentina) are not taking action. While the newly industrializing countries are keeping a low profile, Kazakhstan, which saw the potential for profit in the targets assigned to Ukraine and Russia, volunteered to take on commitments. Kazakhstan's emissions are currently 50% below 1990 levels and if the country can settle for a stabilization target, it can cash in through emission trading (Grubb et al., 1999: 263). Under pressure Argentina too agreed to take on a commitment to reduce the rate of emission growth. But with the financial crises in Argentina, it is unlikely that there will be much follow-up to this commitment. This is an important issue in light of the fact that future commitments will soon have to be negotiated for developed and developing countries. How should such commitments be developed? The existing targets in the regime are but a poor precedent since some very rich countries have been allowed to increase their emissions. There are several proposals in the literature to try and increase the equity in the regime by developing proposals based on certain principles. These include the Triptych approach (Phylipsen et al., 1998), the FAIR approach, the CSE approach (Agarwal, 2000), the Pew Centre approach, etc (Berk et al., 2002). The challenge is that providing developing countries the right to increase emissions needs to be matched by decreased emissions in the developed world. As long as emissions are closely related to economic growth for both groups of countries, there will be reluctance to take measures.

There is also considerable annoyance both in the developed and the developing world about the role of the oil exporting countries. These countries, perceived to be among the richer developing countries, have not invested their new found wealth wisely in their domestic economies. Instead they now fear that any steps to reduce greenhouse gas emissions will lead to a reduction in fossil fuel imports and will thus affect their income. They opposed targets in the Kyoto Protocol until it was made clear to them that the developing countries would proceed without them in future. Since then they have been asking for compensation for the negative effects of policies taken to reduce dependence on oil exports. This is seen by all other countries as unfair. First, these countries are polluters and they are among the richest. From the perspective that they are polluters, they have no right to ask for compensation from the West. From the perspective that they are among the richest developing countries (that have perhaps misused their wealth), it does not appear fair that they want resources made available by the West, since the most vulnerable countries should be the first to be helped. Nevertheless, the strong negotiating tactics of this group of countries has tabled the issue, secured follow-up decisions at Marrakech, and it is unclear what sort of precedent this is likely to set for the future.

As already mentioned, the top twenty countries are responsible for the bulk of the emissions. From an efficiency perspective, these top twenty countries should take action. China and India fall into this category, even though their per capita emissions are quite low. Both these countries have taken the position that as developing countries they will only take action once the North has demonstrated effective leadership. The developed countries find this position as uncompromising and unproductive. However, if one examines the domestic policies in both countries, it is increasingly becoming clear that these countries are, in fact, in the process of transition and such transition may be beneficial in terms of reducing their greenhouse gas emissions.

Let us take the case of India. In a recent project (Gupta et al., 2002), it was estimated that India with 16% of the global population emits about 2.3% of global emissions and has a per capita emission level of 0.2 mt Cy^{-1}. At the same time, India is in the process of liberalizing the electricity and industrial sectors. A number of policies and laws have been adopted since 1990 which promote energy conservation, renewable energy, and efficiency in power plants. Court decisions too are having an impact on the fuels used by public transport. These policies could lead to a drop in the rate of growth of emissions (Gupta, 2001). Estimates show that while India and China's electricity sector is set to grow by 5 times in 2020 relative to 1990, their emissions of greenhouse gases is likely to only grow by 4 and 3 times respectively, indicating a decoupling of emissions from growth (Gupta et al., 2001; see also Zhang, 1999). With the visit of Clinton to India, interest in CDM projects too have increased considerably. There is considerable potential in both countries to realize energy efficiency options. Then why are the governments taking a defensive approach when a constructive approach could have been more productive and opportune? The answer lies in a mixture of reasons. These governments are afraid that policies may not necessarily lead to expected results; the transition period is a painful exercise and very unpredictable; a principled perspective that it is the turn of the North to take action; perhaps a general ignorance about the exact nature of the domestic emissions and sinks and the possible beneficial effects of existing and planned policies whose objective is to prepare the countries for greater integration in the global economy.

Conclusion

A brief overview of the key developments in the climate change regime has been presented and the following conclusions may be drawn from the review. First, there is relentless progress in terms of new science, new negotiations, new decisions and a process of reporting and monitoring. This process is gradually pushing the regime further despite a number of setbacks in the process, and it is possible that the ultimate success of the regime depends on such regular processes.

Second, the regime has faced many setbacks. The original FCCC treaty did not include legally binding quantitative commitments for the developed countries because of US reluctance. Although such commitments were incorporated in the Kyoto Protocol, this was only possible on the condition that the Protocol would also include a number of flexible mechanisms, new gases and sinks. This concession, likewise, was made to accommodate the US and other developed countries. The withdrawal of the US from the regime could have led to a complete breakdown in negotiations but instead helped to cement the remaining countries and their commitments.

Third, while the costs of implementing the regime appear to be the major challenge, these costs amount to only 0.1-1.1% of GNP in 2010. This may be large in actual terms, but is relatively insignificant; further, most of the options to reduce greenhouse gas emissions can be achieved at negative costs. While this is true for the developed countries, for developing countries the opening up of their markets calls for increased competitiveness on their part. This may in itself force developing country companies to try and use energy efficiency opportunities to decrease the costs of production. The case study of India shows that there is considerable potential.

Fourth, the flexibility mechanisms might lead to an inflation of the existing targets of the developed countries. On the other hand, they might lead to accelerated diffusion of modern technologies worldwide and this may be of considerable significance in helping the world move from fossil-fuel-based economies to greenhouse gas friendly economies. By reducing the costs of meeting national commitments, this provides the necessary incentive to keep all countries (except the US) on board in the negotiating process.

Fifth, the key challenge in future will be future commitments for countries. As indicated, existing coalitions and classifications appear to be based on historical circumstances, geographical proximity, and ideological interests. This may not be an adequate basis for negotiating future targets and commitments. The developed country group includes many relatively poor countries. The developing country group includes some rich countries. This undermines the process of negotiation when it is of vital importance that coalitions in the climate change regime be based on real as opposed to perceived common interests.

Acknowledgements

This paper is based partially on the on-going research within the Vrije Universiteit Amsterdom project on the Law of Sustainable Development and Climate Change.

Footnotes

1. A regime is defined as including the principles, rules, norms and procedures in a given area of international relations (Krasner, 1982).
2. An important source of methane is wet rice cultivation which takes place in South and South-East Asia. Other sources of methane are mining, enteric fermentation, manure management and waste disposal.
3. About 82% of total emissions of developed countries is CO_2; methane amounts to about 12%, N_2O about 4% and HFC, PFC and SF_6 about 2% (Oberthür and Ott, 1999: 7).
4. There is uncertainty about how these gases influence the climate. Not much is known about the impacts of organic and black carbon from fossil fuel burning, mineral dust and the indirect effects of aerosols and solar warming. As well, there are cooling effects from the depletion of the stratospheric ozone layer, the accumulation of sulfates and aerosols and biomass burning, which is not well understood.
5. These include the political declarations of the Ministers of Environment in Noordwijk in November 1989 (Noordwijk Declaration, 1989), the Second World Climate Conference (SWCC, 1990) and several others.
6. By an amendment, Turkey is no longer included in the list in Annex II to the Convention.
7. See Freestone and Hey (eds.) (1996) for more details on the precautionary principle.
8. This implies that the resources should be additional to that provided through normal aid channels.
9. For a detailed appraisal of the Kyoto Protocol, see Yamin, 1998.
10. In Luxembourg, on the third day of Whitsunday, celebrations, participants in a procession in honor of Saint Vitus jump three steps forward and two steps backwards to the rhythm of music.
11. The UNECE Convention on the Protection and Use of Transboundary Watercourses and International Lakes 1992 and the European Union's Water Framework Directive include the polluter pays principle.

References

Agarwal, A 2000. Making the Kyoto Protocol Work: Ecological and Economic Effectiveness and Equity in the Climate Regime. CSE Statement, New Delhi.

Agarwal, A., Narain, S., and Sharma, A. 1999. Green Politics: Global Environmental Negotiations. Centre for Science and Environment, New Delhi.

Agenda 21,1992. Report on the UN Conference on Environment and Development, Rio de Janeiro, 3-14 June 1992.UN doc. A/CONF.151/26/Rev.1 (Vols.1-III).

Berk, M., Gupta, J. and Jansen, J. 2003. Comprehensive approaches to differentiation of future climate commitments—some examples. In: Issues in International Climate Policy : Theory and Policy, E. van, Ierland, J. Gupta and M. Kok (eds.). Edward Elgar Publishers. Cheltenham Glos, pp. 171-200.

Climate Change Secretariat, 2001. The Marrakech Accords and the Marrakech Declaration. Climate Change Secretariat, Bonn.

Clinton, W.J. 1997. Remarks by the President on Global Climate Change, National Geographic Society, October 22, 1997.

Dasgupta, C. 1994. The Climate Change Negotiations. In: Negotiating Climate Change: The Inside Story of the Rio Convention. I.M. Mintzer, and J.A. Leonard, (eds.) Cambridge Univ. Press, Cambridge, England, pp. 129-148.

FCCC, 1992. United Nations Framework Convention on Climate Change, New York, 9 May 1992, in force 24 March 1994, I.L.M., no. 31.

Freestone, D. and E. Hey (eds.). 1996. Precautionary Principle: Book of Essays. Environmental Policy and Law Series, Kluwer Law International, The Hague.

Grubb, M. and Yamin F. 2001. Climatic Collapse at The Hague: What Happened, Why and Where Do We Go From Here?, International Affairs, 77, (2): 261-276.

Grubb, M., Vrolijk, C., and Brack, D. 1999. The Kyoto Protocol. Earthscan/RIIA, London.

Gupta, J. 1995. The Global Environment Facility in its North-South Context. Environ. Politics, 4, (1): 19-43.

Gupta, J. 1997. The Climate Change Convention and Developing Countries — From Conflict to Consensus? Environment and Policy Series. Kluwer Academic Publishers, Dordrecht, The Netherlands.

Gupta J. 1998. Leadership in the Climate Regime: Inspiring the commitment of developing countries in the post-Kyoto phase. Review European Community and Int. Environ. Law, 7, (2) 178-188.

Gupta, J. 2001. India and Climate Change Policy: Between Diplomatic Defensiveness and Industrial Transformation. Energy and Environment, 12, (2-3): 217-236.

Gupta, J. and Grubb, M. (eds.). 2000. Climate Change and European Leadership: A Sustainable Role for Europe. Environment and Policy Series, Kluwer Academic Publishers, Dordrecht, The Netherlands.

Gupta, J., Vlasblom, J., Kroeze, C., Blok, K., Bode, J.W., Boudri, C., Dorland, K., and Hisschemöller, M. 2002. An Asian Dilemma. Report No. 410 200 097, Dutch National Research Programme on Global Air Pollution and Climate Change. Bilthoven, The Netherlands.

Houghton, J.T., Jenkins, G.J. and Ephraums, J.J. (1990). Climate Change: The IPCC Scientific Assessment, Cambridge Univ. Press, Cambridge, England.

Houghton, J.T., Ding, Y., Griggs, D.J., Noguer, M., van der Linden, P.J. and Xiaosu, D. (eds.). 2001. Climate Change 2001: The Scientific Basis – Contribution of Working Group I to the Third Assessment Report of the Intergovernmental Panel on Climate Change (IPCC). Cambridge Univ. Press, Cambridge, England.

IPCC 2000. Financing and Partnerships for Technology Transfer. Special Report on Technology Transfer. Inter-Governmental Panel on Climate Change. Cambridge Univ. Press, Cambridge, England.

Krasner, S.D. 1982. Structural Causes and Regime Consequences: Regimes as Intervening Variables, International Organization, 36, (2): 186.

Maya, R.S. and Gupta J. (eds.). 1996. Joint Implementation: Carbon Colonies or Business Opportunities? Weighing the Odds in an Information Vacuum. Southern Centre, Zimbabwe, pp. 43-61.

Metz, B., Davidson, O., Swart R., and Pan J., (eds.). Climate Change 2001: Mitigation – Contribution of Working Group III to the Third Assessment Report of the Intergovernmental Panel on Climate Change (IPCC). Cambridge Univ. Press, Cambridge, England.

Noordwijk Declaration on Climate Change. 1989. In: P. Vellinga, P. Kendall, and J. Gupta (eds.) Ministry of Housing, Physical Planning and Environment, Hague, The Netherlands.

Oberthür, S. and Ott, H.E. 1999. The Kyoto Protocol. International Climate Policy for the 21st Century. Springer-Verlag, Berlin.

Phylipsen, G.J.M., Bode, J.W., Blok, K., Merkus, H. and Metz, B. 1998. A Triptych Sectoral Approach to Burden Differentiation; GHG Emissions in the European Bubble. Energy Policy, 26, (12): 929-943.

SWCC (1990). Ministerial Declaration of the Second World Climate Conference and Scientific Declaration of the Second World Climate Conference. Second World Climate Conference. Geneva.

Watson, R.T., Noble, I.R., Bolin, B., Ravindranath, N.H., Verardo, D.J., and. Dokken, D.J., 2000. Land Use, Land-Use Change, and Forestry, IPCC, Cambridge Univ. Press, Cambridge, England

Watson, R.T., Zinoera, M.C., Moss, R.H., and Dokken, D.J. 1998. The Regional Impacts of Climate Change: An Assessment of Vulnerability. Cambridge Univ. Press, Cambridge, England.

WRI. 2002. Contributions to Global Warming Map. World Resources Institute. http://www.wri.org/climate/contributions_map.html.

Yamin, F. 1998. The Kyoto Protocol: Origins, Assessment and Future Challenges. Rev. European Community and Int. Environ. Law, 7 (2): 113-27.

Zhang, Z.X. 1999. Is China Taking Action to Limit its Greenhouse Gas Emissions? Past Evidence and Future Prospects. In: Promoting Development While Limiting Greenhouse Gas Emissions: Trends and Baselines. J. Glodemberg, and W. Reid (eds.). UNDP and WRI, New York, NY.

Section II
INSTRUMENTS AND INSTITUTIONS OF/FOR KYOTO PROTOCOL

CHAPTER 5

Kyoto Protocol and Desirable Institutions for Mitigation of Global Warming

Akinobu Yasumoto
Adviser, Global Industrial and Social Progress Research Institute (GISPRI)
c/o Mitsui-OSK Line Bldg. 3F, Toranomon 2-1-1, Minato-ku,
Tokyo, 105-0001 Japan   e-mail: yasumoto@gispri.or.jp
Senior Adviser, Research Institute of Market Structure (RIMS),
the Tokyo Commodity Exchange (TOCOM)
c/o Miyamae Bldg. 8F, Nihonbashi-Kakigaracho 1-38-9, Chuo-ku,
Tokyo, 103-0014 Japan   e-mail: a-yasumoto@rims.jp

Introduction

The Kyoto Protocol is an epoch-making protocol not only for setting a quantified emission limitation and reduction commitment of each Annex I country but also for adopting the Kyoto mechanisms. The Kyoto mechanisms in particular present remarkable rationality and it is worth noting the significance of their adoption.

Global warming caused by greenhouse gas is an issue that involves the entire earth, where every country "has a right to, and should, promote sustainable development"[1]. Because of this, institutions to mitigate global warming should be "cost-effective so as to ensure global benefits at the lowest possible cost". Considering the finiteness of global resources including environment, the greater the inefficiency and waste in resource usage, the less hope for attaining sustainable development. Moreover, any burden that may arise is more proper to be borne on the basis of "differentiated responsibility" of each country. This is because the rapid concentration of greenhouse gas since the Industrial Revolution is closely and deeply related to the modernization of countries, which today are called developed countries. Both the Kyoto Protocol and its parent agreement of the UN Framework Convention on Climate Change are based on such basic principles. It is important to note that to fulfill such principles requires the Kyoto mechanisms.

Nevertheless, in view of international negotiation on the details of the Kyoto Protocol, and the review status of domestic institutions ongoing in Japan and European countries, it is regrettable that there has been a deep-rooted move to restrict the use of Kyoto mechanisms. The concept behind such a move is that developed countries must reduce emissions on their own and to use the Kyoto mechanisms to rely on other countries' reduction is a "loophole". Such logic seems quite plausible. Yet it ignores the excellent function of division of labor. As long as we rely on such logic, to realize sustainable development in the world with a balance between environment, society, and economy, is merely a dream.

The United States is the country that generally shows the attitude of earnest compliance with international commitment once agreed upon. There is no way of knowing the true intention of President Bush, and it would be no surprise if one reason why the USA distances itself from the Kyoto Protocol is a concern that, once a restriction is set upon use of the Kyoto Protocol, not only the USA's compliance with it, but that of the entire world also would be endangered.

Global warming is a serious problem. No matter how much we feel its importance emotionally, that will never suffice. Without implementation of a rational institution, the solution to global warming will obviously move farther away. What we must do is allow full utilization of the Kyoto mechanisms and let each country of the world arrange a domestic institution that can guarantee the certain implementation of the Protocol.

Significance of the Kyoto mechanisms

The Kyoto Protocol assigns in Annex B to each of the Annex I countries (developed countries and the countries in economic transition) of the Convention a definite emission limitation or reduction against the year 1990 results to be achieved during the first commitment period. To be specific, it sets a target for Annex I countries as a whole to reduce emissions by 5%, with the assigned amounts for Japan 94%, USA 93%, EU 92%, and Russia 100%, with no commitments to reduce emission assigned to developing countries.

To ensure actualization of the assigned commitments, it is essential to minimize the necessary cost of compliance (emission reduction cost) in each country, and thus in the world as a whole, and thereby ensure sustainable development.[2] Now, to minimize the emission reduction cost in the world as a whole, every country and every economic entity must have an equal marginal abatement cost.[3] The assigned amount for each country, however, was determined as a result of negotiated compromises, and was not set with due consideration of each country's marginal abatement cost curve. Therefore, if, to successfully achieve the quantified commitment, each country were to reduce emissions mainly by domestic measures, such as environmental tax and voluntary agreements, which cannot be operated concomitant with the Kyoto mechanisms, this would necessarily differentiate the marginal abatement costs among countries. In other words, it is not possible to minimize reduction costs in the world as a whole.

In this sense, international emission trading is an ingenious mechanism. If a country with higher marginal abatement cost purchases, in its efforts to comply with quantified commitment, a part of a lower-cost country's assigned amount as an "emission permit" to gain emission additional to its own assigned amount, then it becomes eventually possible to attain an emission permit price that is equal between every country. The result is not only that each country can comply with its own quantified commitment, but also that the total emission reduction cost of international emission trading participant countries (Annex I countries) is minimized as the marginal abatement cost of each country equalizes through the emission permit price.

Now, developing countries that generally have lower production efficiencies and lag behind in energy savings etc., have greater opportunities to reduce greenhouse gas emissions and their marginal abatement costs are significantly less than those of developed countries. However, developing countries have no assigned amounts to be complied with, thus it is not possible to purchase the part of their assigned amount as an emission permit. Therefore, if a developed country (or economic entity) implements a project of energy saving etc. in a developing country, and the amount (credit) verified as the reduction attained by such a project can be added to the assigned amount of that developed country, then it is possible for developed countries to reduce emissions at the emission reduction costs as low as those of developing countries. Such a mechanism is the Clean Development Mechanism (CDM).[4]

If the fungibility between the credits earned by CDM and the emission permits traded in international emission trading can be maintained, and they can be traded freely among countries, then the marginal abatement cost becomes equal worldwide, including developing countries, thereby minimizing the emission reduction cost of the world. In addition, since this would allow reduction in developing countries in which emission reduction obligations based on an assigned amount have not been imposed, the emission reduction of the world as a whole would be significantly larger than the case without Kyoto mechanisms. Moreover, it would bring economic benefits to developing countries and could even expect reduction in environmental pollutants such as SOx and NOx.

Furthermore, the presence of the Kyoto mechanisms totally alters the significance of the assigned amount. Without the Kyoto mechanisms, the assigned amount means an "emission target" for which each country implements domestic reduction measures during a commitment period. With the Kyoto mechanisms, conversely it would lose the meaning of emission target, as each country could add emission volume over its own initial assigned amount by obtaining emission permits or credits from other countries. Some people seem to consider that emission quantity during a commitment period must be lower than the assigned amount, which is definitely a misconception. Under the Kyoto mechanisms, assigned amounts should be considered the initial allocation of emission permits that determine the burden sharing of emission reduction cost in the world as a whole, and should not be considered a target.

Contradictions in the Kyoto Protocol and distortion in negotiation

Considering the significance of the Kyoto mechanisms in the concept of the Kyoto Protocol, their free utilization is clearly essential for sustainable development in the world. Nevertheless, the Kyoto Protocol Art. 17 stipulates that international emission trading must be "supplemental" to "domestic actions" to meet the quantified commitment of each country.

Such a stipulation is based on the concept that the major cause of greenhouse gas concentration build-up is modernization and industrialization since the Industrial Revolution among countries numbered as the developed countries of today, and so such countries must be the ones to reduce greenhouse gas emissions. In other words, it is the "sweep your garden by yourself" concept. It sounds plausible, but, as discussed above, such a concept will not make sustainable development possible. To introduce the Kyoto mechanisms, yet not allow their utilization unless supplementary, is itself a contradiction.

Whether the Kyoto mechanisms must be supplemental or not was the focus of the post-COP3 negotiation on the Kyoto Protocol. The advocating party of "supplementary" was EU. By introducing and utilizing a so-called "bubble" system in which EU's assigned amount is reallocated among its member countries, EU would have a lower reduction cost than would be the case were a definitive amount assigned to each country individually. Recent documents by EU itself indicate that, even when EU is to implement emission reduction of twice as much as its total target of 8% reduction, EU has a marginal reduction cost of 20 euro/CO_2-ton (about $ 70 per per C ton). Compared with Japan's $ 300-900 per C ton, and the USA's $ 100-200 per C ton, this is a significant margin. If Central and Eastern European countries join the EU, EU could incorporate their "hot air" as well.

Furthermore, EU has raised, along with the issue of "supplementary", the issue of "liability" in emission permits[5], for the purpose of preventing the noncompliance caused by overselling emission permits to other countries. As seen in Chair Pronk's proposal, EU has insisted on the "commitment period reserve" system to practically restrict international emission trading. In this system, a certain volume of assigned amount must not be sold to other countries, and the remaining part of the assigned amount must be reserved. This is equal to restricting emission permit exports.[6] Moreover, if compliance with the quantified commitment by the purchase of emission permits must be strictly supplemental to domestic actions, this becomes the same as encouraging an import restriction on emission permits. To allow restrictions on both export and import of emission permits is equal to letting each country adopt a strategic restriction, in which a seller country takes advantage of its dominant position and restricts its export of emission permit to raise permit prices, and a buyer country restricts its import with an aim to pull down permit prices. Such strategic restrictions may make international emission trading practically unusable. Restrictions on the Kyoto mechanisms present no harm to the EU, but deliver a severe blow to Japan with its higher marginal abatement cost and the USA which has a greater reduction quantity. But above all, such restrictions raise the problems of obstructing emission cost minimization in the world as a whole and impeding each country's compliance with its own quantified commitment. In this context the so-called "banking" of an assigned amount for the next commitment period is also a problematical institution since it is difficult to clearly distinguish banking from strategic restriction of emission permits export by the government.

For the trade of goods, there is a principle of free and non-discriminatory import and export practices within the framework of GATT–WTO, given the past unfortunate history of import and export restrictions. The Kyoto mechanisms seek to establish a kind of property right to greenhouse gas emissions, which, until now, have been an asset of no value, and to turn them into tradable resources. For example, steel has been manufactured with materials such as iron ore and coal, but after the Kyoto Protocol it will have to add another material, called emission permits, to its line of raw materials. As we must strictly prohibit strategic restrictions by governments on the import and export of steel, iron ore and coal, emission permits should be subjected to the WTO principle in future. Assuming that some restrictions on the use of the Kyoto mechanisms were imposed, this would be equal to letting the Kyoto Protocol turn its back on sustainable development.

Environmental tax incompatible with the Kyoto Protocol

Assuming that each country will comply with its quantified commitment based on the principle of sustainable development, it is necessary to fulfill at least the following conditions upon designing domestic institutions:

(a) Condition for compliance: comply with each country's own assigned amount.

(b) Condition for reduction cost minimization: minimize the emission reduction cost of each country, in other words, a marginal abatement cost of an economic entity in its own country becomes equal to that of other economic entities.

Nonfulfillment of condition (a) means noncompliance of the Kyoto Protocol. Fulfillment of (a) must accompany the attainment of (b), otherwise a country will have a cost-wise disadvantage in international competition, with increase in problems such as unemployment and the oversea transfer of businesses. Therefore, to fulfill (b), a domestic institution for minimizing the overall reduction cost of that country must be introduced. An institution to equalize the marginal abatement cost of each economic entity in the country is just an institution for it. Such an institution is commonly called an economic instrument and good examples are environmental tax and emission trading.

However, environmental tax such as carbon tax is hardly compatible with the Kyoto Protocol, despite its attractive name. In the institution of environmental tax, it is not clear how much reduction the tax rate applied could realize. Some think that a *post facto* adjustment of tax rate based on performance would suffice. But, such is not practical for two reasons. First, even if a tax rate were adjustable based on, for example, the performance of the past one year, the real state of the economy would change in the next year due to market fluctuation and other factors; thus the tax rate would never be adjusted to a proper level. Secondly, because tax rates are the subjects of congressional resolution in many countries, frequent adjustment would be neither possible nor practical. Therefore, environmental tax would inevitably lead to emission reductions that are either over- or underfulfillment of an assigned amount.

Others may think that the above-mentioned defect could be compensated by a policy mix, using the international emission trading and other Kyoto mechanisms. The Global Warming Mitigation Measure Team of the Central Environmental Council for the Environmental Minister of Japan has adopted this notion. This policy mix is not compatible with the Kyoto Protocol either. The reason is clear. When assuming the world situation of each country adopting its own environmental tax, such a tax in each country would need to be adjusted after completion of a commitment period, since environmental tax cannot grasp exactly how much reduction is to be made during a commitment period. However, if the total of unattained reduction in the world is greater than the total of emission reduction surplus, there must be a country or countries that cannot meet compliance no matter how much money is spent. This is because there is no room for market forces to work through the price mechanism after completion of a commitment period. An institution that embraces such structural risks can hardly be consistent with the Kyoto Protocol.

Which type of emission trading institution is superior

A domestic institution consistent with the Kyoto Protocol is emission trading. One must note that there are many types of emission trading institutions and those satisfying the above conditions (a) and (b) are quite limited. In addition, it is important to fulfill the following condition when designing emission trading institutions, whether international or domestic:

(c) Condition for minimizing implementation cost: minimize the cost of implementing the institution.

In other words, to ensure the implementation of an institution, both the government and private sector must bear some costs of implementation. For example, to implement the emission trading institution, the government must bear the cost of monitoring etc. of emission quantity, whereas the emission source party must bear various costs for compliance, including those for measuring and recording emission quantity. If any trouble in implementation of the institution arises, then judicial costs may need to be considered. Here, those costs of the government and private sector are treated as the implementation costs. Needless to say, it is undoubtedly more desirable to have an institution of low implementation cost.

A desirable type of emission trading in the light of conditions (a) to (c) each is reviewed below.

First of all, tradable permits can be classified into two types:

(i) Set a baseline on a certain level of emission from each type of source and, whenever an actual emission is less than the baseline, recognize this margin as a tradable credit for emission.

(ii) Divide the assigned amount for each country into small units and recognize such units as tradable permits for emission.

Between the above two, (i) is more popular among some people in their hasty conclusion that, to improve unit requirement by energy savings and other measures, one can obtain credits and make businesses using such credits. Whether it is possible to actually sell such credits or not depends on the credit prices, in other words, the costs of energy savings, etc., at the back of credit prices. As in the case of trading conventional goods, trading is done on the basis of a party with higher marginal cost of production buying from a party with lower marginal cost. Were this fact noted, the aforesaid hasty conclusion and misunderstanding would never have arisen. In the first place, (i) presents a problem of making compliance of a quantified commitment difficult, no matter how much the amount is spent as implementation cost, since it is difficult to correlate the assigned amount of each country and the baseline set for an individual economic entity or for a project, especially since credit verification is quite labor consuming. Therefore, (i) is not fully compatible with the Kyoto Protocol.

On the other hand, for (ii) it is easier to relate emission permit issuances to the assigned amount, thus facilitating use of the international emission trading, that is trading of the assigned amount for each country. If control of emission permits is done appropriately, then (ii) may be able to fulfill all the three conditions (a) to (c). Whether this is realized would depend on how the monitoring between the face volume of emission permits and actual emission quantities of their holders is done. Control methods can be classified into two broad types:

(1) control at the downstream (emission) phase,
(2) control at the upstream (import or production) phase.

In discussing emission trading, it is easy to imagine the emission phase control type as a stereotype, uplifted by the inspirational term 'emission' trading. However, the number of emitters is as many as that of consumers and those of households and firms. Considering how labor consuming it is to grasp emission at individual emission sources and to reconcile it with emission permits, one has to admit that the unconquerable wall in terms of implementation costs deters the control at this phase. For this reason some persons say that the control must target major emission sources such as big companies. However, such an institution fails to motivate those not subjected to the institution to own emission permits or to reduce emissions, so there will be a difference in marginal abatement costs between those subjected and not subjected, making it impossible to fulfill the condition of reduction cost minimization. Others may say it can be supplemented by imposing environmental taxes on nonsubjected small- and medium-size companies, but this is hardly a good idea, considering the aforementioned flow of environmental tax. Nonetheless, it is absurd to introduce an institution that needs huge implementation costs, under which consumers and small- and medium-size companies must be troubled about the procurement of emission permits.

On the contrary, control at the upstream phase is remarkably easy. Not only in fossil fuel importing countries, but also in producing countries such as the USA and Russia, the total number of importers and producers is so insignificant compared with the enormous numbers of emission sources. Moreover, upon assuming, for example, the import (or production) of fossil fuels, to reconcile the amount of import (or production) with emission permits[7] at Customs (or at the time of domestic shipment) does not require much in implementation costs. Control at the upstream phase should facilitate and ensure compliance of a quantified commitment at less implementation cost.

But would this satisfy the condition of cost minimization? Some persons are apprehensive that by placing the control point at the upstream phase, the emissions themselves would not be directly regulated nor efficiently reduced. This is a groundless apprehension, however. As mentioned earlier, the introduction of emission permits signifies that they become one of the essential raw materials, like iron ore and coal, in the production of steel. To place the control point upstream demands the procurement of emission permits at the Customs (or domestic shipment) corresponding to the amount of greenhouse gas emission from coal. Since importers themselves are not the users of coal, they need to procure emission permits based upon the forecasted demand for emission permits among their customer steel makers. The demand quantity will be the greenhouse gas emission quantity that can realize the maximization of profits among steel makers. In other words, it is the quantity of demand that corresponds to the emission quantity at the points where the value of steel produced, with marginal increase over a unit emission increase, multiplied by steel price (marginal value productivity of emission) is equal to the market price of emission permits. By selecting the emission quantity at such points as well as the steel production corresponding to that quantity, it is possible to realize the most efficient utilization of greenhouse gas emissions, under the market prices, at that time, of emission permits and steel. A view of this in terms of marginal abatement cost would be as follows. A steel maker per se will reduce emissions in the range where self-reduction is lower in cost than the case of letting importer procure emission permits (marginal abatement cost is lower than emission permit price). If the reduction quantity increases, and the procurement of importers costs less than steel maker reduction (marginal abatement cost is equal or higher than emission permit price), then a steel maker will choose emission permit procurement through importers. Therefore, there must be an equilibrium point where "emission permit price = marginal value productivity per unit of greenhouse gas = marginal abatement cost." Such a relationship exists not only for steel makers but for every economic entity, so that the upstream control can meet the cost minimization condition. If each country were to adopt an institution to make emission permits or CDM credits issued by other governments valid indiscriminately along with those issued by its own government, then emission permits would be traded at the international market price, resulting in global equalization of marginal abatement cost. Therefore, an institution to control the possession of emission permits upstream would fulfill all three conditions of (a) to (c).[8]

Emission trading based on assigned amounts for each country may be of two types depending on the initial allocation of emission permits to each individual economic entity, although it is difficult to assess them since conditions (a) to (c) do not clearly differentiate. Two types of emission permit allocation are allocation *in gratis* (grandfathering) and allocation *in onerous*. Allocation *in gratis* will be given to each economic entity based on some kind of a standard, but emission permits themselves have market values based on the limitation to total allowable emissions, so *in gratis* allocation is the same as a gift or a subsidy to the recipients. Accordingly, allocation *in gratis* may disturb competitive relations among countries.[9] How to implement the allocation of such benefits on what basis is a serious problem.[10]

On the contrary, onerous allocation is to sell emission permits in a market through auctioning etc., and, in this case, the revenues of permit sales will go to a government. If the government uses all such revenues to reduce social insurances and to cut the taxes that have greater distortional effects on existing resource allocation, then there will be so-called double dividends. It may be desirable also to use the revenues for technological development related to greenhouse gas.

A pitfall of emission trading

The foregoing leads to the conclusion that as long as assessing emission trading institutions is based on the conditions of (a) to (c), the best institution would seem to be the emission trading institution founded on assigned amounts with a control point upstream that indiscriminately accepts the validity of emission permits or CDM credits issued by other country governments as well as those issued by the country's own government. However, no such type of institution has been accepted in any country thus far. For instance, the institution adopted in UK differs in many respects.

The gist of the UK type institution is to allow 80% exemption in tax rate on the climate change levy (CCL) for voluntary agreement participants and, in addition, allow participation in emission trading. This institution has severe defects. First of all, it embraces a system called CCL in which emission reduction quantity cannot be predetermined as in the case of environmental tax, and most of the emission permits are generated by improvement in unit requirement, which hardly relates to the amount assigned to each country. This type of institution is unlikely to clear condition (a). Also, between those entered in voluntary agreement and allowed to participate in emission trading, and those on which CCL is imposed, a difference in marginal abatement cost arises, so a problem arises of not being able to clear the condition (b). This can be clearly shown when assuming a case wherein the reduction amount with CCL

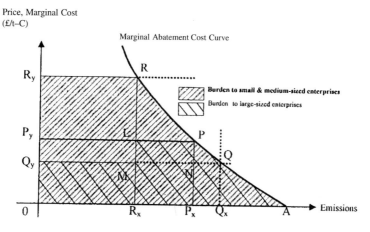

Fig. 5.1:Inefficiency and Inequity of the UK Type Emission Trade Institution

imposed on an economic entity is equal to the commitment amount of the same entity in a voluntary agreement, as in Figure. 5.1. Moreover, one should note that cost burden would vary widely depending on whether CCL or voluntary agreement and emission trade is selected.

Figure 5.1. illustrates the assumed case in which the reduction quantity upon the selection of CCL is equal to the commitment reduction quantity upon voluntary agreement for the same economic entity. The emission quantity is measured along the horizontal axis of the figure and marginal abatement cost and price level of emission permit along the vertical axis. Now let OA be the emission quantity of business as usual (BaU), then the marginal abatement cost curve is described as ascending leftward from A since the greater the emission reduction (the less emission quantity), the more costly measures and instruments must be employed. If the tax rate OR_y is imposed, R_xA is reduction quantity of this economic entity and OR_x is emission quantity for which CCL must be paid. This means the marginal abatement cost at actual emission quantity (OR_x) is OR_y and the total cost burden for this economic entity is the sum of reduction cost RR_xA and tax expense R_yOR_xR. If this economic entity chooses the voluntary agreement in which reduction commitment is R_xA, the tax expense shifts from R_yOR_xR to Q_yOR_xM since the reduced tax rate OQ_y is applied, although the marginal abatement cost at emission quantity (OR_x) remains OR_y. Here let OP_y be the price level of emission permits.[11] Reduction quantity and emission quantity are determined by the intersection of the marginal abatement cost curve and price level line, so P_xA and OP_x are reduction quantity and emission quantity, respectively. This economic entity must purchase emission permits for $R_xP_x(=OP_x - OR_x)$, and marginal abatement cost at actual emission quantity (OP_x) shifts from OR_y to OP_y in response to the change of actual emission quantity from OR_x to OP_x. Therefore, should this economic entity participate in emission trading, it could decrease the marginal abatement cost and the total cost burden, which is the sum of reduction cost PP_xA, tax expense Q_yOR_xM, and emission permits expense LR_xP_xP.

Thus, it might be better to let everyone participate in such voluntary agreement and emission trading, but the control point is placed on the emission phase; hence the greater the number of participants, the greater the problems of implementation cost (especially monitoring cost). As a matter of fact, it would be difficult to allow small- and medium-size firms to participate in emission trading because of low cost-effectiveness of monitoring.

This might lead to a serious equity problem not only in cost burden, but also in competitiveness between large size firms and small- and medium-size ones, since product price, a key element of competitiveness, is dependent on marginal cost, which includes marginal abatement cost.

"EU Emissions Trading Scheme" scheduled to be introduced from 2005 is also a scheme with such problems as follows. The scheme, first of all, does not provide a clear relationship between the tradable allowances under the EU directive and EU's assigned amounts under the Kyoto Protocol, and the direct fungibility with assigned amounts of nonmember countries is denied. Secondly, the participating sectors of the scheme represent only a part of the industry sector. Thirdly, the monitoring points is set at the emission phase, which has higher implementation cost. Therefore, the scheme does not fulfill any of conditions (a) to (c). The reason why the UK or the EU can adopt such inefficient schemes is because the Kyoto Protocol has allocated relatively generous assigned amounts to those countries compared to their energy and economic situations.

Conclusion

Should global warming present a risk of severe disasters in future for the Earth and human beings, one must seriously and sincerely address the Kyoto Protocol. What should be done is clear from the above discussion.

There should be no restrictions on the utilization of Kyoto mechanisms such as international emission trading in Protocol, whatever the name of the restriction. Moreover, WTO rule should guarantee free and indiscriminate implementation of emission trading and CDM.

If every country is to adopt a domestic emission trading system based on assigned amounts with the control point at the upstream phase, then globally it must be possible to realize the reduction of greenhouse gas with certainty at the minimum reduction cost and minimum implementation cost. Among domestic institutions, there are some, e.g.

environmental tax, that may fail if other countries adopt the same institution. Therefore, the Kyoto Protocol should have referred to the framework of domestic institutions right from the beginning.

What is required now is to actively propose a framework that can fully attain the spirit of the Kyoto Protocol and lead to sustainable development of our world. Most of all, it is important to note that the only countries that have assigned amounts can utilize the most beneficial type of emission trading system which meet the aforementioned conditions (a) to (c).

Footnotes

1. "Sustainable development" is possible only when the conservation of environment and natural resources, economic development, and social progress will not adversely affect, but rather complement, each other. Any economic development that merely consumes environment and natural resources will never last, while environmental improvement to be done by low cost effectiveness will increase the burden on the economy and society, and may lead to unemployment and poverty. The effects of environmental and natural resource degradation tend to occur intensively in poorer regions, and the poverty itself may lead to degradation of the environment and natural resources through deforestation and other activities. Therefore, to have sustainable development requires a comprehensive strategy that can mutually enhance the three aforesaid elements.

2. In fact, the Kyoto Protocol defines in its Art. 2.3 "in consideration of article 3 ... need to work on the implementation of policies and measures" and Art. 3 states "policies and measures to address climate change must have larger cost effectiveness that can bring global benefits at the minimum cost possible." Moreover, the Kyoto Protocol and the Convention mention "sustainable development" in many places. These facts can lead to the interpretation that the Kyoto Protocol intends to minimize emission reduction cost of each country and the whole world.

3. The logic behind a certain reduction target based on the principles of "cost effectiveness" and "sustainable development" meaning minimization of reduction costs for each country and the whole world can be explained as follows:

   As reduction quantity increases, room for energy savings and fuel switching decreases, so that for each unit addition of reduction quantity, the cost to attain such reduction inevitably rises. Therefore, to minimize reduction cost of society as a whole, the reduction must start from an economic entity that has the lowest cost needed to reduce one additional carbon unit (marginal abatement cost) at the initial state. With increase in reduction quantity, this particular economic entity's cost increases. When it becomes equal to the cost of the second lowest economic entity, both entities increase reduction quantity at the same rate, and continue the same for the third lowest cost entity and so on, until the total reduction of economic entities reaches the quantified commitment of society as a whole. By doing this, reduction is done in the order of lower cost requirement to reduce each additional unit of carbon; thus society as a whole can attain quantified commitment at minimum cost. At the time of attainment, the marginal abatement cost must be equal among economic entities.

This condition would be the same for one country or the world as a whole, by simply replacing "society as a whole" in the above sentences with "a nation" or "world as a whole". In short, if the marginal abatement cost of each individual economic entity in the world were equal, the reduction costs of a country and the whole world would be minimized.

4. Kyoto mechanisms include Joint Implementation (JI) in addition to the international emission trading and CDM. JI is an instrument to implement a project of energy saving etc. jointly among Annex I countries and to make it possible to convert the reduction quantity derived from such project to credit. Credit must be offset by transfer of the assigned amount of the host country, otherwise the total assigned amounts of both countries will exceed their initial assigned amounts. Therefore the provision of JI in the Kyoto Protocol may not be necessary.

5. The problem is how to maintain the validity of emission permits when the purchaser of the said permits cannot comply with its assigned amount.

6. In COP7 the Parties reached the accord that not only the assigned amount to each country but also the assigned amount and credit transferred from the other countries can be used for the commitment period reserve. This accord relieves the restrictive characteristic of commitment reserve period. But it should be noted that the commitment period reserve would become essentially unnecessary if every country were to adopt such an institution as would ensure compliance with its quantified commitment.

7. If the control point is placed at the upstream phase, "emission permits" must be considered rather as "import permits" and "production permits".

8. Normally, efficiencies are not influenced by initial allocation.

9. The trade distortive effects caused by *in gratis* allocation of emission permits and exemption from environmental tax should be one of the main subjects in the New Round of WTO in the near future.

10. Needless to say, the allocation of assigned amounts for each country in the Kyoto Protocol is an *in gratis* allocation.

11. Normally OP_y is thought to be lower than OR_y; otherwise every economic entity would continue to pay tax for emission.

References

Baron, R. 1999. An Assessment of Liability Rules for International Emission Trading. IEA Information Paper, IEA.

Bohm, P. 1998. Public investment issues and efficient climate change policy, part 1, In: The Welfare State, Public Investment, and Growth. H. Shibata and T. Ihori (eds.) Springer Verlag, Tokyo.

Ellerman, A.D. Obstacles to CO_2 Trading: A familiar problem. In: Climate Change Policy: Practical Strategies to Promote Economic Growth and Environmental Quality. C. Walker et al (eds). American Council for Capital Formation Center for Policy Research, Washington DC.

Goulder, L.H. 1999. Confronting Distribution (and Political) Impacts of CO_2 Abatement Policies: What Does It Cost? Report of Committee on the Compatibility between Economic Development and Environmental Protection, GISPRI.

Hizen, Y. and Saijo T. 2002. Price disclosure, marginal abatement cost information and market power in a bilateral GHG emissions trading experiment. In: Experimental Business Research. R. Zwick and A. Rapoport (eds.). Kluwer Acad. Publ., Dordrecht, Netherlands.

Hizen, Y. and Saijo, T. 2001. Designing GHG emissions trading institutions in Kyoto Potocol: An expermental approach. Environmental Modelling and Software 16(6).

Hizen, Y., Kusakawa T., Niizawa, H., and Saijo, T. 2001. Two patterns of price dynamics observed in greenhouse gases emissions trading experiments: An application of points equilibrium, unpublished manuscript, ISER, Osaka University.

IPCC. Climate Change 2001, Mitigation. (Contribution of Working Group III to the Third Assessment Report of the Intergovernmental Panel on Climate Change), Cambridge University Press, Cambridge, UK.

Kusakawa, T. and Saijo, T. 2001. Liability Experiments: Seller's or Buyer's? Unpublished manuscript, ISER, Osaka University.

Niizawa, H. 2001. Design of domestic policy for preventing global warming. Ann. Rev. Kobe Univ. Commerce, 31, 55-72.

Saijo, T. 2001. The Kyoto Protocol and global environmental strategies of the EU, the U.S. and Japan: A Perspective from Japan, ISER,unpublished manuscript.

Yasumoto, A. 2001. Kyoto mechanism and domestic institutions. J. Japan Inst. Energy, 80 (893).

Yasumoto, A. 2001. Domestic institution and voluntary plan for mitigation of global warming. Energy Forum no. 562.

Yasumoto, A. 2000. Emission trading, a key for mitigation of global warming. Centre of Public Opinion (Chuou-kouron), Tokyo, Japan.

CHAPTER 6

Can the Kyoto Goals be Achieved Using the Oceans as Sinks?

Paul Johnston[a], David Santillo[a] and Bill Hare[b]
[a]Greenpeace Research Laboratories, Department of Biological Sciences
University of Exeter, Exeter EX4 4PS, UK
[b]Visiting Scientist, Potsdam Institute for Climate Impact Research (PIK)
Telegrafenberg A31, P.O. Box 60 12 03, 14412 Potsdam, Germany

Summary

This chapter outlines technological proposals to accelerate the sequestration of the main greenhouse gas, carbon dioxide, into oceanic reservoirs, either through direct disposal of fossil fuel-derived CO_2 in the water column or at the seabed or through attempts to enhance biological uptake through large-scale ocean fertilization programmes. These proposals are evaluated against a number of criteria necessary for acceptance of ocean disposal/ sequestration as a sustainable contribution to the goals and commitments of the Kyoto Protocol to the United Nations Framework Convention on Climate Change (UNFCCC), as well as consistency with other international legal instruments. The chapter concludes that the ultimate objective of the UNFCCC, namely stabilization of greenhouse gas concentrations at levels that would prevent dangerous interference with the climate system, cannot be met through such Although theoretically capable of reducing atmospheric CO_2 levels in the coming century to a similar level as an emission reductions at source strategy, CO_2 disposal in the ocean will result in higher CO_2 levels, and hence larger climate changes and sea level rise in future centuries. Given the threat these technologies present to effective implementation of the Kyoto Protocol and to sustainable energy and waste management more broadly, efforts should be made to ensure that ocean disposal/sequestration of CO_2 cannot be used to offset against commitments to emissions reductions for greenhouse gases.

Introduction

In 1992, following the United Nations Conference on Environment and Development (UNCED, Rio de Janeiro, June 1992), 166 nations signed the United Nations Framework Convention on Climate Change (UNFCCC 1992). The Convention established the objective to achieve *"stabilisation of greenhouse gas concentrations in the atmosphere at a level that would prevent dangerous interference with the climate system"*. The Kyoto Protocol to this Convention, adopted in 1997 (UN 1997), sets legally binding commitments to reductions and/or limitations in emissions of greenhouse gases specific to each of the industrialized countries listed in Annex B to the Protocol. The ratifications of Poland, Canada and New Zealand during December 2002 brought the number of Parties to the Protocol to beyond 100 (see http://unfccc.int), though without the ratification of the Russian Federation, the total of industrialized country CO_2 emissions covered remains short of the 55% required for entry into force of the Protocol.

Although many gases arising from anthropogenic activity contribute to climate forcing, CO_2 is quantitatively by far the greatest contributor (60% of the total radiative forcing of the well mixed greenhouse gases[1] and set to rise further).

Despite the key intention of the Kyoto Protocol being the reduction of emissions at source, a number of proposed approaches to climate change mitigation rely instead on the development of technical mechanisms for the "management" of anthropogenic CO_2 in the environment. In broad terms, such approaches endeavour to limit the magnitude of atmospheric increases in CO_2 concentration through attempts either to influence the partitioning of carbon between different environmental compartments or to isolate generated CO_2 from the atmosphere over long time-scales. Beyond the manipulation of terrestrial carbon stocks through management of land-use changes, proposed strategies include direct disposal of liquid or solid CO_2 at sea (both above and beneath the seabed) and the enhancement of uptake of CO_2 by natural biological processes in the oceans (through ocean "fertilisation"). In each case, these sea-based techniques involve a high degree of intervention with ecosystems and with the global carbon cycle, involving manipulations on the scale of "planetary engineering".

At best these proposals are subject to enormous uncertainties regarding their likely effectiveness in stabilizing atmospheric CO_2 concentrations, while having the potential for adverse impacts at local, regional and global level. Some nevertheless regard them as necessary contributions towards combating climate change (e.g. Hoffert et al. 2002). At worst they are portrayed not merely as measures to mitigate climate change to which we are already committed but also as permitting continued exploitation of fossil fuel reserves, with the attendant environmental problems.

This chapter focuses on issues of technical feasibility, environmental impact and legality under existing international law of ocean related storage proposals for CO_2. It asks whether such interventions with the oceanic component of the climate system should be employed as a contribution to meeting the goals and commitments laid down in the Framework Convention on Climate Change and its Kyoto Protocol.

The UNFCCC and its Kyoto Protocol

In order to put this latter question in context, it is first necessary to understand what the ultimate objective of the UNFCCC and its Kyoto Protocol is, and to review the mechanisms envisaged in order to meet this goal. Article 2 of the Convention sets the objective of stabilizing greenhouse gas concentrations at levels that would prevent dangerous interference with the climate system, with the latter being defined as "the totality of the atmosphere, hydrosphere, biosphere and geosphere and their interactions". The specific reduction target of 5% against 1990 emissions for industrialised countries set for the first commitment period of the Kyoto Protocol (2008-2102) is generally recognised as only the beginning of what will need to be much more substantial cuts in emissions in future commitment periods if the objective of the Convention itself is to be met.

Under Articles 3.3 and 3.4, the Protocol permits the limited use of terrestrial carbon offsets (i.e. verifiable changes in terrestrial carbon stocks) against emissions commitments, though this has been, and is likely to remain one of the most controversial aspects of the Protocol. Significantly, these provisions mean, for example, that an ocean fertilisation "sink" cannot be counted towards Annex B Parties' obligations to limit emissions.

Articles 3.1 and 3.2 of the Protocol specify that industrialised Parties listed in Annex B must limit or reduce their emissions of greenhouse gases according to an agreed schedule. The manner in which sinks can be used to offset emissions is set out in subsequent Articles 3.3 and 3.4. Article 3.3 limits this to:

"direct human-induced land-use change and forestry activities, limited to afforestation, reforestation and deforestation since 1990".

Moreover, removal by such sinks must be amenable to quantification and reporting in a "transparent and verifiable manner" (in a legally binding regime developed under Articles 5, 7 and 8 of the Protocol). Article 3.4 provides for additional sink activities to be agreed by the Conference of the Parties (COP) to the FCCC (serving as the meeting of the Parties to the Protocol), but these are still limited to:

"additional human-induced activities related to changes in greenhouse gas emissions by sources and removals by sinks in the agricultural soils and the land-use change and forestry categories".[2]

Theoretically, certain other provisions of the Protocol could allow the use of oceanic storage of carbon for credits. These include Article 6 of the Protocol, which permits Joint Implementation projects between industrialized countries for credits and the Clean Development Mechanism under Article 12, which allows for developed countries to purchase credits from projects done in developing countries. In each case, however, specific enabling decisions of the Parties would be required, and this is unlikely without a full review of the issues. An IPCC special report on CO_2 storage technologies is now in preparation, including a section on the implications of carbon capture and storage for emission reporting and monitoring provisions of the Convention and Protocol.

An increasing body of research has focused on strategies to reduce the scale of further atmospheric increases in CO_2 concentration through large-scale human intervention in oceanic carbon cycles. As noted above, these schemes involve either direct introduction of fossil-fuel derived CO_2 or enhancement of biological uptake of CO_2 from the atmosphere to the oceans. In the most part, proposals have been developed largely as engineering concepts, with a focus on technical and economic feasibility and with relatively little regard either for adverse impacts on the receiving environment or for the existing legal instruments within which they would necessarily operate. This legal regime includes not only the provisions of the Kyoto Protocol outlined above, but also those of Conventions established to ensure protection of the marine environment from human activities (specifically the 1972 London Convention on dumping of wastes at sea and, more generally, the UN Convention on Law of the Sea (UNCLOS)).

The attractiveness of ocean disposal and/or sequestration strategies for climate change mitigation is based primarily on the scale of the oceans as a reservoir of carbon and, therefore, as an apparent sink for further emissions. Indeed, the scale of the ocean carbon reservoir relative to terrestrial and atmospheric reservoirs is often cited as a justification for the pursuit of such technological approaches (Table 6.1). The importance of the oceans as a reservoir and ultimate sink of carbon is recognised by both the UNFCCC and the Kyoto Protocol, and is not in dispute. What is questionable, however, is whether human interventions of the nature proposed, ostensibly designed to accelerate uptake of anthropogenic CO_2 emissions by the oceans, are consistent with the requirements under Article 2 of the UNFCCC to prevent dangerous interference with the climate system and Article 4.1 (d) to "sustainable management...conservation and enhancement" of the ocean carbon reservoir:

"promote sustainable management, and promote and cooperate in the conservation and enhancement, as appropriate, of sinks and reservoirs of all greenhouse gases not controlled by the Montreal Protocol, including biomass, forests and oceans as well as other terrestrial, coastal and marine ecosystems;"

Whilst proponents of ocean carbon sequestration see this as a mandate for pursuing their research under the Convention (e.g. Adams et al. 2002), it could well be argued that ocean disposal/sequestration of CO_2 is neither sustainable in concept or practice nor consistent with the conservation or enhancement of marine ecosystems. Under Article 2, two issues arise:

- the deliberate direct addition of CO_2 to the oceans constitutes a potentially dangerous source of interference to the climate system;
- a sizeable fraction of this CO_2 will reappear in the atmosphere over centuries, leading to higher CO_2 concentrations than would have occurred had emissions been reduced at source[3]:

Table 6.1. Estimates of carbon reservoirs of different biosphere compartments and order of magnitude estimates of potential capacities for carbon sequestration (adapted from Herzog 2001)

Reservoir size	Gt (billion tonnes) carbon
Oceans	44 000
Atmosphere	750
Terrestrial	2 200
Sequestration potential	**Gt (billion tonnes) carbon**
Oceans	1000s
Deep saline formations	100s-1000s
Depleted oil and gas reservoirs	100s
Coal seams	10s-100s
Terrestrial	10s

From a more technical viewpoint, it is also questionable that the capacity of ocean sinks can be related in such a simplistic way to the estimated scale of the global carbon reservoir the oceans represent. Most proposals under consideration necessarily involve introductions of large quantities of CO_2 at specific sites, or fertilisation to increase atmospheric drawdown within a particular region, such that the specific characteristics of those locations will have a substantial impact on the fate of the carbon so introduced.

Moreover, despite increasing research and knowledge, our understanding of oceanic carbon cycles, and therefore the manner in which such interventions would interact with and impact upon them, remains limited. In short, it is difficult to see how claims that ocean disposal/sequestration of CO_2 merely represents an acceleration of the natural uptake of CO_2 from the atmosphere can be verified and justified given the current state of knowledge of ocean carbon fluxes.

Outline of the ocean disposal and sequestration technologies proposed

Characteristically the technologies proposed involve very large scale projects designed either to prevent carbon dioxide reaching the atmosphere (CO_2

disposal and or storage) or to sequester carbon from the atmosphere. Proposals fall into three distinct categories:

- The fertilisation of open waters to increase primary production and hence to absorb more carbon in fixed form which will eventually be incorporated into the ocean sediments
- Disposal of captured carbon dioxide directly into oceanic waters.
- Injection of captured CO_2 into sub-seabed geological formations.

All of these technologies would need to be implemented on a global scale to achieve their designed effect on the working of the planet as whole (Marland, 1996) as they are designed to modify, through human intervention, the operation of global scale biogeochemical cycles, in this case the carbon cycle. For the purposes of this chapter, which specifically addresses the applicability of the oceans as sinks under Kyoto, discussion is focused primarily on the first two of these concepts, involving interaction with the ocean carbon reservoir itself.

Ocean Fertilisation

Fertilisation with iron

Iron is a limiting micro-nutrient for algal growth in some large areas of the worlds oceans. Ocean fertilisation with iron has been theoretically and, to an extent, practically explored. A series of experiments conducted in the Equatorial Pacific Ocean (IRONEX) were designed to assess the scope for increased algal production through supply of the limiting nutrient over an area of 64 km^2. In response to a single introduction of iron to these high nitrate/low chlorophyll waters, a biological response was observed. A second experiment in which multiple iron additions were made showed that both particulate and dissolved organic carbon did increase, while dissolved CO_2 concentrations decreased. Significantly, the researchers stated that the experiments were "not intended as preliminary steps to climate manipulation". Inevitably, however, these studies have been discussed in relation to their applicability to climate change mitigation (Ormerod & Angel, 1998) and modelling studies have been carried out based upon the concept.

Models have shown (see: Ormerod & Angel, 1998) that the largest effect of iron fertilisation would be in the Southern Ocean, where a doubling of primary production could take place (albeit concentrated over a small proportion of the total area). However, the draw-down of atmospheric carbon dioxide would be less than that implied by the amount of new primary production generated since some of the carbon dioxide would be supplied by changes in the seawater bicarbonate equilibrium. Efficiency may also be impacted by unpredicted secondary effects on planktonic community structure and/or food webs as a result of fertilisation.

Moreover, the nature, scale and permanence of the fertilisation operation required to sustain increased carbon assimilation likely render it infeasible in practical terms. It has been estimated, for example, that for this approach to sequester 0.5 Gt carbon *per annum* would require 2700 ships or 200 aircraft to treat 1.7×10^7 km^{-2} of sea surface with around 470,000 tonnes of iron per year (Sarmiento & Orr, 1991) and on a continuous basis.

Fertilisation with macro-nutrients

Addition of the macro-nutrients, nitrate and phosphate to seawater has also been proposed (e.g. Shoji & Jones 2001). The concept has been investigated in practical terms through the Norwegian MARICULT project, geared primarily to increase supplies of food and natural raw materials (Ormerod & Angel, 1998). Nonetheless, the idea that this might increase oceanic uptake of CO_2 has not gone unrecognised. Most of the potential impacts of this strategy are common to the iron fertilisation concept. In the case of macro-nutrients, however, many of these impacts have been observed in natural systems impacted by nutrients introduced *via* sewage or agricultural sources. Ecosystem changes, anoxic waters and the appearance of nuisance species have all been documented in coastal waters subject to enhanced macro-nutrient loading (see: Johnston et al., 1998). Ultimately, the effectiveness of such approaches in enhancing carbon sequestration in the longer term are highly uncertain, while the risks of undesirable secondary impacts are substantial (Trull et al. 2001).

Disposal of CO_2 into ocean waters

Mitigation of climate change impacts by means of the direct introduction of fossil fuel-derived carbon dioxide into marine waters was first proposed in 1977. Thus far, proposals for ocean disposal of carbon dioxide involve one of three methods of introduction:

- Introduction by pipeline into deepwater followed by dissolution (see Drange et al. 2001)
- Dispersion following discharge of dry-ice blocks or liquid CO_2 from a ship
- Formation of a lake of liquid CO_2 in the deep ocean (see Brewer, 2000)

These concepts are illustrated in Figure 6.1a-c.

In practice, the discharge of dry-ice blocks is likely to prove much more expensive than production of liquid carbon dioxide. It costs around twice as much to produce solid as opposed to liquid CO_2 (Golomb, 1993). The solid material would tend to sink; it has been calculated that cubes of 3-4m dimension would sink below 3000m before 50% of the CO_2 dissolved. One idea which has grown from this is to use shaped part solid/part liquid penetrators, 4-5m in diameter and 20-40m in length which sink and embed

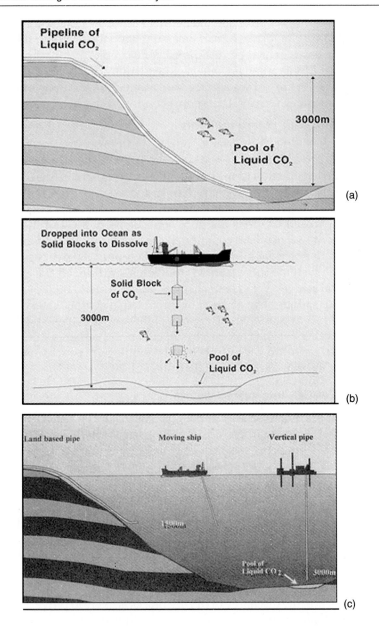

Fig. 6.1a-c: Concepts for the deep-sea disposal of carbon dioxide captured from power and industrial plants. These involve the discharge of liquid carbon dioxide from land based pipelines into deep water (Figure a) or the disposal of blocks of dry ice directly from vessels (Figure b). Figure (c) shows the proposed disposal of liquid carbon dioxide from a towed pipeline into midocean waters and the discharge of liquid carbon dioxide onto the seabed from a fixed platform.

themselves in the sea-floor (Guevel et al., 1996). Once again, the energy penalty is likely to be high and a large infrastructural cost is implied given that capacity will need to exist to handle the approximately 500 tonnes of CO_2 produced by a 500MW coal fired power plant each hour.

Most research effort has focused on the introduction of CO_2 by pipeline. At depths of less than 500m, introduction of CO_2 will create bubble plumes since, at ambient pressures and temperatures, carbon dioxide exists as a gas. These bubbles will dissolve in the seawater and, it is suggested, become trapped below the ocean thermocline. Although in engineering terms this is a relatively simple option, the retention time of the gas would be relatively short (around 50 years).

At prevailing temperatures and pressures between 500m and 3000m, carbon dioxide exists as a positively buoyant liquid. Hence, it will form a droplet plume, perhaps with the droplets covered in film of hydrate which could slow dissolution. Below 3000m, it is anticipated that CO_2 will form a dense liquid plume which could ultimately form a lake of liquid CO_2 on the sea bottom (Adams et al., 1995) with a clathrate surface. Retention times in these cases should be much greater since deep water exchanges with surface water at a much lower rate than surface water interacts with the atmosphere in the upper mixed layer of the sea. Calculations have suggested that a lake of CO_2 on the deep ocean floor derived from 1GW of coal fired power plant over ten years of operation could cover $654,500m^2$ of sea floor to a depth of 80.9m and contain 58 million tonnes of the liquefied gas (Wong & Hirai, 1997). The assumption that the pool would remain 80.9m in depth is merely a hypothetical construct to operate the mathematical model. Unless constrained by local topography, it seems reasonable to assume that, in practice, the liquid CO_2 would spread out to form a thinner layer over a much wider area, with the consequence of more widespread impacts.

Ocean disposal/sequestration: a sustainable strategy?

A diversity of concerns arise in relation to the proposals outlined above to exploit the oceans as a "storage" or disposal site for continued anthropogenic emissions of CO_2. To be justifiable as measures contributing to the attainment of Kyoto commitments, ocean disposal and sequestration programmes would need to satisfy a number of key criteria, which may be summarised as below:

1. **consistent** with Article 2 of the UNFCCC - the measure must not contradict or prejudice the stabilization of greenhouse gas concentrations at levels that would prevent dangerous climate change;

2. **predictable and effective** – they would need to be founded upon an understanding of marine circulation patterns and carbon cycles sufficient to allow reliable prediction of behaviour and effects

following intervention and to guarantee containment over effective timescales;

3. **sustainable** – such programmes should not conflict with basic principles of sustainability, including intergenerational responsibility, avoiding systematic changes in chemical composition and maintenance of the bases of biological diversity and productivity (including at the abyssal seafloor);

4. **verifiable** –the amounts of carbon sequestered would need to be quantifiable and verifiable;

5. **legal** – all such sea-based activities would clearly need to be consistent with other legislative instruments governing the use and protection of the marine environment;

6. **efficient** – the additional costs of such operations, in terms of CO_2 generated, must not in themselves significantly add to the equilibrium concentrations of CO_2 in the atmosphere by contrast with other measures, either in the short or the long-term; and

7. **complementary** – the pursuit of any such measures must not restrict the availability or effectiveness of other measures to address climate change, especially the reduction of CO_2 generation and emissions at source through development of alternative technologies and energy sources.

These criteria provide a framework against which the acceptability of current proposals for the direct ocean disposal and/or enhanced biological uptake of CO_2 through ocean fertilisation can be evaluated. It must be said, however, that to date, the development of the ocean disposal/sequestration concept has taken place largely within a technological nimbus, seemingly insulated from these more holistic considerations, despite the magnitude of the questions and uncertainties that surround them.

Do we know enough about ocean carbon cycles for ocean disposal/ sequestration of CO_2 to be predictable and effective?

An understanding of the role of the oceans in the carbon cycle is central to any evaluation of the likely impacts of ocean disposal of CO_2 and associated storage techniques. Dominating global carbon reserves, the oceans play a large part in the carbon cycle. In simple terms, it is the equilibrium which exists between ocean carbon dioxide concentrations and those in the atmosphere, and the way in which this equilibrium is driven, that governs atmospheric carbon dioxide concentrations in the long term. The uncertainties and limitations to knowledge of the global carbon cycle, and of the specific role of the oceans, have recently been highlighted by Falkowski et al. (2000). In large part these limitations are due to poor spatial and temporal resolution of understanding of these processes in the ocean despite the data generated by large ocean research programmes on this topic and to the resultant inadequacies of the mathematical models applied to such studies (Follows et al., 1996; Ormerod, 1996). CO_2 uptake from the oceans has largely been inferred from calculations based upon sparse total inorganic carbon measurements. Improving the precision of these estimates by suitably intense sampling programmes is likely to be some decades into the future (Goyet et al., 1997).

Uncertainties in the capacity of the ocean "sinks"

Even with the current uncertainties which exist in relation to the size of carbon reservoirs and fluxes, simple calculations based upon the actual concentration of CO_2 and the theoretical saturation achievable in deep ocean waters indicate a notional capacity for several million Gt of CO_2. In practical terms, this is constrained by the amount of carbonate ion available to neutralise the carbonic acid formed by dissolved carbon dioxide if environmentally significant excursions in overall oceanic pH values are to be avoided. Estimates for carbonate available in deep ocean sediments range between only 1,600 Gt and 4,920 Gt, with dissolved carbonate in waters deeper than 500m at 1,320 Gt. While it must be stressed that these values are highly speculative due to the paucity of empirical data (see: Wong & Hirai, 1997), the actual capacity is clearly likely to be orders of magnitude less than notional values.

Recently, Rau and Caldeira (1999) and Caldeira and Rau (2000) have proposed carbon capture/sequestration schemes which employ additional reactions with carbonates (e.g. limestone) prior to ocean disposal in an attempt to circumvent this constraint. Although reducing the potential for pH changes resulting from direct CO_2 injection, such approaches (described simply as "accelerated weathering") fail to solve many of the more conceptual and ethical problems surrounding human intervention with the ocean carbon cycle on this scale, aside from the almost unimaginable scale of the infrastructure necessary in order to put such measures in to practice.

Uncertainties in the timescales of containment

In addition to uncertainties and unknowns relating to overall capacity, fluxes and even immediate fate of artificially introduced CO_2, major questions surround the timescales over which sequestered carbon will remain isolated from the atmosphere. These questions of retention time are central to the viability of ocean disposal schemata, with timescales of 500-1000 years considered necessary for "effective" containment (ignoring longer-term considerations).

Broadly speaking, the retention time for CO_2 disposed of to the ocean is assumed to increase with the depth at which it is discharged (Wong & Matear, 1993; GESAMP, 1997), though with considerable regional variations related to ocean circulation patterns (Bacastow et al., 1995; Ormerod, 1996). Of course, the disposal of liquid CO_2 in the water column would be subject to ocean circulation processes to a far greater extent than that deposited on the sea floor in solid or liquid form. Eventually, however, ocean processes will dissolve all forms of dumped CO_2 and transport it towards the surface. It has been estimated that at a depth of >3000m a pool of CO_2 containing 58.4 Mt would dissolve into the deep water over approximately 240 years. Various values for the dissolution and movement of CO_2 in the deep sea have then been derived through the use of observations and models, though the models have acknowledged limitations (Ormerod, 1996) and hence these predictions are subject to a high degree of uncertainty.

Indeed, some previously held assumptions regarding residence times for deep ocean waters have recently been challenged. Furthermore, the potential for climate change (even to the extent to which we are already committed) to modify these residence times through impacts on circulation drivers and patterns have been largely overlooked.

For example, assumptions that the age of the water in some deep ocean basins of the North Pacific was in excess of 1000 years have been challenged through the use of passive tracers such as chlorofluorocarbons, tritium and radiocarbon. Turnover time for deep water in the basins around the margins of the North Pacific (the Okhotsk, Japan and Bering Seas) is now estimated at only 100 years or less as opposed to the previously accepted values of up to 1300 years. Similarly, the basins of the Arctic Ocean may ventilate within only 75-300 years (see: Wong & Matear, 1996). This obviously throws into serious question the presumed utility of these basins for CO_2 disposal which had previously been assumed to be possible. Although other regions may turn over more slowly than previously thought, the general picture is one of limited understanding of circulation patterns, hardly a reliable basis for assuming predictable and effective containment of CO_2.

In addition, there appears to be extreme uncertainty about how rapidly vertical mixing takes place, with widely divergent values cited in the literature (Wong & Matear, 1996). No deep measurements are available for the crucial parameter (diffusivity) which may, in any case, vary according to ocean floor topography on a site specific basis. Areas with steep structures on the ocean floor (e.g. seamounts) have diffusivities much higher than accepted "background" values, lending weight to the hypothesis that such topography-induced turbulence may dominate deep ocean mixing (Munk & Wunsch, 1998, Chiswell & Sutton, 1998). Wind driving is also considered to be highly important (Wunsch, 1998). The rate of vertical mixing of deep Pacific water is, accordingly, a subject of debate. Based upon ^{13}C and ^{14}C tracer studies, a mixing rate of 30 m y^{-1} has been inferred, implying that dumped CO_2 could reach surface waters in 140-200 years instead of the 1000-1500 years suggested in earlier literature. Overall, a great deal of uncertainty exists concerning the mechanisms of ocean circulation and heat exchange. Mixing is unlikely to be uniform, but may take place at localised active regions, such as well stirred bottom slopes (see: Killworth, 1998).

It appears to date that none of the analyses of the fate of the CO_2 have considered the fact that the ocean circulation patterns could change either as a result of natural variability or as a result of climate change. Indeed, it has been suggested (Sarmiento et al., 1998) that ocean circulation changes modelled under a global warming scenario could have very significant implications for oceanic CO_2 uptake in the future. As one possible example, while the basic phenomena associated with the El-Niño-Southern Oscillation are known, the deep oceanic changes associated with it are much more poorly understood. Moreover, complex land-ocean interactions in carbon budgets which occur during ENSO events are only now beginning to be documented. For example, the Amazon Basin, which normally acts as a substantial sink of carbon, may, in ENSO years, act as a significant carbon source (Tian et al., 1998).

Similarly, the modelled predictions that North Atlantic thermohaline circulation, the process whereby surface water is drawn into deep water in the North Atlantic, may weaken or stop under changed climatic conditions, are also of significance (Rahmstorf, 1995; DETR, 1998; Schlosser et al., 1991). There appears to be an intricate and intimate relationship between the North Atlantic Oscillation (NAO) and the physical properties and behaviour of deep currents flowing in the Denmark Strait (between Greenland and Iceland). The NAO is the dominant recurrent mode of atmospheric behaviour in the North Atlantic Sector (Dickson, 1998; Dickson et al., 1999), and yet none of the CO_2 disposal studies appear to have considered the consequences of impacts on this process which could arise from changes in sea-ice cover under a changed climatic regime (Yang & Neelin, 1993; Hunt et al., 1995).

Indeed, there is some evidence in the palaeorecord to suggest that changes in deep ocean circulation could have acted as a climate "switch" in the past (MacLeod & Huber, 1996). Changes in deep water circulation affecting waters where large quantities of carbon dioxide have been disposed of could, therefore, lead to a much more rapid return of carbon dioxide to the atmosphere than is predicted in the various studies which have not considered these factors. At the very least a rapid efflux of disposed CO_2 from deep waters would render the supposed rationale for ocean disposal completely irrelevant. Even relatively small changes in the pattern or location of oceanic upwelling could lead to disposed CO_2 reaching the ocean surface much faster than anticipated.

Finally, it must be remembered that under conditions of changed climate there could be changes in the frequency and intensity of storms and in the strength of prevailing winds (IPCC 1996). A direct coupling exists between wind and ocean circulation. According to modelling exercises supported by observations (Wunsch, 1998), much of the work done by wind occurs in the Southern Ocean. Changes in wind intensity and consequent impacts upon ocean-atmosphere CO_2 exchange could be of considerable importance in determining the residence time of CO_2 in the ocean. Similarly, other research has shown that the depth of convection in the oceans is increased as more heat is removed from surface waters by wind action (Curry et al., 1998).

Clearly, the significance of all these phenomena for oceanic CO_2 cycling is not known. Such studies, however, are making it clear that chemical and physical properties of deep water masses are far from immutable and that the signatures of sea surface conditions are translated, in time, to deep waters and vice versa. The uncertainties which arise as a result of these findings do not appear to have been taken into consideration in analyses of CO_2 disposal (see: Wong & Hirai,1997).

Is ocean disposal/sequestration of CO_2 sustainable?

Perhaps the most generic formulation of the concept of sustainability is that captured within the Rio Declaration arising from the UN Conference on Environment and Development (UNCED 1992):

> "the right to development must be fulfilled so as to equitably meet developmental and environmental needs of present and future generations"

From the preceding discussion, it is already apparent that proposals for large-scale human intervention in oceanic carbon cycles could be seen to fall foul of this concept of intergenerational equity, especially as assumptions regarding effectiveness in space and time are highly questionable. The timeframes over which some interventions, especially fertilisation programmes, would need to be conducted carry considerable trans-generational responsibilities to continue the process. In the case of iron fertilisation of the Southern Ocean, for example, models suggest that if the exercise was stopped after 50 years, a significant proportion of the carbon fixed in this way would simply return to the atmosphere as CO_2 over the subsequent half century (Ormerod & Angel, 1998).

Issues such as energy penalties incurred in implementing such interventionist approaches and their potential to inhibit progress towards carbon-neutral, renewable alternatives may be seen to reduce further the consistency of these schemata with sustainable practice. These additional concerns are addressed further below.

An alternative description of sustainability, though by no means inconsistent with the concept above, is provided by the list of first order principles set out by Cairns (1997). Of these four principles, three may provide a useful metric against which the sustainability of CO_2 disposal/

sequestration proposals can be evaluated, namely that neither i) man-made nor ii) natural substances should be permitted to accumulate systematically in the biosphere and that iii) the fundamental bases for productivity must not be systematically depleted. The potential for CO_2 disposal or ocean fertilisation operations to impact systematically on the diversity and productivity of natural systems is of concern. Although the stated purpose of ocean disposal/sequestration is to reduce catastrophic impacts on marine and terrestrial systems resulting from CO_2 otherwise being released directly to atmosphere, it must be recognised that the schemata proposed can lead to unpredictable and essentially irreversible impacts of their own.

All proposed oceanic CO_2 disposal/sequestration schemes have potential to cause considerable ecosystem disturbance. Considerable uncertainty surrounds the impacts of ocean fertilization on marine ecosystems. A series of experimental introductions of iron into the Southern Ocean promoted a bloom of phytoplankton (Boyd et al, 2000) but in doing so, produced significant changes in community composition and the microbial food web (Hall & Safi, 2001). These changes are disturbances of the marine ecosystem. If ocean fertilization was implemented on a larger scale, there are concerns that these changes could alter the oceans' food webs and biogeochemical cycles (Chisholm et al., 2001) causing adverse effects on biodiversity. There are also possibilities of nuisance or toxic phytoplankton blooms and the risk of deep ocean anoxia from sustained fertilization (Hall & Safi, 2001).

In the case of CO_2 introduced directly at depth, seawater pH values will be altered, with potential adverse consequences for marine organisms (Ametistova et al., 2002). It has been predicted that the fall in pH associated with a CO_2 plume could disrupt marine nitrification and lead to unpredictable phenomena at both the ecosystem and community level (Huesemann et al. 2002). Organisms unable to avoid regions of low pH because of limited mobility will be most affected, layers of low pH water could prevent vertical migration of species and change the composition of particles, affecting nutrient availability (Ametistova et al., 2002). Deep-sea organisms are highly sensitive to changes in pH and CO_2 concentration (Seibel & Walsh, 2001). Thus, even small changes in pH or CO_2 could have adverse consequences for deep-sea ecology and hence for global biogeochemical cycles that depend on these ecosystems (Seibel & Walsh, 2001). CO_2 introduced upon seamount ecosystems raises further concerns. Benthic fauna in these ecosystems exhibit high degrees of endemism and where fish congregations are of highly adapted (K-selected) species (Koslow et al. 2000; de Forges et al. 2000).

The overall ecological and biodiversity implications of ocean CO_2 disposal are very likely to be negative, especially to benthic systems, although there is a lack of detailed knowledge of the faunal assemblages likely to be impacted, and the extent of the areas affected (Seibel and Walsh 2001). There is a need, for example, to obtain precise information on the responses of deep sea animals to elevated CO_2 exposure over their whole life-cycle under realistic conditions (Omori et al., 1996).

In order generally to justify large scale ocean dumping activities, it has been suggested (Angel & Rice, 1996) that large scale in situ experiments should be conducted to investigate impacts. The suggestion has been repeated in connection with the CO_2 ocean disposal option by Ormerod & Angel (1996), who state that the biological impacts of large-scale discharges (of CO_2) can only be adequately assessed by a careful manipulation of a total ecosystem. This may be true, but the ecosystem concerned needs to be fully understood at the outset of the experiment. Under current circumstances, therefore, it is difficult to see what such an experiment would achieve since the ecosystems in question are so poorly characterised. Hence only the grossest effects are likely to be detected and the more subtle, but equally important, changes forced upon ecosystem structure and function are likely to be overlooked. Such an experiment, conducted without rigorous preparation over many years and followed up by many years of subsequent observation, would be scientifically worthless.

Could quantities of carbon sequestered in oceans be verified?

It is noted above that, while it is true that Articles 3.3.and 3.4 of the Kyoto protocol explicitly allow for the offsetting of sequestration into "sinks" against commitments to emission reductions, the clear intention is that this should apply to management of terrestrial carbon reserves only. Moreover, Article 3.3 requires that any such offsetting must be "reported in a transparent and verifiable manner". As Caldeira (2002) notes:

> "such a system of credits or debits cannot rely on trust alone; there must be some independent way of verifying carbon stored by a sequestration project"

This reporting requirement, vital to the environmental effectiveness, proper governance and evaluation of Kyoto Protocol implementation, will nevertheless undoubtedly be subject to enormous operational difficulties and measurement uncertainties. Extending such a concept to the marine environment, the transparent and verifiable reporting of increases in oceanic carbon reserves seems inevitably to run up against insurmountable problems. Aside from the scale of uncertainties and lack of understanding of the fate of introduced carbon (or indeed the effectiveness of fertilisation schemes in enhancing draw-down), Caldeira (2002) stresses that reabsorption from the atmosphere will add further complexity, since it will be essential that such reabsorbed carbon is not included in offset calculations. Even if the other substantive issues surrounding ocean disposal/ sequestration of CO_2 could be resolved, therefore, it would appear that the fundamental barriers to quantification and verifiability for carbon so sequestered would render its offsetting against emission reduction commitments difficult in the extreme, if not impossible.

Do ocean disposal/sequestration proposals conflict with other existing legal instruments?

Before governments commit themselves to continue to invest in research into the technical feasibility of disposing of fossil fuel generated CO_2 to the oceans there is a need seriously to consider the practicability and feasibility of such options under elements of international law other than the Kyoto Protocol and the FCCC.

For example, unless the international regime governing the dumping of wastes at sea, both above and under the seabed (the 1972 London Convention), is made the subject of drastic revisions, the direct disposal of fossil fuel-derived CO_2 from vessels and platforms would violate international law. The possibility of such revisions is neither realistic nor desirable, particularly as acceptance of CO_2 dumping could open the door to reconsideration of dumping as an option for other industrial wastes. In the words of the UN Group of Experts of Scientific Aspects of Marine Environment Protection (GESAMP 1997):

> "For technical and financial reasons CO_2 (dry ice) disposal appears to be an unattractive option unless it is to be dumped from vessels. Dumping from ships, however, comes under the aegis of the Convention on the Prevention of Marine Pollution by Dumping of Wastes and Other Matter (London Convention, 1972). In 1993 Contracting Parties to this Convention adopted a prohibition on dumping of industrial wastes...at sea that took effect on January 1st 1996. It therefore seems unlikely unless the Convention can be amended to permit the dumping of CO_2, from ships, that any of the current Parties to the Convention...could give approval to such a practice. It should be further noted that the same conclusion would apply to liquid CO_2 disposal from vessels and platforms which would also fall within the purview of the London Convention 1972".

GESAMP (1997) was correct to point out that that the dumping of CO_2 at sea is a concept that is not in accordance with existing international law. In addition to dumping at sea from ships, aircraft, platforms and other man-made structures at sea, the Convention also covers the disposal of wastes and other matter under the seabed. This was agreed by the Thirteenth Consultative Meeting of Contracting Parties to the London Convention (1990) which adopted by vote Resolution LDC.41 13 to this effect. Six years later, in adopting the 1996 Protocol to the London Convention, the Contracting Parties unanimously agreed that:

> " 'Dumping' means: any deliberate disposal into the sea of wastes or other

matter from vessels, aircraft. Platforms or other man-made structures at sea;
[....as well as] any storage of wastes or other matter in the seabed and the
subsoil thereof from vessels aircraft platforms or other man-made structures
at sea." (Article I, 4.1.1 and I, 4,1,1.2 of the 1996 Protocol to the London
Convention)

According to Article XV 2 of the London Convention 1972, any amendments to the Convention require a two thirds majority of those Contracting Parties in attendance at the time of the vote and:

"will be based on scientific or technical considerations."

The prohibition on the dumping of industrial waste at sea on or under the seabed is legally binding on all Contracting Parties to the London Convention (78 countries by the end of 2002) and also upon all Contracting Parties to the United Nations Convention on the Law of the Sea (UNCLOS, 1982). This is in accordance with Article 210.6 ("Pollution by Dumping") of UNCLOS, which states that:

"National laws regulations and measures shall be no less effective in preventing
reducing and controlling such pollution than the global rules and standards".

The acceptability under international law of disposing of CO_2 via a pipeline from the shore into the marine environment remains an open question. In the preambular text to Resolution LC. 49 (16), banning the dumping of industrial waste at sea, Contracting Parties to the London Convention:

"Reaffirm[ed] further the agreement that a better protection of the marine
environment by cessation of dumping of industrial waste should not result in
unacceptable environmental effects elsewhere."

This statement was primarily addressing the need to give priority to waste avoidance and clean production and also to avoid wastes previously dumped at sea being placed in land-based dumps in a manner which could not be considered to be environmentally sound. Nevertheless, the need to avoid wastes that could be dumped at sea entering the marine environment via land based sources and activities was also very much present in the minds of the negotiators.

Moreover, in adopting the 1996 Protocol to the London Convention, Contracting Parties reiterated in Article 2 of the Protocol their obligation to:

"individually and collectively protect and preserve the environment from ALL
SOURCES of pollution".

Further they also agreed in Article 3.3 of the Protocol that:

"In implementing the provisions of this Protocol, Contracting Parties shall act so as not to transfer, directly or indirectly, damage or likelihood of damage from one part of the environment to another or transform one type of pollution into another".

The most recent expression of the universal trend to eliminate the dumping and land based discharge of wastes to the marine environment is contained in the 1998 Sintra Statement which was adopted unanimously by the Environment Ministers of the countries bordering the North East Atlantic, together with the European Commissioner for the Environment (OSPAR, 1998). The signatories to this statement:

"reemphasise[d] [their] commitment to prevent the sea being used as a dumping ground for waste, whether from the sea or from land based activities".

In common with GESAMP (1997) the Independent World Commission on the Oceans (IWCO 1998), established to consider thoroughly all issues of ocean science, ocean conservation and ocean governance, noted that the dumping of CO_2 at sea is banned by the London Convention and that, furthermore:

"The Framework Convention on Climate Change and its Kyoto Protocol do not provide for Parties to dump or store CO_2 in international waters and thereby to offset their emissions".

The provisions of existing Conventions, coupled with the current political reality, would seem, therefore, to rule out the consideration of oceanic disposal of anthropogenic CO_2 as a potential strategy to address current and future emissions. Indeed, it was questions concerning the legality of ocean disposal of CO_2 under the more regional (North East Atlantic) OSPAR Convention (1992) which led the Norwegian Environment Ministry to refuse a permit for an experimental release planned in the Norwegian Sea during the summer of 2002 (Adams et al. 2002, Burke 2002).

Is the energy penalty for capture and disposal or for fertilisation acceptable?

Many proposals to dispose of CO_2 in to deep ocean waters have been predicated upon the assumed possibility to capture clean CO_2 from the majority of the world's power stations and transport it for disposal. This strategy clearly carries with it a significant energy penalty from the outset. A typical capture scheme involves absorbing CO_2 from flue gases into a solvent (e.g. commonly monoethanolamine, MEA) followed by steam stripping. Disposal of the complete flue gas is unlikely to be viable since the carbon dioxide content ranges from about 3% in the case of gas turbine plant (Langeland & Wilhelmsen, 1993), similar to a natural gas combined cycle plant (3.6%), to flue gas concentrations of around 14% for a coal fired power plant. In the case of some specialised petrochemical processes this may be as high as 80% (Bailey & McDonald, 1993) while coal based plant with oxygen fuel and CO_2 recycle could reach 92%.

Current energy penalties, in other words the reduction in utility output as a result of using the process, range from between 15 and 24% for gas-fired plant to between 27-37% for conventional coal fired plant. Advanced coal plant attracts an energy penalty of 13-17% (Herzog et al., 1997). These figures do not include liquefaction and disposal penalties. These estimates are broadly supported by other estimates (Leci, 1996) suggesting absolute reductions in efficiency of up to 35%. Effectively, capture alone would reduce the thermal efficiency of a typical power plant from around 35% to 25% (GESAMP, 1997) accounting for around 30% of the total energy content of the coal used as fuel. Some improvements envisaged for the future suggest penalties could be brought down to between 9% and 15%. This energy penalty effectively results in a reduced efficiency of the plant.

A further 10% of the fuel value of the coal would be required to liquefy and compress the carbon dioxide to a pressure enabling deep-water discharge (150 Atm) (Golomb, 1993). Other estimates (Haugen & Eide, 1996) include liquefaction and disposal penalties with those from capture. These suggest that for a 500MW power station the electricity consumption for the MEA process amounts to between 18 and 24MW. In addition the steam requirement for stripping the solvent equates to between 70 and 140 MW and is smallest for gas fired power stations. In order to prepare the stripped CO_2 for disposal it must be liquefied, pressurised and purified. Assuming injection into an aquifer, this will require an additional 55-60MW for a coal fired plant and about 30-40MW for a gas fired plant. Transport of CO_2 from a pulverised coal plant over 100 km would require a further 3MW and, if

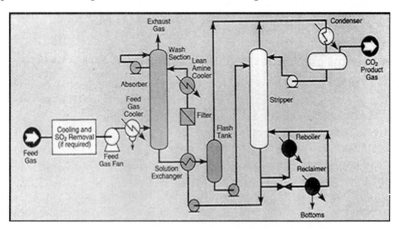

Fig. 6.2. Schematic diagram for a process for the capture of carbon dioxide from power station/industrial process flue gases. This indicates the energy demanding components of the process and the waste streams generated as a result.

the full capacity of a large diameter pipe were used, the total energy required for this phase would be 12-14MW.

Accordingly, the total electricity loss from a 500MW facility capturing and disposing of CO_2 can be expected to amount to at least 80MW (16%) while heat losses due to steam consumption account for at least 70MW. A total penalty of 25%-45% can, therefore, be expected on the basis of current technology. This will inevitably result in greater quantities of carbon dioxide being generated for a given unit of energy produced.

This reduction in efficiency has considerable long-term implications for the concept of ocean dumping of carbon dioxide in the context of Article 2 of the UNFCCC. As noted above, a dumping strategy would reduce the transient peak of CO_2 over the next century (as would an emission reduction strategy aimed at the same CO_2 concentrations within that timeframe) but ultimately would result in CO_2 levels approaching those expected had no emissions reductions been undertaken. If the emissions from energy used in the capture and disposal processes are taken into account, the long term stabilisation value could perhaps even be significantly higher than under a "business as usual" scenario (depending upon the precise mix of fossil fuels used in energy generation and their relative carbon intensities). Hence, the assertion (Herzog et al., 1997) that ocean disposal represents simply a short circuit of natural processes is somewhat disingenuous since it does not appear to accommodate the fact that adoption of the strategy will lead in the longer term to higher CO_2 concentrations than would otherwise have been the case. This has been recognised in modelling exercises which have shown that, where marine disposal systems are used, the long term atmospheric concentrations ultimately exceed those from systems without controls using disposal to ocean (Flannery et al., 1993).

Will the pursuit of ocean disposal/sequestration impact negatively on the development of source-oriented and carbon neutral solutions?

It has been observed that technical progress in carbon management is outpacing consideration of its limitations and potential risks. Large-scale adoption of such strategies would constitute a major technological and societal choice that would be difficult to reverse. There is a strong risk that this choice will be made without adequate reflection upon its implications (Parson & Keith, 1998) given the current slow political progress on genuine abatement through *inter alia* the development of renewable energy resources.

In turn, this suggests the potential for a serious conflict within corporations and governments regarding allocation of resources for onward investment in energy technologies. There are finite resources available in the corporate and public domains to direct at development of sustainable (renewable) energy technologies and the associated infrastructure. Carbon management strategies designed to maintain fossil fuel consumption and use into the future are diametrically opposed in philosophy and intent to the development of renewable energy resources and of highly efficient technologies and transport systems. Given that a highly significant component of investment in renewable energy development is being made by multinational oil companies, there is the very real prospect that such investment will be held hostage to the development of expensive carbon management technologies. There also appears to be a presumption in the energy industry that the relatively low cost "least regrets" options which include improvements to energy supply, end-use efficiency, substitution with less carbon intense fuels and inexpensive renewables will not solve the medium to long term problem. Accordingly, it has been argued that technologies to capture and remove CO_2 could have a central role to play in the future. Crucially, they are seen by some in the energy industries as a long term option which allows for continued large scale use of fossil energy resources (Herzog et al., 1997).

In the light of the numerous legal and political uncertainties that would surround any proposal to dump CO_2 in the marine environment or artificially to enhance uptake through fertilisation on a massive scale, along with the considerable scientific and technical problems and uncertainties outlined elsewhere in this document, it would appear that further investment in such options would represent a wasteful drain on available resources. These resources might be much better utilised in the research, development and promotion of renewable energy alternatives and the implementation of effective energy efficiency programmes. Such an approach would clearly be more consistent with the conclusions expressed by the IWCO (1998)

Summary and Conclusions

The first commitment period reduction targets of the Kyoto Protocol are widely seen as a significant first step in reducing CO_2 and other greenhouse gas emissions. The 5% reduction target for industrialized countries is by no means adequate to achieving the ultimate objective of the Convention (WRI, 1998) and further and deeper emission reduction commitments by the industrialized countries will be essential in the second commitment period (2013-2017) and beyond. Developing countries will need to switch more rapidly to energy efficient technologies and environmentally sound renewable energy supply sources. In this context ongoing development of unsustainable ocean disposal/sequestration strategies will continue to draw vital resources away from the research, development and implementation of energy efficiency programmes and renewable energy alternatives.

The potential for unpredicted, and possibly serious or even irreversible, effects on current or future generations resulting from the various ocean disposal strategies outlined above is indisputable and threatens to violate the principle of sustainability established as a result of the first UN

Conference on Environment and Development (UNCED 1992). At the same time, the assumption that ocean sequestration or disposal of CO_2 will be effective in combating climate change and in particular meeting the ultimate objective of the UNFCCC, stabilization of greenhouse gas concentrations at levels that prevent dangerous interference with the climate system, remains very much open to debate. Indeed, the use of such strategies to meet the provisions of the Kyoto Protocol would undermine its environmental effectiveness and, by loading longer term increases of CO_2 into the atmosphere in future centuries, contradict the principal of intergenerational responsibility. In other words, use of "ocean sinks" in meeting short-term emission targets in the coming decades would be at the expense of further climate change in the future.

The optimistic view (Herzog et al., 1997) that CO_2 disposal strategies allow for wide scale use of fossil fuels into the future can be challenged on fundamental grounds. It makes little difference to the final long-term level of CO_2 achieved in the atmosphere. Furthermore, the seemingly widely held assumption that:

> "From a long term perspective (i.e. at the time scale of a millennium) the question of ethics of using the ocean does not arise because the fate of most CO_2 discharged to the atmosphere will be to finish up in the deep ocean" (Angel, 1998)

...is quite wrong. The ocean disposal of CO_2 raises very deep ethical questions indeed, particularly since the strategy likely carries with it a commitment for future generations to a worse equilibrium state than would be the case by reducing emissions at source or even in some case continuing to emit on a "business as usual" basis. As carbon cycle modellers have noted, the CO_2 does not disappear and that "ignoring the truly long-term effects is a political and ethical choice" (Tans, 1997).

Although CO_2 disposal in the oceans may diminish the extent of the transient peak in atmospheric CO_2 concentrations over the coming century it certainly would bequeath to future generations most likely unacceptable levels of climate change. Such an outcome could only be averted via technologies that extracted CO_2 direct from the atmosphere, to soak up the flux of anthropogenically injected CO_2 outgassing from the oceans in future centuries. In effect, the strategy simply delays the full impact of climate change to a point in the future. This key trans-generational failing is not addressed by proponents of the strategy (e.g. Herzog et al., 1997). Since the final CO_2 stabilisation value will be key determinant of the ambient climate of the time, the ocean disposal option effectively shifts a substantial burden of the responsibility (perhaps 30-50%) into the future at which point it will not be recoverable or avoidable.

In 1998, the Independent World Commission on the Oceans (IWCO) considered the implications of CO_2 disposal at sea, concluding that:

"In the context of global warming, the importance of ocean-atmosphere interactions, the role of the oceans as a CO_2 sink and the precautionary principle, it is imperative that people and governments exploit as a first priority, the manifold opportunities that exist for reducing carbon emissions and consider only with circumspection the potential of the oceans as a site for CO_2 disposal". (IWCO 1998)

From the analysis presented above, it seems clear that the position of the IWCO on this issue has substantial merit. It can only be hoped that the "people and governments" to whom their concluding statement was addressed will take due note and will not look to the oceans as a convenient "sink" to allow Kyoto commitments to be met in accounting terms while failing to address the underlying problem of the unsustainable fossil-fuel economy.

Footnotes

[1] See Table 6.1 of IPCC TAR WGI for summary of radiative forcing from 1750-1998 of main greenhouse gases.

[2] Essentially these were defined in the Bonn and Marrakech Accords of 2001 to included restricted volumes of forest management credits, cropland and grazing land management and revegetation (see FCCC/CP/2001/13/Add.1 at *www.unfccc.int*) and are beyond the scope of this chapter to elaborate).

[3] This could, for example, render impossible the future attainment of climate protection goals such as those set by the European Union of limiting warming below 2oC increase above pre-industrial levels.

References

Adams, E.E., Golomb, D.S. & Herzog, H.J. (1995) Ocean disposal of CO_2 at intermediate depths. Energy Conversion and Management 36 (6-9): 447-452.

Adams, E., Akai, M., Alendal, G., Golmen, L., Haugan, P., Herzog, H., Masutani, S., Murai, S., Nihous, G., Ohsumi, T., Shirayama, Y., Smith, C., Vetter, E., Wong, C.S., (2002) International field experiment on ocean carbon sequestration. Environmental Science and Technology 36 (21): 399A.

Ametistova, L., J. Twidell, & J. Briden, (2002) The sequestration switch: removing industrial CO_2 by direct ocean absorption. Science of the Total Environment **289**: 213-223.

Angel, M.V. (1998) The use of the oceans and environmental impact. In: Report of the Advisory Group on R&D on Ocean Sequestration of CO_2, 26th–27th March 1998, Heathrow, UK. Publ. International Energy Agency, Cheltenham, UK. Report No: PH3/2.

Angel, M.V. & Rice, A.L. (1996) The ecology of the deep sea and its relevance to global waste management. Journal of Applied Ecology 33: 754-772.

Bacastow, R.B., Cole, K.H., Dewey, R.K. & Stegen, G.R. (1995) Effectiveness of CO_2 sequestration in the oceans considering location and depth. Energy Conservation and Management 36 (6-9): 555-558.

Bailey, R.T. & McDonald, M.M. (1993) CO_2 capture and use for EOR in Western Canada 1. General Overview. Energy Conversion and Management 34 (9-11): 1145-1150.

Boyd P.W., Watson A.J., Law C.S., Abraham E.R., Trull T., Murdoch R., Bakker D.C.E., Bowie A.R., Buesseler K.O., Chang H., Charette M., Croot P., Downing K., Frew R., Gall M., Hadfield M., Hall J., Harvey M., Jameson G., LaRoche J., Liddicoat M., Ling R., Maldonado M.T., McKay R.M., Nodder S., Pickmere S., Pridmore R., Rintoul

S., Safi K., Sutton P., Strzepek R., Tanneberger K., Turner S., Waite A., Zeldis J. (2000) A mesoscale phytoplankton bloom in the polar Southern Ocean stimulated by iron fertilization. *Nature*, **407**, (6805) 695-702.

Brewer, P.G., (2000) Contemplating action: storing carbon dioxide in the ocean. Oceanography, **13** (2), 84-92

Burke, M. (2002) Sequestration experiment is drowning. Environmental Science & Technology, News, November 1 2002

Cairns, J. (1997) Defining goals and conditions for a sustainable world. Environmental Health Perspectives 105 (11): 1164-1170.

Caldeira, K. & Rau, G.H. (2000) Accelerating carbonate dissolution to sequester carbon dioxide in the ocean: geochemical implications. Geophysical Research Letters 27(2): 225-228

Caldeira, K. (2002) Monitoring of ocean storage projects. IPCC Workshop on Carbon Separation and Storage, 18-22 November 2002, Regina, Canada

Chisholm S.W., Falkowski P.G., Cullen J.J. (2001) Oceans - Dis-crediting ocean fertilization *Science*, **294** (5541), 309-310

Chiswell, S.M. & Sutton, P.J.H. (1998) A deep eddy in the Antarctic intermediate water north of the Chatham Rise. Journal of Physical Oceanography 28: 535-540.

Curry, R.G., McCartney, M.S. & Joyce, T.M. (1998) Oceanic transport of subpolar climate signals to mid depth sub tropical waters. Nature 391: 575-577.

de Forges, B.R., Koslow, J.A. & Poore, G.C.B. (2000) Diversity and endemism of the benthic seamount fauna in the southwest Pacific. *Nature*, **405**, 944-947

DETR (1998) Climate Change and Its Impacts: Some Highlights from the Ongoing UK Research Programme: A First Look at Results from the Hadley Centre's New Climate Model. UK Meteorological Office and UK Department of Environment Transport and The Regions, November 1998: 12pp.

Dickson, B. (1998) All change in the Arctic. Nature 397: 389-390.

Dickson, B; Meincke, J., Vassie, I., Jungclaus, J. & Osterhus, S. (1999) Possible predictability in overflow from the Denmark Strait. Nature 397: 243-246.

Drange, H., Alendal, G. & Johanessen, O.M. (2001) Ocean release of fossil fuel CO_2: a case study. *Geophysical Research Letters*, **28** (13) 2637-2640.

Falkowski, P., Scholes, R.J., Boyle, E., Canadell, J., Canfield, D., Elser, J., Gruber, N., Hibbard, K., Hogberg, P., Linder, S., Mackkenzie, F.T., Morre, B., Pedersen, T., Rosenthal, Y., Seitzinger, S., Smetacek, V., Steffen, W. (2000) The Global Carbon Cycle: A Test of Our Knowledge of Earth as a System. *Science* 290: 291-296.

Flannery, B.P., Kheshgi, H.S., Hoffert, M.I. & Lapenis, A.G. (1993) Assessing the effectiveness of marine CO_2 disposal. Energy Conversion and Management. 34 (9-11): 983-989

Follows, M.J., Williams, R.G. & Marshall, J.C.(1996) The solubility pump of carbon in the subtropical gyre of the North Atlantic. Journal of Marine Research 54: 605-630.

GESAMP (1997). (IMO/FAO/UNESCO-IOC/WMO/WHO/IAEA/UN/UNEP Joint Group of Experts on the Scientific Aspects of Marine Environmental Protection) Report of the twenty-seventh session of GESAMP, Nairobi, Kenya, 14-18 April 1997. GESAMP Reports and Studies No. 63: 45pp.

Golomb, D. (1993) Ocean disposal of CO_2: Feasibility, economics and effects. Energy Conversion and Management 34 (9-11): 967-976.

Goyet, C., Healy, R., McCue, S.J. & Glover, D.M. (1997) Interpolation of TCO_2 data on a $1^\circ \times 1^\circ$ grid throughout the water column below 500m depth in the Atlantic Ocean. Deep-Sea Research I 44 (12): 1945-1955.

Guevel, P., Fruman, D.H. & Murray, N. (1996) Conceptual design of an integrated solid CO_2 penetrator marine disposal system. Energy Conversion and Management 37 (6-8): 1053-1060

Hall, J.A. & Safi, K., (2001) The impact of in situ Fe fertilization on the microbial food web in the Southern Ocean. *Deep-Sea Research II*, **48**, 2591-2613.

Haugen, H.A. & Eide, L.I. (1996) CO_2 capture and disposal: The realism of large scale scenarios. Energy Conversion and Management 37 (6-8): 1061-1066

Herzog, H., Drake, E. & Adams, E. (1997) CO_2 Capture, re-use and storage technologies for mitigating global climate change. White Paper Final Report, publ. Energy Laboratory, Massachusetts Institute of Technology, US Department of Energy Order No: DE-AF22-96PC01257.

Herzog, H. (2001) What future for carbon capture and sequestration? Environmental Science & Technology April 1 2001: 149A-153A

Hoffert, M. I., K. Caldeira, et al. (2002). "Advanced Technology Paths to Global Climate Stability: Energy for a Greenhouse Planet." Science 298(5595): 981-987.

Huesemann, M.H., Skillman, A.D. & Crecelius, E.A. (2002) The inhibition of marine nitrification by ocean disposal of carbon dioxide. Marine Pollution Bulletin, 44, 142-148

Hunt, B.G., Gordon, H.B. & Davies, H.L. (1995) Impact of the greenhouse effect on sea-ice characteristics and snow accumulation in the polar regions. International Journal of Climatology 15: 3-23.

IPCC (1996) Climate Change 1995. The Science of Climate Change. Contribution of Working Group I to the Second Assessment Report of the Intergovernmental Panel on Climate Change. Cambridge University Press: 572pp

IWCO (1998) The Ocean-Our Future. Report of the Independent World Commission on the Oceans. Publ. IWCO.pp 89-90

Johnston, P., Santillo, D., Stringer, J., Ashton, J., McKay, B., Verbeek, M., Jackson, E., Landman, J., van den Broek, J., Samsom, D., Simmonds, M., (1998) Report on the World's Oceans. Greenpeace Research Laboratories Report May 1998. Greenpeace International: 154 pp. ISBN: 90-73361-45-1

Killworth, P. (1998) Something stirs in the deep. Nature 396: 720-721

Koslow, J.A., Boehlert, G.W., Gordon, J.D.M., Headrich, R.L., Lorance, P., & Parin, N. (2000) Continental slope and deep-sea fisheries: implications for a fragile ecosystem. ICES Journal of Marine Science, 57, 548-557.

Langeland, K. & Wilhelmsen, K. (1993) A study of the costs and energy requirement for carbon dioxide disposal. Energy Conversion and Management 34 (9-11): 807-814

Leci, C.L. (1996) Financial implications of power generation costs resulting from the parasitic effect of CO_2 capture using liquid scrubbing technology from power station flue gases. Energy Conversion and Management 37 (6-8): 915-921

Marland, G. (1996) Could we/should we engineer the Earth's climate? Climatic Change 33: 275-278

MacLeod, K.G. & Huber, B.T. (1996) Reorganization of deep ocean circulation accompanying a late Cretaceous extinction event. Nature 380: 422-425.

Munk, W. & Wunsch, C. (1998). Abyssal recipes II: Energetics of tidal and wind mixing. Deep Sea Research 45: 1976-2009.

Omori, M., Norman, C.P., Maeda, M., Kimura, B. & Takahashi, M. (1996) Some considerations on the environmental impact of oceanic disposal of CO_2 with special reference to midwater organisms. In: Ormerod, B & Angel, M. (1996) Ocean Storage of Carbon Dioxide: Workshop 2: Environmental Impact. International Energy Agency Greenhouse Gas R&D Programme, Cheltenham, UK: 83-98.

OSPAR (1998) Sintra Statement, Summary Record of the Ministerial Meeting of the OSPAR Commission, Sintra, Portugal, July 1998, OSPAR 98/14/1: Annex 45.

Ormerod, B. (1996) Ocean Storage of Carbon Dioxide: Workshop 1: Ocean Circulation. International Energy Agency Greenhouse Gas R&D Programme, Cheltenham, UK: 113pp.

Ormerod, B & Angel, M. (1996) Ocean Storage of Carbon Dioxide: Workshop 2: Environmental Impact. International Energy Agency Greenhouse Gas R&D Programme, Cheltenham, UK: 131pp.

Ormerod, B., & Angel, M., (1998) Ocean Fertilisation as a CO_2 Sequestration Option. International Energy Agency Greenhouse Gas R&D Programme, Cheltenham, UK: 50pp.

Parson, E.A. & Keith, D.W. (1998) Fossil fuels without CO_2 emissions. Science 282: 1053-1054

Pfannkuche, O. (1996) The deep sea benthic boundary layer. In: Ormerod, B & Angel, M. (1996) Ocean Storage of Carbon Dioxide: Workshop 2: Environmental Impact. International Energy Agency Greenhouse Gas R&D Programme, Cheltenham, UK: 99-111.

Rahmstorf, S. (1995) Bifurcation of the Atlantic thermohaline circulation in responses to changes in the hydrological cycle. Nature 378: 146-149.

Rau, G.H. & Caldeira, K. (1999) Enhanced carbonate dissolution: a means of sequestering waste CO_2 as ocean bicarbonate. Energy Conversion and Management 40: 1803-1813.

Sarmiento, J.L. & Orr, J.C. (1991) Three dimensional simulations of the impact of Southern Ocean nutrient depletion on atmospheric CO_2 and ocean chemistry. Limnology and Oceanography 36: 1928-1950.

Sarmiento, J.L., Hughes, T.M.C, Stouffer, R.J., & Manabe, S. (1998) Simulated response of the ocean to anthropogenic climate warming. Nature 393: 245-249.

Schlosser, P., Bonisch, G., Rhein, M. & Bayer, R. (1991) Reduction of deepwater formation in the Greenland Sea during the 1980s: Evidence from tracer data. Science 251: 1054-1056.

Schulze, E.-D., Valentini, R. & Sanz, M.-J. (2002) The long way from Kyoto to Marrakesh: Implications of the Kyoto Protocol negotiations for global ecology. Global Change Biology 8: 505-518

Seibel, B, A. & Walsh, P.J. (2001) Potential impacts of CO_2 injection on deep-sea biota. Science, 294, 319-320.

Shoji, K; Jones, I.S.F. (2001) The costing of carbon credits from ocean nourishment plants. The Science of the Total Environment, 277, 27-31

Tans, P.P. (1997) The CO_2 lifetime concept should be banished. An editorial comment. Climatic Change 37: 487-490.

Tian, H., Melillo, J.M., Kicklighter, D.W., McGuire, A.D., Helfrich, J.V.K. (III), Moore, B. (III) & Vorosmarty, C.J. (1998) Effect of interannual climate variability on carbon storage in Amazonian ecosystems. Nature 396: 664-667.

Trull, T., Rintoul, S.R., Hadfiled, M. & Abraham, E.R. (2001) Circulation and seasonal evolution of polar waters south of Australia: implications for iron fertilization of the Southern Ocean. Deep-Sea Research II, 48, 2439-2466

UN (1997) Kyoto Protocol to the United Nations Framework Convention on Climate Change (UNFCCC 1992), UNFCCC Document FCCC/CP/1997/7/Add.1 [available at http://unfccc.int]

UNCED (1992) Rio Declaration on Environment and Development, United Nations Conference on Environment and Development [http://www.un.org/documents/ga/conf151/aconf15126-1annex1.htm]

Wong, C.S. & Hirai, S. (1997) Ocean Storage of Carbon Dioxide: A Review of Oceanic Carbonate and CO_2 hydrate chemistry. Publ. International Energy Agency Greenhouse Gas R&D Programme, Cheltenham, UK: 90pp.

Wong, C.S. & Matear, R. (1993) The storage of anthropogenic carbon dioxide in the ocean. Energy Conservation and Management 34 (9-11): 873-880.

Wong, C.S. & Matear, R.J. (1996) Use of active and passive tracers to deduce the effects of circulation and mixing on the disposal of CO_2 in the North Pacific. In: Ormerod, B. (1996a) Ocean Storage of Carbon Dioxide: Workshop 1: Ocean Circulation. International Energy Agency Greenhouse Gas R&D Programme, Cheltenham, UK: 105-113.

WRI (1998) 1998-99 World Resources: A Guide to the Global Environment. Publ. The World Resources Institute/United Nations Environment Programme/United Nations Development Programme/The World Bank. Oxford University Press, Oxford: 369pp.

Wunsch, C. (1998) The work done by the wind on the oceanic general circulation. Journal of Physical Oceanography 28: 2332-2340.

Yang, J. & Neelin, J.D. (1993) Sea-ice interaction with the thermohaline circulation. Geophysical Research Letters 20 (2): 217-220.

The Idea of Social Progress and Emissions Trading

Jon Rosales
University of Minnesota, 180 McNeal Hall, 1985 Buford Avenue,
St. Paul, MN 55108, USA   email: rosaφφ31@umn.edu

Introduction

As the Kyoto Protocol comes into force, many in the general public are surprised to learn of the actual outworkings and ramifications of the treaty. The mechanisms that carry out the Protocol, the so-called flexibility mechanisms, allow the heaviest polluting countries of the world a way out of making emission reduction cuts by purchasing credits to offset their emissions, thereby mitigating the necessity for expensive changes in production and energy use. The three mechanisms—joint implementation, the clean development mechanism, and emissions trading—allow those countries that emit greenhouse gases (GHG) over their emission targets the option to purchase credits or to produce credits in other countries to offset their own emissions and to meet their Protocol commitments.[1] In effect, these three mechanisms create an international market for GHG credits that can be bought or sold. As it is the most ambitious policy option of the three, this chapter focuses on emissions trading and the philosophic motivations for pursuing such a policy. The idea of social progress, as conceived in the late 18th century, will be laid out as a motivating factor of this policy.

In interviews with key policy-makers, analysts, and advocates of emissions trading attending the last three rounds of negotiations of the Conference of Parties (COP) to the United Nations Framework Convention on Climate Change (UNFCCC)[2] two related themes emerged:

- ❑ Emissions trading is a means to achieve our environmental goals at the least cost;
- ❑ Emissions trading is a win-win proposition allowing for both more efficient economic growth and environmental improvement.

These two points are the most common way proponents of emissions trading frame the policy within the UNFCCC. What is the critical thinker to make of emissions trading and of the way it is being framed?

The policy can be put in its economic context. Gorman and Solomon (2002) do such an exercise explaining the economic history of emissions trading. They trace the theoretical foundation of emissions trading to Coase (1960) who suggested that environmental degradation was the result of a lack of bargaining between parties with clearly defined private property rights. According to Coase, negotiation was the most efficient course for social actions and any action by the state that inhibits negotiation between private parties should be discarded. In other words, pollution is the result of the absence of a market for that pollution. In the 1960s and early 1970s economists started to apply this principle to air and water pollution. Emissions trading schemes appeared in the United States to manage leaded gasoline and, internationally with the Montreal Protocol, to phase out chlorofluorocarbons or CFCs. Emissions trading, however, really took root with the U.S. Environmental Protection Agency's sulfur dioxide (SO_2) program that was part of the 1990 Clean Air Act Amendments and is a model upon which the UNFCCC's GHG emissions trading regime is patterned. Like then, emissions trading is now touted as the least-cost method to achieve GHG emission reductions by taking advantage of the differential costs of mitigating emissions around the globe and trusting that a market will find these cost differentials.

This explanation is useful in understanding how the economic rationale of emissions trading originated, but to understand the philosophy behind emissions trading, of why least-cost approaches and economic efficiency itself are important, a deeper analysis is needed. The approach taken here outlines emissions trading as a production of the philosophical orientation of the decision-makers and the institutions involved. This chapter proposes a philosophical interpretation of why this flexibility mechanism was put forth and accepted.

My underlying assumption is that sociopolitical institutions, such as emissions trading, are negotiated as they are reproduced. I take a theoretical proposition that the motivations for pursuing a GHG emissions trading policy through the UNFCCC can be best exemplified by the writings of the French positivist philosophers of the late 18th century. The nature of thought that supports emissions trading came to fruition during the 18th century in Europe as philosophers debated the significance and direction of changing social conditions that accompanied the French, Industrial, and American Revolutions. New ways of thinking were introduced during these periods of social upheaval replacing old structures and extending others, and the idea of social progress, that civilization moved steadily toward increased perfection, emerged in Western Europe.

This approach is important because *the idea of social progress overvalues the future at the expense of the present with no regard for the inequity of the past.* It focuses the attention of, in this case, climate change policy-makers on what can be achieved at some future date as opposed to how present social and environmental conditions were achieved. Such a future orientation diminishes other potentially more effective ways of mitigating climate change. This chapter offers one interpretation of the tremendous obstacles other voices are facing within the UNFCCC. How this way of thinking spread in Western society as the idea of social progress took root is explained. The emissions trading policy is an attempt to continue this rendition of social progress in the face of climate change.[3]

Original and other sources pertaining to three principal thinkers of the idea of social progress—Turgot and Condorcet of the French tradition, and Smith of the Scottish—are referenced. There are many other philosophers of the time (e.g. Saint-Simon, Comte, and Hutchinson) that could be referenced as well, but these three captured the idea of social progress and the role of the economy that is seen in the current climate change policy. Influenced by the theories of evolution and the law of gravity they suggested broadly that society moved forward, progressing toward something better, from infancy to maturity. Their ideas coalesced into the idea of social progress that has become ubiquitous in the West, especially in the United States. Condorcet used Turgot's ideas to formulate a theory of social progress that, it is suggested, supports the discourse on emissions trading. Turgot also influenced the work of Adam Smith and subsequent economic thought that pervades the discourse on emissions trading.

The idea of social progress has been compared to fate, providence, and immortality (Ekirch, 1951). In this way the idea of social progress is less a belief or political or economic directive, but regarded as an autonomous process. "[T]he supreme law of progress… carries along and dominates everything; men are but its instruments… it is no more in our power to withdraw ourselves from its influence or to control its action than it is to change at our pleasure the primitive impulse which makes our planet circle the sun" (Kumar, 1978, p. 31). Noble (1983: 22) adds: "[Y]ou can't stand in the way of progress, nor should you – even if it kills you." We shall consider how and why this idea became so ingrained as to propel us in our routines and actions of civil society, and to account for the emissions trading policy.

Historical Context

The underlying factors influencing the idea of social progress predate the French and American Revolutions. The factors can be seen in Asia Minor and Arab text and later in Greek and Roman thought. Greek philosophy started in Asia Minor in the early 6th century B.C. and became the foundation of Roman thought. "[I]n all theoretical subjects such as philosophy and the sciences, the Romans were completely dependent on the Greeks, their teachings and writings, their doctrines and terminology" (Kristeller, 1979: 221-2). Plato, for example, was held in high regard in Roman antiquity, especially in the areas of mathematical sciences. In fact, the philosophers and intellectuals we focus on later in this chapter also held Greek thought in high regard, according it with the utmost in refinement in the areas of art and civics. Medieval Europe did not explode into its Renaissance with innate ideas; it was built upon the shoulders of these Asian and Arab civilizations that lie to its east, and later its own Greek and Roman traditions.

The Roman agglomeration of Greek, Arab, and Asia Minor traditions in particular captivated medieval Europe. Europe in the Middle Ages lived for many centuries under the direct tradition of Roman institutions of learning, business, literature, and language. This period developed "a common source of material, a common terminology, a common set of definitions of problems and a common method of discussing... problems" (Kristeller, 1979: 38), and flourished in close alliance with Catholic theology and authority.

The imprint of theology permeated medieval European society. For many, the purpose of this life was to secure happiness in the world to come. "The idea of a life beyond the grave was in control, and the great things of this life were conducted with reference to the next" (Bury, 1924: viii). With theology held in such high regard, the authority was naturally the Church with its theological tradition.

The medieval Catholic Church was Western civilization's legitimizing agent at the time. For many, the Church guided life questions, deep and ordinary, on a day-to-day basis. Answers were definitive, although sometimes debated within the church body. The Church had the scriptures, tradition, and momentum to guide its proclamations, declaring that life on earth is a pilgrimage to the next and only good deeds, fasting, prayer, or grace, channeled through the Church, allowed humans to escape the material world. "During the early Middle Ages, learning became to a large extent a monopoly of the clergy since the most important schools were those attached to the monasteries of the Benedictine order. Their instruction was based on the scheme of [what was] considered the sum total of secular learning" (Kristeller, 1979: 111). The wording of the period was almost entirely in the possession of the Catholic clergy. The Middle Ages were religious, and the Church held great authority and legitimacy for the aristocracy, but Europe's sole attention was not preoccupied by things religious or theological. The Church's position of authority was never entirely complete. Other more secular modes of knowledge existed that regained visibility in the period to come.

The Renaissance

There were strong pagan influences that increasingly affected life outside of medieval Europe's religious and theological pursuits that resulted in only a nominal adherence to the doctrines of the Church. Paganism was influential in the growth of all non-religious intellectual interests; it was not so much opposed to religious doctrine or competing for attention but ran through European culture as a less visible current. As Europe entered the Renaissance, these nonreligious pursuits and paths to knowledge gathered momentum. "If an age where the nonreligious concerns that had been growing for centuries attained the kind of equilibrium with religious and theological thought, or even began to surpass it in vitality and appeal, must be called pagan, the Renaissance was pagan, at least in certain places and faces" (Kristeller, 1979: 68). These strong pagan influences eventually allowed for an era of humanism that took over Europe and sent it on a trajectory that eventually challenged the authority of the Church.

The diffusion and access to Arab, Asia Minor, and Greek texts during the Renaissance in Europe also helped usher in an era of humanism. Humanist Europe sought cultural guidance in these texts of classical antiquity. Their access infused Renaissance Europe with forgotten and previously inaccessible ideas that could be called an information revolution. With their translations into Latin, Arab science and Greek philosophy was made accessible.

The entire Greek literature, for example, was made available for scholars and laity with the invention of the printing press. The copying and editing of such a vast amount of information would not have been possible without the ability to reproduce text. The Renaissance extended its source material to the entire range of extant sources that incited the original ideas to follow in the Enlightenment period. Humanism sought to explore and apply these classical influences in all walks of medieval life. As opposed to the religious focus of the Church, there was an emphasis on humanity, on a person's feelings, opinions, experiences, and surroundings and how these could be explored through the humanities. Rhetoric, with a classicist orientation, was of particular interest. It allowed for the expression of ideas the imagination could ponder. Humanism eventually influenced other disciplines such as logic, mathematics, astronomy, medicine, and law–even theology.

The persistent and systematic focus of the ensuing Renaissance period was on the dignity of humanity and the exceptionality of human ability. Humanism was both a revival and rebirth of the ability for learning in Europe after a long period of decay when it had been contested and subdued by the Church, and it is within this period that we find embryonic notions of social progress.

Like the UNFCCC, the Renaissance was concerned with humans and their problems. The Renaissance cannot be called a complete reversal of focus from that of a Church-dominated culture of the Middle Ages, but rather a rearrangement of many medieval characteristics. It was a period likened to the revival of ancient learning, literature and art, thus the name Renaissance. The Renaissance was a cultural and literary movement that had profound philosophical implications but in itself was not solely philosophical. It cut across national, religious, and professional divisions of the period in Europe. It spread from Italy to all areas of Renaissance culture including the sciences. It removed secular culture and learning from the grip of the Church.[4]

Kristeller (1979: 132) summarizes the period as follows: "The thought of the Renaissance, considered as a whole, worked with some of the old material supplied to it by medieval thought but produced out of that something new and different which expressed its own insights and aspirations." As these ideas confronted the old social arrangement, a new brew of social arrangements emerged in Europe, ushering in the Enlightenment.

The Renaissance melded European civilization around dignity and respect for human ability. Descartes proclaimed "I think, therefore I am." It was an age of human opportunity, where all that was wanted was possible through human ability. By means of reason and rationality, humanism provided the ability to control the physical world, to know all its secrets, to master its ways, and to turn them to satisfy desires. Truth, it said, was manifest here and now on earth for us to find. The answers lie within human ability focused on this life rather than on life eternal. The larger meaning of life, even morality, became a terrestrial pursuit. The Renaissance was also an age of fermentation that saw old ideas come forth to renegotiate the role of social institutions, especially the Church.

The Enlightenment

The European Enlightenment, roughly mid 15th century–18th century, ordered, sifted, developed, and clarified the Renaissance, breaking down and replacing the older forms of knowledge and constructing new forms based on reason and rationality. It was a period that unshackled the restraining authoritarian system of the Church. It was a period characterized as opposing the "power of convention, tradition, and authority in all fields of knowledge... removing the rubble of the ages in order to make visible the solid foundations of the structure of knowledge" (Cassirer, 1951: 234). It, like other structures of thought, claimed to have possession of the permanent and universal truth.

The promise of this new mode of knowledge was that each human being carries within him/herself sources of knowledge to be exercised by their ability for reason. For some, the abilities made creators of humans –

an emancipation of the mind associated with the natural order. For others, the Creator endowed these abilities. Descartes explains, "What we clearly and distinctly see to be true must indeed be true; for otherwise God would be deceiving us" (quoted in Popper, 1965: 7).

Reason and rationality delivered science as the new legitimizing agent and an era of "unparalleled epistemological optimism" (Popper, 1965: 5) began. Scientific experiment became the key for discovering the secrets of nature and augmentation of knowledge. This method binds science to a never-ending pursuit of knowledge to discard, revise, or develop new theories. By discarding or revising old ideas and developing new ones, the scientist's role was to clear up conflicting ideas, organizing them into a systematic body of ever-increasing, objectively reasoned, and confirmed knowledge.

Scholars began to notice the results of these new modes of inquiry and noticed the pace of such discoveries compared with classical antiquity. It was the scientific discoveries of the 17th and 18th centuries "that gradually brought about a neat separation between genuine and false sciences and put an end to the fantastic cosmology of the Middle Ages and Renaissance" (Kristeller, 1979: 123). It was then that the Enlightenment took hold and its promises permeated European culture and eventually diffused to help frame the current topic, climate change policy.

It is from these structures of knowledge acquisition, as they coalesced in Europe during the Enlightenment, that the current social systems evolved within the UNFCCC. *It is therefore important to consider the nature and potential of thought itself, as it became known in this period.* A mechanical and analytic approach to knowledge acquisition became indispensable, as Cassirer (1951: 25) writes of the period:

> The mind neither creates nor invents; it repeats and constructs. But in this repetition it can exhibit almost inexhaustible powers. It expands the visible universe beyond all bounds; it traverses the infinity of space and time; and yet it is increasingly engaged in the production of ever new shapes within itself.

The new mottoes of the Enlightenment were *libido sciendi* and *sapere aude* – lust for knowledge and dare to know! The dominant approach to knowledge acquisition became commensurable with these new compulsions.

Western Europeans began to look at phenomena in this context and propelled by the confidence in their ability, determined the construction of not only the phenomena in question, but its context as well. This represented an infusion of confidence in the ability of humans to delve into the questions afflicting society. The confidence, no matter how scant, was viewed as being better than the old structures of knowledge acquisition. "Not doubt, but dogma, is the most dreaded foe of knowledge; not ignorance as such, but ignorance which pretends to be truth and wants to pass for truth, is the force which inflicts the mortal wound on knowledge" (Cassirer, 1951: 161). Human desire and capability for knowledge thus became embodied in the processes of social inquiry. Human capacity for science and desires for economic comfort now frame the climate change debate. We shall return to this later.

Reason

Underlying the new way of knowledge acquisition was a confidence in the human ability for reason. "[I]nstead of being moved by instincts and governed by force, men were capable of being moved and governed by reason" (Mumford, 1934: 182). Reason was reliable, something permanent, and ever present like, it was thought, the permanence of natural systems. But reason was dissimilar in that it lay in the human mind as something that could be controlled. With increased control over reason humans also gained ability over mental disturbances, or what was referred to as the passions. Reason, it was thought, "subsists in spite of all the passions which make war on it, in spite of all the tyrants who would drown it in blood, in spite of imposters who would annihilate it by superstition" (quoted in Bury, 1924: 150). Reason represented a powerful stimulant in faith in humankind.

By uncovering what is truly permanent, reason unified independent efforts at the discovery of truth. "Reason is the same for all thinking subjects, all nations, all epochs, and all cultures" (Cassirer, 1951: 6). This certainty was to be spread as a light to all places as the basis for knowledge and all ways of knowing. As Cassirer (1951: 95-6) continues:

> Reason, as the system of clear and distinct ideas, and the world, as the totality of created being, can nowhere fail to harmonize; for they merely represent different versions or different expressions of the same essence... It is only through this medium that we recognize and act upon external objects... There is no true knowledge of things except in so far as we relate our sense perceptions to ideas of pure reason.

Leibnetz took it further by saying, "There is nothing without a reason" (quoted in Partington, 1993: 204).

The faith in this ability stems from the perception that reason uncovers what is truly timeless and permanent. An order is found through the sphere of pure reason, it was thought, that predates all human power and is valid independent of any power. This way of knowing allows understanding to "unfold gradually, with ever-increasing clarity and perfection, as knowledge of the facts progresses" (Cassirer, 1951: 9). By the ability to understand its order, reason separated humans from the mystery that is Nature. The act of reason represented our participation in Nature and gave access to an intelligible world. Our task was to throw reason at Nature and truth would cascade upon us.

European philosophers and intellectuals in this period sought universal order and law in Nature using natural science as the model to reveal these truths. Nature was viewed as the trove of order that only applied reason could exhume. It is "nothing but a force implanted in things and the law by which all entities proceed along their proper paths" (quoted in Cassirer, 1951: 44). The assumption that there is a fundamental law behind natural processes became the hypothesis of the Enlightenment. Nature is not the organization of things but the origin and foundation of truths. Nature represented truth, it was thought, for it is always in harmony with itself.

The position that truth is found in Nature threatened the very foundations of the Church. It shook the throne upon which traditional authority resided as it contested the Church's stranglehold on claims to the truth. The truth in Nature was revealed but only to those that employed their abilities using a proper methodology. As Cassirer (1951: 43) again explains: "This truth is revealed not in God's word but in his work; it is not based on the testimony of Scripture or tradition but is visible to us at all times. But it is understandable only to those who know nature's handwriting and can decipher her text."

The method of inquiry, based on reason, rationality and observation, held the key to understanding the writing in Nature. The ability now beholden in the Enlightenment allowed humans to read the truth in Nature. "The supremacy of reason shook the thrones from which authority and tradition had tyrannised over the brains of men. [It] was equivalent to a declaration of the Independence of Man" (Bury, 1924: 65).

Not all Enlightenment thinkers discarded divine authority. St. Thomas Aquinas, for example, called "both natural and divine law radiations of the divine being" (Cassirer, 1951: 241). Such calls defended powerful interests. A faith in human ability to reveal, not simply to discover or conjure or interpret, original truths loosened the Churchs grip as the sole decipherer and provides of truth. With a renewed faith in the moral abilities for reason truths were not solely revealed to organized religion as a mediator between the creator and the masses; original ideas were simply the expression for the first time of what should be evident to all – the truth embedded since the beginning of time in Nature. In other words, the authority of truth had always been there, since the beginning of time, and was not constructed along with the history of the Church.

Reason thus reaffirmed the potential of humanity that our destiny lies within our abilities. "It follows that the end of mankind upon earth is to reach a state in which all the relations of life should be ordered according to reason, not instinctively but with full consciousness and deliberate purpose" (Bury, 1924: 251). We could take control of our own destiny and no longer be subject to the whims of Nature and passion. Applied purposeful reason would now guide humanity.

The parallels with the science-policy process adopted by the Intergovernmental Panel on Climate Change (IPCC), which informs the UNFCCC, are almost too obvious to mention. The IPCC and the UNFCCC uphold that policy must be based on good science, or what scientists have reasoned from observing the climate to be true.

Observation, as the operational display of reason, accepts that there is an objective reality that can be known. It is a reality we can encounter with our senses and decipher with our ability to reason. If reality is tangible, visible, and consistent, then its order can be revealed. The form of the phenomena arises from the facts themselves gathered by observation. The logic of the facts is then fully realizable by employing reason and rationality. Once the form or causality is revealed, it offers the opportunity for prediction and control.

Prediction and Control

Driving the pursuit of knowledge and explanation was an impulse to control.[5] Of particular intent was control over Nature. As Sir Francis Bacon, often called the first philosopher of the modern age, advocated, if we can understand Nature, let's make it work for us. Newton exposed the regularities of Nature and it was thought, if it could be understood, it could be governed. Similarly, the UNFCCC is an endeavor, among other things, to manage the global carbon cycle at the optimal concentrations of atmospheric CO_2 concentrations.

As Nature is controlled, it was argued, so could the social realm be. By gaining knowledge of Nature, humanity was no longer subject to its forces. Rather, with knowledge of natural forces, humanity could "guide them toward a goal... [and] bring about an equilibrium which assures the preservation of the community" (Cassirer, 1951: 214). Humans can escape the will of Nature and create a secure social world as a rationally structured arrangement. The hope was to control the social environment for increased harmony and fulfillment of needs and desires. As Montesquieu asserted: "Man is not simply subject to the necessity of nature; he can and should shape his own destiny as a free agent, and bring about his destined and proper future" (in Cassirer, 1951: 215).

Natural Law

The philosophers and intellectuals of the idea of social progress abided by an interpretation that natural and social phenomena are both determined by permanent laws; there is a unified natural order to social and natural phenomena that operates according to natural law. They contemplated the human causes of production, such as the physical causes of motion or heat, for a theory of society. This natural law would be revealed if the impeding social dialectics, such as undue government interference or predisposition to prejudice, were pruned. Natural law, in other words, was the discovery of the cause of social events and discontents. If the optimal law were found, it would represent perfection and society would become balanced in a natural order. Natural law represented the "mechanical justice of the social economy" (Strong 1932: 53). The goal was to find the equivalence to the permanence that could be observed in Nature.

A starting assumption had to be established to make the idea of social progress work. That assumption is the constancy of Nature. The constancy, or permanent and undiminished capacity of Nature, was not original to the Renaissance or Enlightenment period. Humans have always looked to Nature for knowledge and to learn how to organize themselves. This assumption is a foundation of the acquisition of knowledge. Knowledge will progress only if inquiry is on a sure foundation, and this foundation is not sturdy unless it is submitted that the laws of Nature are constant. There has to be something to learn from. If this hypothesis is not accepted, if we consider it possible that Nature is not uniform and changes its properties with time, then we cannot be ensured that social progress will continue since the foundation of progress may change. This point is essential to the theory of social progress.

With this foundation, a new method of inquiry sought to explain why events happen, seeking causality in relationships. The method sought out patterns for prediction wherein certain events regularly follow other events. The acceptance that Nature is permanent and that reasoned objectivity would reveal causality encouraged those engaged in inquiry to reveal the code of events in question. This was done in pursuit of objective universal laws similar to those ascribed to the natural world. Physical and social events were reduced to a formulaic analogue. The planets rotate in ellipses governed by laws of gravity. Politics is engineering with arbitrary conjecture or emotion. Universal explanations can be applied to all types of events, conditions, culture, and humans. Once the true explanations are found, humanity would do best by resigning itself to acceptance of these causalities and their rationality. Some readers may see similarities with market arguments for environmental and other problems.

Science as the Conduit of Social Progress

Memory was the easiest means to store these new explanations as long as the amount was small enough for the human brain to handle. When human memory became too small for the amount of knowledge being generated, out of necessity, writing and classification evolved. Logic was born out of classification and gave order to diffuse processes. Scientific theory stemmed from this basic sorting mechanism. Therefore, science, its methods and theory, are simply a "logically ordered description" (Georgescu-Roegen, 1971: 25). This method of inquiry became institutionalized in the sciences with the Scientific Method. The method became the medium to apply reason and arrive at the truth embedded in Nature. A simple but rigorous application of the Scientific Method would open up unimaginable frontiers for humanity.

This approach was embraced for knowledge acquisition of many topics in the Western tradition, not only in the sciences but also in the humanities. In art, for example, there was a strong movement in the Enlightenment to interpret, test, and measure art by using the laws of reason. The otherwise pleasurable experiences associated with art were cast off as momentary, for true art, like physics, had to represent what was permanent. As there are laws of nature, there must be similar laws of the imitation of nature. Many other aspects of culture and individuality were also affected by the new application of reason and its methods. For example, a poem of the Enlightenment period asks (quoted in Cassirer, 1951: 281-2):

> What is virtue?
> It is reason put in practice.
> And talent?
> Reason brilliantly set forth.
> Spirit?
> Reason well expressed.

Taste is simply refined good sense, and genius is reason sublime.

Religion, of course, was no exception; it too was swayed by the new mode of inquiry and knowledge acquisition. This period did not reject the beliefs of religion; rather it offered a new basis of faith. As Cassirer (1951: 39-40) explains:

> Knowledge of 'nature' is synonymous with knowledge of creation... The 'realm of nature' is thus opposed to the 'realm of grace.' The one is communicated to us through its sense perception and its supplementary processes of logical judgment and inference, of the discursive use of the understanding; the other is accessible only through the power of revelation.

The use of the Scientific Method to discover truth relegated religion, along with other claims not based on reason, to the heap of dogma. The concern now was with the certainty allowed by inquiry into Nature. It no longer dealt with a system of beliefs but with a process to uncover certainties. These certainties lead to a firm belief in the progressive acquisition of knowledge and the extension of a firm basis for future social progress.

The achievements of science converted humanity's imagination toward the promise of social progress. Science would allow humans to understand the structures of physical phenomena and apply the organization to society. Saint-Simon (1975: 29) suggested that science would reveal the "laws of social organization." It must be remembered that the idea of civilization itself was not certain and was deeply debated at this time.

Social theorists, envious of this perceived power to uncover permanence and certainty, applied the same concept to the social realm. The promise of science was to govern so precisely to do away with social dialectics; the system run by science would be its last, everlasting. The true methods of the physical sciences would be applied to the social sciences, for this was their most efficacious task. Precisely organized social organization would unleash the ordered manufacturing of social desires. It would produce a made-to order social system based on what was perceived to be everlasting in Nature. Nature continued on its own, therefore why couldn't society be organized correspondingly?

The human dimensions of social organization were thus diminished and labeled inferior, even polluting. Any nonprogress was attributed to human error such as barbarism, inefficiency, or ignorance. The task then became easily defined – to rid society of these human errors, bask in the growth of universal reason and enjoy the "happy destiny of humanity" (Bury, 1924: 128).

The assumption that science would continue indefinitely to discover permanent structures was what ascended and would keep science in its position as the guiding conduit of social progress. The promise of science is to guide nations of people and their social institutions, as long as we follow the call and let science correct and augment itself. Again, hints of these beliefs are evident in the science-policy process adopted by the IPCC and UNFCCC.

The idea of social progress is thus held with science as its tireless organizing drone that offers solace and hope for the future. Our fears were laid to rest in our ability for further understanding and an application of a rational order. Science, and its method, took their position and soothed the irrational passions of humanity. A passage from Noble (1983: 20) captures the idea nicely: "[S]cience was heard and the savage hearts of men were melted, the scabs fell from their eyes, a new life thrilled through their veins, their apprehensions were ennobled, and as science spoke, the multitude knelt in love and obedience."

These became the underlying operating principles of Western organization that are particularly evident in current technological and economic systems. They negotiated their way to become the standards of legitimacy and are visible in the policy choices now coming online with the Kyoto Protocol and in the organizational structure of the IPCC and UNFCCC.

Technological and Economic Imperatives of Social Progress

Thus history can now be seen in two areas that influence climate change policy – technical rationality and market logic.

With the Exhibition of 1851 in London and its display of science and technology shown to the masses, belief in human progress swept through

European society. The Exhibition showed the power of steam, railroad, and telegraphy. It showed inventions that "abridged space, economised time, eased bodily suffering, and reduced in some ways the friction of life" (Bury, 1924: 332). As Marshall (1938) pointed out, technology also produces new opportunities, it stimulates the creation of new jobs. Technology also allows labor to be more productive, thus increasing the value of labor. Technology takes over routine tasks, and doing that which humans cannot. Technology through machinery relieves manual stress, thus the monotony of life. Management, engineering, and design, that foster fine character, replace arduous jobs. Technology becomes an implement of social welfare. The idea of progress has now become realized in the form of technology through applied science.

A "technology as progress" doctrine became ingrained as machines became more efficient with time becoming tangible evidence of progress. The momentum of technology as progress dominates our mindset from the past, and the promise of technology dominates us from the future (Noble, 1983: 10). At all stages of this progression, efficiency improvement is the technological imperative and crowning evidence of progress.

Industrialism also became progress. The quantitative attribute is linked to the qualitative virtue where an increase in the quantity results in an increase in quality of life. Technological progress through increased efficiency and productivity leads to the economic growth of material goods, which leads to social progress. Economic growth is nourished by technology loosening supply-side constraints. Progress is realized through increased efficiency or productivity. Primary to both is technology, the result being increased output per unit of input. Industrialism thus becomes progress by producing more. More being preferred to less, a monotonic arrangement so engrained in our experience, it has become ubiquitous.

This tradition is seen in the choice of emissions trading as a viable policy for climate change. Emissions trading offers decision-makers an option to internalize environmental conditions within the doctrine of progress. Emissions trading is an option that applies the technological imperative of efficiency while allowing for the continuance of economic growth.

* * *

The Renaissance ushered in an era of respect for human ability. Humanism took hold in Western Europe as an infusion of information from Greek and other sources were translated and disseminated for mass consumption via the printing press. The Enlightenment period that followed revived confidence in human ability with the application of reason. The period is characterized as embracing the exploration of the sources of knowledge that lie within humans and shunning the traditional authority of the Church. Faith now lies in human ability. The foundation of enlightened thought resides in Nature as the stock of knowledge. Our ability for reason and rationality, applied through observation, allowed for the facts embedded in Nature to unfold with ever-increased clarity and augmentation. This direction of knowledge acquisition directly influenced the perception in society of social progress.

An analysis of the context of the ideas of social progress up to this point can be summarized as follows:

- ❑ Although contested and negotiated, reason became the dominant way of knowing in Western culture during the Enlightenment period.
- ❑ Reason delivered science as the new legitimizing agent and ushered in an era of optimistic confidence in human abilities.
- ❑ The Scientific Method became the mode of discovering the secrets of Nature and the augmentation of knowledge.
- ❑ Faith in this method of knowing stems from the perception that reason uncovers what is truly timeless and permanent. This way of knowing allows understanding to "unfold gradually, with ever increasing clarity and perfection, as knowledge of the facts progresses" (Cassirer 1951: 9).
- ❑ By its cumulative nature, reason develops an impression of social progress as we move forward in time and gather more knowledge.
- ❑ This knowledge provided by science became incarnate through technological industrial application.
- ❑ The linear progression of social progress was realized through technological and economic systems based on efficiency and growth.
- ❑ This linear progression, this cumulative cascade of knowledge led many to accept the idea of social progress and to search for its underlying characteristics.

Social progress was viewed as a "synthesis of the past and prophecy of the future" (Bury, 1924: 5). In order to fully realize the doctrine of social progress, one had to assume its continuance in the future, not just focus on the achievements of the past. If it were known that the world is coming to an end in 100 or 200 years, for example, the meaning of progress would be lost. The full conception of the doctrine of social progress was more than just transmitting the all that has been achieved to posterity, but to augment those discoveries for an even better future. As Leibnetz stated, "the present is pregnant with the future" (quoted in Bury, 1924: 194).

Law of Social Progress

The idea of social progress took hold in the second half of the 18ᵗʰ century through French and Scottish philosophic and intellectual movements.

Several French renditions existed of the idea but all were shaped by and adhered to a faith in humanism that developed earlier (Table 7.1). In this section we turn to the specific ideas of the French and Scottish Enlightenment that developed a theory of social progress. The aim of philosophers of these traditions was to discover the law of social progress analogous to, for example, Newton's discovery of the law of gravity.

Table 7.1. 18th century schools of thought on progress in the French tradition

School	Position on Progress
Perfectabilists	Progress is a continuous and unlimited extension of social perfectability
Physiocrats	Progress is realized through economic liberalization (i.e., free trade)
Encyclopedists	Progress is realized through the unification of all knowledge
Ideologues	Progress is realized by basing moral and political ideas on physical properties to avoid perception and judgment errors

These schools of social progress largely held to the ideals of (Strong, 1932):
- ❑ Liberalism: to use humanist abilities to solve social dialectics;
- ❑ Positivism: to base social choice on what could be positively observed;
- ❑ Utilitarianism: a focus on human needs and desires.

In the last decade of the 18th century, these renditions of the idea of social progress embodied the ideas of the Enlightenment, particularly in the tumult of the French Revolution. Before the French Revolution, social change in Europe happened very slowly, almost imperceptibly, giving the notion of permanence. Change, in fact, was not sought after as it was equated with calamity – plagues, natural disasters, invading armies (Schapiro, 1934). Propelled by the ideas coming out of these groups the French Revolution challenged this conventional thinking, applying the ideas of the Enlightenment toward what they thought would change the world. The French Revolution gave "immense acceleration to human hopes" (Schapiro, 1934: 254).

The French Revolution was predicated on the ideas of individual freedom, representative government, popular sovereignty, self-determination, religious tolerance, popular education, individual civil rights, the nation-state, and *laissez faire* (Schapiro, 1934). These social institutions changed as France, and especially England, transformed from feudal societies to mercantilism and capitalism. With the rise of a bourgeois class and free enterprise, a focus on markets, rather than the restricted commerce of the guilds, became a national priority. It is in this era that we find the beginnings of the current market-based orientation to environmental policy, and the roots of emissions trading.

It was also in this period that French and Scottish philosophers began work on a unified theory of social progress. What they searched for was the Law of Progress. Bury (1924) accredits Abbe de Saint Pierre with first formulating the idea of social progress in the early 18th century. Tracing the idea to its beginning is not the task at hand, or that important. What is important is to understand how the idea took hold. For this reason we turn to the French Physiocrats, Turgot and Condorcet, and to Adam Smith of the Scottish tradition. Although many other thinkers have pondered and written on the idea of social progress, these three captured the spirit and forward thinking hope that is evident in the discussions and interpretations of emissions trading. Smith is directly connected to the technical rationale for the policy as the father of economics.

Turgot (1727-1781)

Turgot was the Finance Minister for Louis XVI, but his influence spanned the Atlantic influencing such notables as Adam Smith and Benjamin Franklin. Meek (1973) explains that he was influenced by Christianity; scientific and technological discovery; and anthropological studies of American Indians, and like many contemplating social affairs by Locke, Montesquieu, and Hume. These broad interests led him to write on language, history, economics, love and marriage, political geography, natural theology, morality, and science. Turgot, the economist, wrote on the distribution of wages, profit, rent, labor, unemployment and capital, taxation, arguments against the Corn Laws, and to overthrow the guild system.

Turgot gained fame as an economic historian who, consistent with the ideas permeating the French Enlightenment, focused on permanent principles: "To him, reason was the fundamental fact of economic interpretation and application. His appeal by voice and pen, and public effort was to reason" (Shephard, 1903: 247). He aligned with the Physiocrats of 18th century France who suggested that free trade and free enterprise, coupled with land-based labor and capital were the sources of wealth. These sources of wealth offered society the wherewithall to achieve gradual social progress, even as the economy ebbed and flowed through business cycles. Turgot wrote: "The whole human race, through alternate period of rest and unrest, of weal and woe, goes on advancing, although at a slow pace, towards greater perfection" (in Meek, 1973: 41).

Turgot gave two highly influential addresses that were later duplicated in book form mid-18th century, entitled *A Philosophy Review of the Successive Advances of the Human Mind* and *On Universal History*. His seminal work on the subject of social progress, however, was *Reflections on the Formation and Distribution of Wealth* (the text that would heavily influence Adam Smith). These are discussed in turn.

Advances of the Human Mind (1750)

In this address, Turgot reasoned, probably influenced by the work of Locke, that all humans have the same sensory organs, receiving stimuli from the same universe that broadcasts the same message to all people, and since the needs of humans are universal, all humans have proceeded in roughly the same way, from barbarous to refinement (Meek, 1973). Varying and localized circumstances impeded this process accounting for the unequal progress of nations. So Turgot focused on what these circumstances were and why humanity exhibited this apparent general long-run process of refinement. Growth of societies, he suggested, is like that of humans or plants – from infancy to maturity.

Turgot surmised (translated in Meek, 1973: 41):

The phenomena of nature, governed as they are by constant laws, are confined within a circle of revolutions which are always the same. All things perish, and all things spring up again; and in these successive acts of generation through which plants and animals reproduce themselves time does no more than restore continually the counterpart of what it has caused to disappear.

The succession of mankind, on the other hand, affords from age to age an ever-changing spectacle. Reason, the passions, and liberty ceaselessly give rise to new events: all the ages are bound up with one another by a succession of causes and effects which link the present state of the world with all those that have preceded it. The arbitrary signs of speech and writing, by providing men with the means of securing the possession of their ideas and communicating them to others, have made of all the individual stores of knowledge a common treasure-house which one generation transmits to another, an inheritance which is always being enlarged by the discoveries of each age. Thus the human race, considered over the period since its origin, appears to the eye of a philosopher as one vast whole, which itself, like each individual, has its infancy and its advancement.

He proposed in this address that the main vehicle for the process toward perfection is economic surplus. Surplus leads to increased time, leisure, and resources to spend on refining science and art. Commerce, to him, was the catalyst and underpinning to social progress.

Universal History (circa 1751)

In this address Turgot put forth a stage theory of history. He suggested that societies progress through stages, from hunters to shepherds and then on to husbandry (what he later equated with agriculture). He considered hunters savages (like the American Indian, whom he studied). Shepherds, he suggested, expressed the social institutions of private property and the concept of wealth. With husbandry the agriculturalists began producing surplus, which led to the development of towns and trade, and a leisure class. These stages explained the differing positions nations held on the development continuum depending on, he suggested, the population size, natural dowry, geography, communication with the outside, and intermingling with the outside (Meek, 1973).

Turgot here touched on an important idea that the Scottish Enlightenment picked up on, namely, the idea that social development is a by-product of concerted human actions and intent. Turgot wrote, these "gentle passions... led them on their way" (in Meek, 1973: 69-70) which is very similar to Smith's idea of the invisible hand idea. These ideas led others to follow with a general theory of social development, which is interesting, but beyond the subject at hand.

Reflections (1766)

Written for two Chinese foreign students studying in France and who would return to China with a report on French society, *Reflections on the Formation and Distribution of Wealth* would become Turgot's most significant and influential work. In *Reflections* Turgot added a fourth stage of development, that of commerce, characterized by the division of labor and class, and the production of surplus. *Reflections* broadly considered the economic history of a society and what happens to that society with the introduction of this new class of capitalist entrepreneurs. His thesis was to isolate the principles of a universal progression and successive improvement on the constitution of the French economy. He favored free competition versus the regulation of prices because free competition would lead to a price that would satisfy sellers. This was an early version of *laissez faire*. It included a general theory of the returns on capital and on the crucial role of the entrepreneurial class.

Turgot was a contemporary of Adam Smith, the father of economics, and one of the originators of the logic behind emissions trading. Although controversial, according to Lundberg (1964) there is evidence that *Reflections* was the germ for Smith's *Wealth of Nations*. There is ample evidence that Smith had Turgot's *Reflections*, but Lundberg goes further to suggest that it was Smith who translated *Reflections* into English (Lundberg, 1964: 57). Smith was in Paris in the 1760s mingling with Voltaire, Hume, Condorcet, and Turgot taking part in the Parisian economic intellectual society. Smith states no direct reference to Turgot, but there are many phrases and uses of words common to both works. It is known that Turgot corresponded with Smith starting in the early 1760s and that Smith owes much to the Physiocrats and Turgot for the ideas behind the *Wealth of Nations* (Shephard, 1903).

The two also differed. "Turgot would posit reason as the sole determinative factor in construing economic relations" (Shephard, 1903:

257). Smith's base was moral philosophy, arguing what ought to be and what benefits could be derived from the new capitalist system. Smith illustrated and instructed; Turgot critiqued and reasoned.

Smith (1723-1790)

Smith was influenced by the classics, Hobbes, Locke, Montesquieu, Voltaire, Rousseau, Hutchinson, and especially Hume and the Physiocrats – Turgot and Condorcet. But his perspective was from moral philosophy applied to political economy – a "theologian of the 'natural school'" (Shephard, 1903: 255). At the eve of industrialization in Scotland, Smith deduced that the basis of wealth was going through a change from the ownership of land to that of capital and the control of labor. On the infamous pin factory that Smith observed, Strong (1932: 40) writes: "Adam Smith may be accorded the gratuitous honor of being one, if not the first, of the philosophers and prophets of social progress, characteristically concealing the moment of this concept for his system, as he did, in a technology which, at first blush, appears to be indifferent to any such bias." Capital, science, and technology made incarnate, became the lure for riches and prosperity, not only for individuals but also nations. It is here where social progress resides.

Commerce and industrialism brought opportunity for wealth and the acceptance of a "dynamic idea, such as progress" (Schapiro, 1934: 234). Smith considered the nature and causes of such progress as being imbued with what he thought to be the natural propensity for the good to win out in society (Strong, 1932). His aim was to uncover the laws of cause-and-effect operating in society by observing the Scottish economic process in the later part of the 18th century and draw theory rather than developing abstract assumptions from pure reason.

Wealth of Nations (1776)

Smith had two grand theories of social progress—the ethical and the economic (Strong, 1932). His ethical theory was laid out in *The Theory of Moral Sentiments* (1759). This book described a doctrine of harmonious social order guided by God.

His economic theory was laid out in *An Inquiry into the Nature and Causes of the Wealth of Nations* (1776). This book considered the causes of material prosperity and well-being. These came about, he suggested, not through human wisdom or aspirations of opulence, but by the propensity of humans to engage in commerce. Society gradually improves with time on its own, following the innate propensities of its members. Therefore, progress happens even in the absence of human planning.

Smith's ideas on progress were that it comes from free enterprise unrestricted by government policies, resulting in the greatest advantage for each and for all. This was the mode of producing social progress. *The*

Wealth of Nations was "much more than a treatise on economic principles; it contains a history of the gradual economic progress of human society, and it suggests the expectation of an indefinite augmentation of wealth and well-being" (Bury, 1924: 220).

Condorcet (1743-1794)

Condorcet was a mathematician, philosopher, and economist, influenced by Newton with the prospect of universal laws that could be uniformly applied to social organization. He posited that since the laws of nature were constant, the laws of human activity and behavior were also constant. Since humans have the ability to process experience and observation with reason and by Descartes' logical deduction, we can adjust for the better. Reason is independent and supreme. To Condorcet, all progress happens in the mind, it is intellectually determined.

Like Smith, Condorcet was also greatly influenced by Turgot, but it was Condorcet who put the theory behind the law of progress together. Condorcet put these ideas together as events coalesced around him and intellectual thought itself began to accept the idea of gradual refinement of social organization through the human abilities, now believed inherent, of reason and rationality. He applied this orientation to history and the future.

Outlines of a Historical View of the Human Mind (1795)

Ekirch (1951) suggests that this was the first book to fully consider the idea of progress. Condorcet began the task by cataloguing history into epochs, "from the primitive times to the French Revolution" (Schapiro, 1934: 240). The study of history, he suggested, enables the fact and direction of progress to be identified and thereby to accelerate the rate of progression. Condorcet attempted to discover the laws of human behavior based on the history of progress to that point for prediction and control in the future. He posed a hypothesis that each succeeding "instant exerts on the succeeding instant, and thus, in its successive modifications, the advance of the human species towards truth or happiness" (quoted in Bury, 1924: 208). These successive modifications advance humanity increasingly toward the refinement of truth or happiness.

His work and pronouncements were to be the grand narrative of all history to foretell the future. This progress was what underlined human development for all time, and within which lie the hope for humanity in future. This was a dissection of humanity, as a natural scientist would uncover a natural law, to reveal a permanent and all-encompassing explanation of why the past looked so ignorant and miserable, and the future so hopeful. Malthus' *Essay on the Principle of Population* was a response to Condorcet.

For the future, Condorcet sought equality between nations, between individuals and the overall perfectability of humankind. He was a great advocate for liberalism as he set out to sweep away the monarchy that was a "fatal clog on progress" (Frazer, 1933: 13). He foresaw the hope that equality would come from the distribution of commerce as the vehicle for the equal distribution of wealth. He foresaw the end of sex and education discrimination and the increase in life expectancy. Condorcet called for a league of nations with a universal language to establish these institutions (Schapiro, 1934). Religion, inheritance, inequality, and prejudice, he suggested, are errors that retard human progress (Frazer, 1933: 15). This, he wrote, would be learned from the French Revolution.

Condorcet "accepted with enthusiasm and conviction the doctrine of the infinite perfectability of human nature… nature has set no bounds on the improvement of human faculties" (Frazer, 1933: 13). He left four main points (Schapiro, 1934):

❑ Social progress is a natural law;
❑ Social progress could be discovered by studying history;
❑ Humans advanced through changing institutions and conditions; and
❑ Humanity is future determined as we march toward perfectability.

On to Other Nations

Social progress has been an animating and focusing idea of Western civilization. Bury (1924) suggested it was the most important idea in the 18th century. In the 1870s and 1880s the idea of social progress was becoming an article of faith, and into the 1900s was considered a general axiom. The idea of progress has evolved with the growth of science, politics, economics, society, and culture, and is ubiquitous in the West.

In the Enlightenment period inquiry into social progress involved the study of history and revealed that human history had indeed led somewhere. As Condorcet advocated, the study of history "enables us to establish the fact of Progress, and it should enable us to determine its direction in the future, and thereby to accelerate the rate of progression" (Bury, 1924: 211). Of this period, faith in progress became a social religion, a dream of secular immortality.

Nowhere was this so evident as in the United States. The founders of the United States were influenced by the ideas of Bentham, Kant, Locke, and these same French and Scottish philosophers. In fact Condorcet's *Outlines of an Historical View of the Human Mind* (1795) was published in the U.S., in Philadelphia, in 1796 for the U.S. to adopt. After Napolean fell in 1812, America only looked forward. "With the continent opened to the enterprise of its citizens, the faith of a Condorcet… in the power of science and of human reason to effect progress, was shared by the American philosophers in the first decades of the Republic" (Ekirch, 1951: 26). These American philosophers were Benjamin Franklin, Thomas Paine,

John Adams, and Thomas Jefferson. The United States started off with an attitude that it was unique, that others would strive to be like it, and that other countries would discard the old European structures that held them down.

The idea of progress carried over in the formation of the United States' political system; its conception of the West with room to roam and expand; its land of opportunity and material abundance; its faith in science, technology and economic growth; and its conception of the future. "With the concrete evidence of material advancement on every side, progress was the faith of the common man as well as of the philosopher... As defined in Europe by generations of intellectuals, [progress] came to be treated as a purely philosophical idea. However, in America the unique experience and concrete achievements of the people helped to give the concept a dynamic reality" (Ekirch, 1951: 36-37). This led to the mass acceptance of the idea of progress as the *modus* of civil society. It is to these natural laws that the truths are self-evident in the US Declaration of Independence.

The idea of social progress is also seen in the development literature wherein societies work through stages of development, moving incrementally closer to wealth, stability, and freedom. Nations of people from this viewpoint are designated to hierarchical scaling – First, Second, or Third Worlds, or developed and developing countries. Civilization, it is proposed, had progressed from savagery in some parts of the world; the lesser groups of people must then follow the lead of the civilized. The idea of social progress gives hope to all peoples to rise from their condition. As Schapiro (1934: 236) writes: "No people could be permanently barbarous or inferior; existing differences in the degree of civilization among peoples were due to circumstances which could be changed by enlightened laws." Such is the mindset within the UNFCCC and its Kyoto Protocol with coalitions or annexes of industrialized countries (Annex I or B) and nonindustrialized countries (Annex II).

* * *

The contemporary mindset about the future takes these ideas of the French and Scottish philosophers of social progress for granted. "Formerly the subject of much philosophical theorizing by intellectuals, the idea [of progress] has also penetrated to the generality of the people. It has been withdrawn from the exclusive scrutiny of the philosophers to become accepted as a part of the ideology of industrial civilization" (Ekirch, 1951: 7).

These theorizing intellectuals commented and attempted to make sense of what they saw in the social upheaval around them. What they captured in their theories, although it may sound utopian when spelled out, also underlies the efforts to deal with climate change. We accept the

fact that we can absorb climate change into the realms of int
economy and management and continue the faith in a better t
The efforts employed by the UNFCCC nudge the international community
to absorb the threat that climate change poses to social progress.

Emissions Trading as Social Progress

Emissions trading continues in this tradition. As this new trading regime
comes into force through the Kyoto Protocol it represents still another
achievement of civil society in the progression of perfectability. This policy
mechanism represents a deliberate application of reason to modify social
systems to include a mode of controlling climate variability. Faith in the
idea of social progress also represents confidence in human abilities to
take care of problems, no matter how large, as they arise. Emissions
trading is such an effort to adjust the international economy for continued
progress. This is most visible in two of its most central qualities—it is a
market solution and relies on technological improvement. Technological
improvement represents an older rendition of social progress and market
solutions the contemporary modes of social progress. Both are fed by
scientific inquiry.

The UNFCCC, as a social institution, also exemplifies the ideas of
social progress by focusing humanist principles of human desires
(economic growth) and human ability (science). The policy mechanisms
of the Kyoto Protocol insulate economic growth from the mitigation of
climate change, thereby protecting the progress of material well-being.
The science-policy process adopted by the UNFCCC casts reason at Nature
to reveal what is true about climate change and then bases social action
on these truths.

The policy-makers, analysts, and advocates of emissions trading
interviewed at the latest COP meetings aptly described the emissions
trading in this context. It can now be shown that their explanations have
more depth than appeared at first brush.

❑ Emissions trading is a means to achieve our environmental goals at
   the least cost.

There are two parts to this statement. First, in the case of climate
change, environmental improvement required deliberate negotiation in a
contested social environment to establish what environmental goals would
be sought. In this case the social environment was the UNFCCC. The
Kyoto Protocol was the result of this contested negotiation. Delegates at
the third COP meeting in Kyoto, Japan debated in the years preeding the
meetings about what caps or quotas to greenhouse gas emissions
industrialized countries would accept. The Kyoto Protocol was a goal-
setting agreement made with the confidence we share with one another
not only in our abilities for reason, but in its deliberate application to

social affairs. The caps resulted from a scientific analysis of Nature to gather knowledge about its state of affairs, a determination of how much the climate had varied from some level of permanence (pre-Industrial times), and then to set out a game plan to ameliorate nature's condition. Emissions trading emerged as the most ambitious policy mechanism to achieve the environmental goals set by those caps in emissions.

Second, the economy, as the medium of social progress, must not be held back, or at least be minimally affected, or else we affect the amount of social progress awarded to us. Emissions trading is framed as giving us the most environmental improvement per unit expenditure. Environmental improvement is now a commodified condition of social progress. Emissions trading extends to include environmental amenities into the more traditional set of social progress indicators, such as scientific and technological achievement, and economic growth. In fact environmental improvement is a result of a more extensive application of these social institutions.

Once the environmental goals are set, emissions trading relies on a market rationale that centers maximizing social progress at the least cost. Markets are now seen as the engines of efficiency for social progress if they are allowed to find optimality. If there are problems that lie outside the economic process (called externalities), it is suggested that they arise because the market has not been fully applied. Environmental degradation is not the fault of the logic of the market itself (economic growth); rather, it is the fault of actions that impede the full application of the market. In this way economic thought mirrors the Scientific Method. As with science, economic answers are found in ever-increasing iterations of market logic.

In other words, climate change is the result of market interference and ineffectual policies and actions taken by government.[6] Emissions trading follows this logic and is the direct extension of the ideas of social progress put forth by Adam Smith, and later applied by Coase (1960), as the rationale for markets as social progress doctrine. In arguing for least-cost solutions to achieve environmental goals, those interviewed were advocating for a new sort of economic scientific sociology wherein the world will be guided by a general theory of market optimality for environmental improvement. Market economic rationales are now akin to natural law–a balance and harmony of social and environmental goals grounded in perceived truths of least cost and market efficiency. The UNFCCC, through emissions trading, acts out natural law to manage and control the global carbon cycle for continued social progress.

❑ Emissions trading is a win-win proposition allowing for both more efficient economic growth and environmental improvement.

As with the first statement, the economy is the medium for social progress. Progress is measured by the amount of things we produce; this is how we achieve ultimate happiness, measured in material terms. As Francis Hutchinson in the 18th century and Jeremy Bentham in the 19th advocated, theoretical pursuit was to pursue the best social organization that would bring the greatest happiness to the greatest number of individuals. Many philosophers of the Enlightenment also regarded social progress in material terms, but with a progression toward an equitable distribution within society. "[T]he greatest happiness possible for us consists in the greatest possible abundance of objects suitable to our enjoyment and in the greatest liberty to profit by them" (Bury, 1924: 173). This is the optimal make-up of social progress—freedom from material want for all in society. Economic growth as progress, in this regard, must not be impacted or restrained by climate change policy. Emissions trading allows for the economic growth as progress doctrine to continue.

These statements also capture the ideas of the philosophers highlighted in this chapter (Table 7.2). Both Turgot and Smith saw the economy as the source of social progress. Condorcet took a more academic look at the history of scientific and technological achievement as standards of social progress. These three standards of social progress—scientific, technological, and economic—are captured in the emissions trading policy and in the statements above. Within this deeper analysis the critical thinker can start to make sense of the emissions trading policy.

Table 7.2. Enlightenment period philosophers renditions of social progress

Philosopher	On Social Progress
Turgot	Progress through commerce
Smith	Progress from innate goodness to engage in uninhibited trade
Condorcet	History is evidence of the law of progress

Conclusion

This chapter has argued that emissions trading has a history that predates and is deeper and more ubiquitous than the economic efficiency rationale initially posed by Coase (1960) and his followers. Emissions trading has a structured epistemology and ontology that becomes evident with deeper analysis. When we consider this longer and broader take on emissions trading, we obtain a better understanding of how and why emissions trading was negotiated and constructed within the UNFCCC. When we consider the philosophical underpinnings of the institutions surrounding emissions trading we can better understand why certain policy options or mechanisms are offered, supported, and implemented.

We can also use this analysis to identify dissent, as we recognize that these structures are negotiated and temporal. The systems of thought that were ushered in with the Enlightenment, stemming from paganism, were once subordinate to the dominant or hegemonic Catholic Church in Europe. Similarly, there are systems of thought today that lie subordinate or invisible but might gain support as the UNFCCC climate change system negotiates itself into new forms.

The Kyoto flexibility mechanisms—joint implementation, the clean development mechanism, and emissions trading—are temporary on a historical scale. They make sense under the current dominant/hegemonic structures pervading the UNFCCC, but these too are temporary. What we can learn from this analysis is that dissenters, or those supporting the subordinate positions, should continue to work and refine their positions because some day conditions may take hold that propel their positions into dominance.[7]

It is not suggested here that social progress is a grand narrative set forth as an explanation of all for all time. The idea of progress has always been contested, yet it has survived, and continues to influence the thinking of policy-makers, analysts, and advocates shaping the international climate regime. There are other policy options with other histories that must be told. This story has been the story of the dominant, hegemonic policy option coming into force with the Kyoto Protocol. For those interested in change, this is the institutional momentum we face. We better do our homework and tap into histories just as deep to be able to face such a behemoth. Only then can alternatives be heard. As Bury (1924: 160-1) delineated the 18th century, we can similarly head the possibilities of change:

> Displaced, along with his home, from the center of things, he [humanity] discovers a new means of restoring his self-importance; he interprets his humiliation as a deliverance. Finding himself in an insignificant island floating in the immensity of space, he decides that he is at last master of his own destinies; he can fling away the old equipment of final causes, original sin, and the rest; he can construct his own chart and, bound by no cosmic scheme, he need take the universe into account only insofar as he judges it to be to his own profit.

The idea of social progress has been a disaster for millions of people. *It overvalues the future at the expense of the present with no regard for the inequity of the past.* It also diminishes other potentially more effective ways of addressing climate change. The mechanisms coming into force with the Kyoto Protocol will not satisfactorily address equity or mitigate greenhouse gases to a level that achieves climate stability. For these reasons, other ways of knowing climate change must be considered, and advocates must continue working to gain space and be heard within the UNFCCC. This chapter has attempted to lay out what these other voices are facing.

Footnotes

1. Only countries that ratify the Kyoto Protocol can participate in and take advantage of the flexibility mechanisms.
2. The UNFCCC is an international legal framework established in 1992 at the "Earth Summit" in Rio de Janeiro to address climate change with over 180 signatory countries.
3. To describe the context of the idea of social progress, sources referred to in the well-respected Routledge Encyclopedia of Philosophy (1998) will be used. This encyclopedia is an anthology of philosophy categorized by topic and lists the leading authors in the field. Three widely regarded as pre-eminent historians of this period – Bury (1924), Kristeller (1979), Cassirer (1951) – are the primary sources used to describe the events and changes in the mode of knowledge that evolved into the concept of social progress.
4. It is too sweeping a position to state that all social institutions and disciplines were displaced. For example, the old conception of theology remained whereas new constructions of physics emerged. Theology lost its stature, as the new physics initiated by Galileo, applied the scientific method of applied reason, rationality, and objectivity through experimentation.
5. Much has been written on the pursuit of control found in Western civilizations. See for example: *The Chalice and the Blade: Our History, Our Future* (1988) by Riane Eisler; *Close to Home: Women Reconnect Ecology, Helath and Development Worldwide* (1994) by Vandana Shiva; or *Science, Development and Violence: The Revolt against Modernity* (1992) by Claude Alvarez.
6. It is interesting to note that there is no market for carbon; this market is being created, framed, and originated by the state actors involved with the UNFCCC. The state's role is to generate the carbon market then let go and refrain from interference.
7. For example, critics of emissions trading at COP 8 were commenting that an equity-based climate change regime is gaining momentum.

References

Bury, J.B. 1924. The Idea of Progress: An Inquiry into its Origin and Growth. Macmillan and Co., Ltd., London.

Cassirer, Ernst. 1951. The Philosophy of the Enlightenment (translators — Fritz C.A. Koelln and James P. Pettegrove). Princeton Univ. Press: Princeton, NJ (USA).

Claeys, Gregory and Sargent, Lyman Tower. 1999. The Utopia Reader. New York Univ. Press: New York, NY.

Coase, Ronald H. 1960. The Problem of Social Cost. J. Law. Econ. pp. 1-44.

Craig, Edward (ed.) 1998. Routledge Encyclopedia of Philosophy. Routledge: New York, NY.

Ekirch, Arthur Alphonse, 1951. The Idea of Progress in America, 1815-1860. Peter Smith : New York.

Frazer, James George, 1933. Condorcet on the Progress of the Human Mind. Clarendon Press: Oxford, England.

Georgescu-Roegen, Nicholas. 1971. The Entropy Law and the Economic Process, Harvard Univ. Press: Cambridge, MA (USA).

Gorman, Hugh S. and Solomon, Barry D. 2002. The origins and practice of emissions trading. J. Policy History. 14: 293-320.

Lundberg, I. C. 1964. Turgot's Unknown Translator: The Reflections and Adam Smith, Martinus Nijhoff, Hague, Netherlands.

Kristeller, Paul Oskar. 1979. Renaissance Thought and its Sources. Columbia Univ. Press, New York, N.Y.

Kumar, Krishan, 1978. Prophecy and Progress: The Sociology of Industrial and Post-Industrial Society, Penguin Books, New York, NY.

Marshall, Alfred 1938. Principles of Economics. 8th ed. Macmillon, London.

McLean, Iain and Hewitt, Fiona 1994. Condorcet: Foundations of Social Choice and Political Theory. Edward Elgar Publ. Ltd., Hants, England.

Meek, Ronald L. 1973. Turgot on Progress, Sociology and Economics. Cambridge Univ. Press, Cambridge, England.

Mumford, Lewis 1934. Technics and Civilization. Harcourt, Brace and Co., New York, NY.

Noble, David 1983. Present Tense Technology: Technology's Politics. Democracy Vol. 3(2): 8-24.

Partington, Angela (ed.). 1993. The Concise Oxford Dictionary of Quotations. (3rd ed.) Oxford Univ. Press, Oxford, UK.

Pickering, Mary. 1993. Auguste Comte: An Intellectual Biography. Cambridge Univ. Press, Cambridge, England. Vol. I.

Popper, Karl. 1965. Conjectures and Refutations. Harper and Row Publ., New York, NY.

Saint-Simon, Henri. 1975. Selected Writings on Science, Industry and Social Organization. Croom Helm: London, UK.

Schapiro, J. Salwyn. 1934. Condorcet and the Rise of Liberalism. Harcourt, Brace and Co., New York, NY.

Shephard, Robert Perry. 1903. Turgot and the Six Edicts. Studies in History, Economics and Public Law. Columbia Univ. Press, New York, NY, Vol. 28, No. 2.

Strong, Gordon Bartley. 1932. Adam Smith and the Eighteenth Century Concept of Social Progress, Ph.D. diss., University of Chicago, Chicago, IN (USA).

CHAPTER 8

Status Report on CO_2 Emissions Saving through Improved Energy Use in Municipal Solid Waste Incineration Plants

Johnke B.
Federal Environmental Agency of Germany, Bismarck Plate1 Postfad 330022,
D-14191 Berlin, Germany   e-mail: bernt.johnke@uba.de

Introduction

The reduction of climate-relevant emissions of CO_2 is an ambitious goal of the national climate protection program. One way to contribute to this goal is to produce and use energy more efficiently.

Apart from treating waste, German municipal solid waste (MSW) incineration plants also produce and supply waste-derived energy. Possibilities are described herein for additional energy production and recovery in MSW incineration plants to substitute for energy from fossil energy sources and thus reduce climate-relevant CO_2 emissions. Furthermore, the results of this study constitute a recommendation to pay greater attention to energy use arguments in connection with the saving of fossil fuels at new waste incineration sites.

Aim, Subject

Fulfillment of the waste management function means that inevitably a certain amount of climate-relevant emissions of carbon dioxide (CO_2) will always be generated in MSW incineration. In addition, the process produces heat. Consequently, the optimal use of energy production from MSW – coupled with an optimal use of this energy – is of special interest.

The question examined here is whether and to what extent the available energy potential in MSW incineration plants can be utilized and improved in terms of plant technology, in order to thereby contribute to a reduction of climate-relevant carbon dioxide (CO_2) emissions. This analysis is geared to the current state of energy production technology in existing installations.

The results of this study are also intended to demonstrate what contribution utilization of existing energy potential in MSW incineration plants can make to a reduction of CO_2 emissions within the framework of the targets specified in the national climate protection program (see point 5 of the status report). The national climate protection program (BMU, 2000) sets out measures and targets for CO_2 emission reduction until 2010. An emission reduction of the order of 18 – 20% (about 180,000 – 200,000 Gg of CO_2) is expected by 2005 based on the CO_2 emission reduction measures taken since 1990. According to BMU (2000), from 1998 onward an additional CO_2 reduction potential of 50,000 – 70,000 Gg had to be tapped if the target of a 25% reduction by 2005 were to be achieved.

Methodology

- Determination of the fossil carbon fraction in residual MSW (**0.473 Mg of CO_2 per Mg of waste**) and climate-relevant CO_2 emissions (**6,144 Gg of CO_2 per year**) from MSW incineration according to the IPPC guidelines for waste incineration (based on an amount of incinerated waste of approx. **13 million Mg/a**).

- Determination of the gross fuel heat potential in incinerated residual MSW from 56 plants (about **37.448 x 10^6 MWh/a**). Data on calorific value and waste throughput were obtained from an ITAD study (status: 2000).

- Determination of the amount of energy (electricity, heat) supplied by the 56 MSW incineration plants, drawing a distinction between plants supplying electricity only, heat only, and electricity/heat (total of about **14.6 x 10^6 MWh/a**).

  The plants' own energy consumption (about 2.3 x 10^6 MWh) is not known, nor is it known whether this energy was produced internally or externally. Therefore, this proportion was roughly estimated at 1.2 x 10^6 MWh/a for electricity and 1.1 x 10^6 MWh/a for heat, but was not included in subsequent calculations.

- The fuel-related emission factors used are a crucial determinant for the amount of fossil carbon related CO_2 emissions replaced by waste incineration. The CO_2 emission factor of the so-called power plant mix (CO_2 emissions of all power plants/net electricity production) is widely used for comparative computations.

  By substituting for energy produced from fossil fuels, the total energy supplied in 2000 by all 56 plants with grate firing systems resulted in a CO_2 emission savings of about **4,000 Gg CO_2/a**. Of this total, electricity supply accounted for **1,040 Gg CO_2/a,** and heat supply **2,960 Gg CO_2/a**.

- For the computations presented here, the relevant fuel-related emission factors were used for the fuels to be replaced, i.e., **lignite** for base-load electricity (349 kg CO_2 per MWh) and a **fuel mix of 80% fuel oil/20% natural gas** for **heat** (254 kg CO_2 per MWh).
- Determination of the additional usable fuel heat potential in the 56 existing plants, assuming that all are converted to CHP and optimized in terms of energy production and supply capability (additional usable energy potential, heat/power: about **13.15 x 10⁶ MWh/a = 13.15 TWh/a**); this is based on the feasible state of energy production technology (total energy (electricity/heat) production: about 2 – 2.25 MWh/Mg waste) at existing installations.
  The additional usable energy potential determined would have to be corrected to include the proportion for the plants' own energy requirements (**about 2.3 x 10⁶ MWh/a**). As it is not known how high this value really is, and whether this energy is self-supplied or supplied by external sources, own requirements were not taken into account in the computation.
- A CO_2 reduction contribution (of **up to 3,416 Gg CO_2**) due to the substitution of energy production from fossil sources was determined for the additional usable energy potential of about 13.15 x 10⁶ MWh/a from the optimization of energy supply (CHP). This CO_2 reduction contribution does not include CO_2 from own energy requirements. The inclusion of this share would mean that the CO_2 reduction contribution would have to be corrected by about 695 Gg CO_2 (415 Gg CO_2 for own consumption of electricity and 280 Gg CO_2 for own heat requirements).
- The possible contribution of MSW incineration to energy production would bring a **0.5% CO_2 saving** relative to total CO_2 emissions from energy production (641,000 Gg CO_2).
- The additional contribution to energy which could be made by using the 56 plants' additional energy potential of up to 3,416 Gg CO_2 was related to the **CO_2 reduction targets** for the **CO_2 reduction potential** of 50,000 – 70,000 Gg of CO_2 by 2005, as cited in the **national climate protection program**. MSW incineration could make an additional contribution of about **5%**.

Determination of climate-relevant emissions from MSW incineration

General remarks

It is known from computations to determine climate-relevant emissions (CO_2, N_2O, NO_x, NH_3, C_{org}, CO) from waste incineration (Johnke, 1999) that CO_2 emissions are the main component of climate-relevant emissions from MSW incineration, exceeding significantly, by at least 10^2, other climate-relevant emissions in terms of the emission load (computed in CO_2 equivalents).

Determination of emission factor from fossil carbon fraction in waste: Uncertainty is involved in determining the fraction of climate-relevant CO_2 from waste incineration for the purpose of calculating total climate-relevant CO_2 emissions, since the range of variation of CO_2 of biogenic origin allows a high level of decision latitude. One has to rely on literature data to a large extent. Data on results of analyses for the carbon content in waste, its distribution among the various waste fractions and an assumed waste composition are given by Reimann and Hämmendi (1996).

If the default data given in Chapter 5, "Waste Incineration", of the IPPC guidelines entitled "Good Practice Guidance and Uncertainty Management in National Greenhouse Inventories" were used for Germany, a figure of 0.557 Mg CO_2 per Mg waste would be obtained for the share of CO_2 from MSW waste incineration. The IPPC default data for MSW are: 0.4 for the carbon content x 0.4 for fossil carbon as % of total carbon x 0.95 for the efficiency of combustion x 44/12 for conversion from C to CO_2. Because of the composition of residual waste in Germany, the IPPC default for the carbon content appears to be too high. This value would be acceptable for Germany only when used for comparative purposes and when in using it, it could be assumed that all other Parties to the Climate Convention apply it as well and do not exploit the ranges that are listed in addition to the defaults.

A figure of **0.473 Mg CO_2** was used for Germany to estimate emissions of the climate-relevant CO_2 fraction from MWS incineration (CO_2 from waste of fossil origin). (This figure for MSW was determined as follows: 0.34 carbon content of waste x 0.4 fossil carbon as percentage of total carbon x 0.95 efficiency of combustion x 44/12 conversion from C to CO_2.) This specific value of **0.473 Mg CO_2 per Mg waste** is used in the considerations presented below. Based on this value, an **emission factor** of **45 Mg CO_2 per TJ** was computed for residual MSW, equivalent to **164 kg CO_2 per Wh**.

For the waste incineration sector, the IPPC guidelines concern only CO_2 and N_2O as climate-relevant gases and provide equations only for these two substances (Tables 8.1 and 8.2).

Table 8.1: Calculation example for CO_2

Calculation example - CO_2 (amount of MSW incinerated in 2000: 12.99 x 10^3 Gg waste/year (original substance)):
IPPC equation
Emission CO_2 = 12.99 x 10^3 Gg waste/year x 0.473 Gg CO_2/Gg waste
Emission CO_2 = 6,144 Gg/year
Calculation of CO_2 equivalent
Climate-relevant equivalent CO_2 emission = 6,144 Gg CO_2/year

Table 8.2: Calculation example for N_2O

<u>**Calculation example - N_2O**</u> (amount of MSW incinerated in 2000: 12.99 x 10³ Gg waste/year (original substance)):
<u>IPPC equation</u>
Emission N_2O = 12.99 x 10³ Gg waste/year x 11 kg N_2O/Gg waste
Emission N_2O = 0.143 Gg/year
<u>Calculation of CO_2 equivalent</u>
Climate-relevant equivalent CO_2 emission = 0.143 Gg N_2O/year x 310 Gg CO_2/Gg N_2O
Climate-relevant equivalent CO_2 emission = 44.3 Gg CO_2 equivalents/year

Comparison of CO_2 emissions

Energy production accounts for **641,000 Gg** of **total CO_2 emissions** in Germany (1999), amounting to **859,000 Gg,** and the transport sector for **191,000 Gg CO_2.** The contribution of MSW incineration is **6,144 Gg CO_2** (fossil CO_2 fraction) and is thus **< 1%** of energy production's contribution to CO_2.

Table 8.3: Comparison of CO_2 emissions from MSW incineration and total CO_2 emissions in Germany (Source: Daten zur Umwelt 2000)

Emission	Total emissions 1999	Global Warming Potential GWP 1999	Total emissions from MSW incineration (fossil fraction)	Contribution of MSW incineration to total emissions
	$(Gg\ y^{-1})$	$(Gg\ y^{-1})$	$(Gg\ y^{-1})$	(%)
Carbon dioxide CO_2	858,511	858,511	6,144	
Nitrous oxide N_2O	141	43,710	0.14 *(44)**	
Total GWP		**902,221**	**6,188**	**0.7**

* Value in parentheses gives the emission level converted to CO_2 equivalents to enable a comparison with the GWP

Energy use in MSW incineration plants – Evaluation of an unpublished internal survey carried out by ITAD (association of thermal waste treatment plant operators in Germany, ITAD, 2000).

In the area of thermal treatment of MSW, 56 plants with grate-firing systems, 2 pyrolysis plants, 1 pyrolysis/gasification plant (Thermoselect) and 1 gasification plant (SVZ) were in operation in Germany in 2000. The following only considers the energy supply data of the 56 plants with a grate-firing system as available from the ITAD survey, as sufficient energy data are not available for the other four plants.

Existing energy recovery potential

The existing usable energy recovery potential in residual waste from the 56 MSW incineration plants operated in Germany (grate-firing systems) can be estimated as follows:

Amount of residual waste for incineration: 12.99 x 10^6 Mg/a (status: 2000, ITAD survey)

Average calorific value: 10.377 x 10^6 kJ/Mg waste

Gives: 134.8 x 10^{12} kJ/a, or 134.8 x 10^6 GJ/a

Computed gross fuel heat potential: 37.448 x 10^6 MWh/a (conversion via 3.6 GJ/MWh).

> This means that the amount of incinerated residual waste of **12.99 million Mg/a** contains a theoretically available, so-called **gross fuel heat potential** of **37.448 x 10^6 MWh/a.**

Conversion of the heat potential in waste into energy takes place in the combustion chamber of the incineration plant. As well as the gross fuel heat potential, the so-called net fuel heat potential of the steam generated during combustion can be used as a parameter for energy-related considerations. This allows the efficiency of combustion to be taken into account. The combustion efficiency of MSW incineration plants is in the range of 75 – 90%. *(In accordance with the requirement laid down in the Closed Substance Cycle and Waste Management Act, the minimum combustion efficiency in the case of energy recovery was assumed to be 75% (Article 6 (2) of the Act).)*

Table 8.4 shows **net fuel heat potentials** resulting – in contrast to the gross fuel heat potential – from different **combustion efficiencies**:

Table 8.4: Net fuel heat potential of MSW incineration plants as a function of the efficiency of combustion

Combustion efficiency	Net fuel heat potential
75%	Ca. 28 x 10^6 MWh/a
80%	Ca. 30 x 10^6 MWh/a
85%	Ca. 32 x 10^6 MWh/a
90%	Ca. 34 x 10^6 MWh/a

Actual energy use

Total amount of energy supplied (electricity and heat) : All 56 MSW incineration plants use the heat generated during incineration. Depending on the geographical location of the plant, and season, the net fuel heat potential is converted into electricity, district heat, process steam and/or heating steam, made available to external users and used to meet own energy requirements. The total amount of energy actually supplied in 2000 by all incineration plants with grate-firing systems together was **14.6 x 10^6 MWh/a**, of which **electricity** accounted for **2.99 x 10^6 MWh/a** and **heat** for **11.65 x 10^6 MWh/a** (ITAD, 2001). The supply efficiency can be determined by comparing actual energy supply (not including own consumption) and the gross fuel heat potential, as shown in Table 8.5.

Table 8.5: Actual energy supply in relation to the gross fuel heat potential

Gross fuel heat potential	Actual energy supply	Energy supply efficiency
37.448 x 10^6 MWh/a	14.6 x 10^6 MWh/a	Ca 39%

The electricity which all the plants require for their own use must be added to the **net electricity** supplied to the grid by 48 of the 56 plants, amounting in total to **2.99 x 10^6 MWh/a.** The plants' own consumption for pollution control measures (e.g. flue gas treatment) is assumed to be 25 – 50% of their electricity production. If own consumption is set at 40%, **gross electricity** production from MSW comes to about **4.18 x 10^6 MWh/a**.

For comparison, this is equivalent to
- **0.71% of gross electricity production in Germany**.

Of the 56 plants, 49 supplied heat to downstream users (district heat, process steam and/or heating steam). The total amount supplied as heat energy was about **11.65 x 10^6 MWh/a (net heat production)**. Assuming the plants' own consumption for pollution control measures (e.g. flue gas treatment) is 10%, or **1.1 x 10^6 MWh/a**, of their heat production, a figure of about **12.75 x 10^6 MWh/a** is obtained for **gross heat production** from MSW.

> The inclusion of assumptions with respect to own electricity and heat requirements brings gross electricity/heat production by the plants to **16.93 x 10^6 MWh/a**, which—based on the gross fuel heat potential—gives an actual **total energy recovery efficiency of 45.2%.**

Plants supplying electricity only: In 2000, 48 of the 56 plants produced electricity, of which **7 plants** supplied electricity only. Electricity supply by these 7 plants ranged from 18 GWh/a to 185 GWh/a, supply efficiencies varied between 11.9% and 24.1% (theoretical mean: 15.2%). The (net) electricity production of these plants was between 0.33 MWh/Mg waste and 0.7 MWh/Mg waste (Table 8.6).

Table 8.6. Actual electricity supply of MSWI plants supplying electricity only

Number of plants	Amount of waste incinerated (Mg/a)	Gross fuel heat potential (MWh/a)	Amount of electricity supplied
7	1,053,033	2,860,711	480,136

Plants supplying heat only: In 2000, 49 of the 56 plants supplied heat (district heat, process steam and/or heating steam) to downstream users. Of these, **8 plants** produced exclusively heat. Heat supply by these 8 plants ranged from 70 GWh/a to 1,025 GWh/a, supply efficiencies varied between 24.6% and 75% (theoretical mean: about 65%). The (net) heat production of these plants was between 0.6 MWh/Mg waste and 2.25 MWh/Mg waste (Table 8.7).

Those plants which are at the upper end of this band all supply their heat directly to a power station, and usually exclusively as process steam. The extremely high heat recovery efficiencies of these plants are diminished firstly, by the efficiency of the power plants downstream (0.35 – 0.4), and secondly, by the fact that they use electricity from external suppliers to meet their own electricity requirements. (*Note: In the case of plants with very high supply efficiencies, it is necessary to check to what extent the produced energy is actually ascribable to the input waste and whether other factors, such as discrepancies in calorific value, use at low temperatures, soiling of the boiler, use of fuel oil and divergent steam parameters, may have led to miscalculation of efficiencies.*)

Table 8.7: Actual heat supply of MSWI plants supplying heat only.

Number of plants	Amount of waste incinerated (Mg/a)	Gross fuel (MWh/a) heat potential	Amount of heat supplied (MWh/a)
8	2,451,301	5,541,618	4,849,162

Plants supplying electricity and heat: In 2000, 41 of the 56 plants supplied electricity and heat to downstream users (electricity, district heat, process steam and/or heating steam). Electricity supply by these plants ranged from 2.1 GWh/a to 256 GWh/a, supply efficiencies varied between 0.2% and 17.5% (mean: 9.5%). Heat supply by these plants was between 9.37 GWh/a and 1,097 GWh/a and supply efficiencies were between 1.2% and 67.1% (mean: 24.3%). The (net) electricity production of these plants was between 0.006 MWh/Mg waste and 0.47 MWh/Mg waste, (net) heat production between 0.035 MWh/Mg waste and 1.78 MWh/Mg waste. (Table 8.8).

Table 8.8: Actual electricity/heat supply of MSWI plants with CHP.

Number of plants	Amount of waste incinerated (Mg/a)	Gross fuel heat potential (MWh/a)	Amount of electricity (MWh/a)	Amount of heat supplied (MWh/a)
41	9,490,151	27,346,710	2,507,613	6,797,648

Total energy supply efficiencies in MSW incineration

Based on the amount of energy produced by waste incineration plants and the electricity and/or district heat, process steam and heating steam supplied by them, the following total energy supply efficiencies were calculated (Table 8.9).

Table 8.9: Total energy supply efficiencies of waste incineration plants (not including own requirements and based on the gross fuel heat potential)

Total supply efficiency for 7 plants supplying electricity only 11.9 – 24.1% (theoretical mean: 15.2%)
Total supply efficiency for 8 plants supplying heat only: **24.6 – 75% (theoretical mean: ca. 65%)**
Total supply efficiency for 41 plants supplying electricity and heat: **12.5 – 82.5% (theoretical mean: 33.8%).**
This includes **Electricity supply with 0.2 – 17.5% (mean: 9.5%)** and Heat supply with 1.2 – 67.1% (mean: 24.3%).

Discussion

The ITAD data on total supply efficiencies are net figures—not including own consumption, which means an upward change is possible for many plants (for otherwise the plants' own electricity consumption would have to come from fossil energy sources). So, thus far, consideration of net total supply efficiencies (see above section) – not including own consumption – gives only an incomplete, very conservative picture of energy production in MSW incineration.

Determination of credits for climate-neutral energy (electricity/heat) from MSW incineration

CO_2 emission factors of fossil fuels

Depending on which fossil fuel is to be replaced, different specific CO_2 emission factors must be used. These emission factors have been determined and adapted by the Federal Environmental Agency (Sections II 6.4 and I 2.5) for many years. They are obtained by dividing CO_2 emissions by the fuel input, given in Mg/TJ.

Production in a lignite-fired power station at base load, with an emission factor of 349 kg CO_2 per MWh, was assumed for the case of the substitution of electricity by MSW incineration. For the case of heat substitution, an emission factor of 254 kg CO_2 per MWh was assumed, which is obtained by mixing the respective emission factors for 80% fuel oil, of 266.4 kg CO_2 per MWh and 20% natural gas, of 201.6 kg CO_2 per MWh. Table 8.10 lists the fuel-specific emission factors used for computations carried out for comparison with waste.

For electricity, the emission factor of the power plant mix as commonly used for ecological comparisons was not used for the purposes of the computations presented here, because in light of the reduction target of the national climate protection program, the priority goal should be the replacement of fossil energy sources.

Table 8.10 also gives the CO_2 equivalence factors generally applied for conversion to CO_2 equivalent emissions.

Table 8.10. List of Fuel-specific emission factors

Emission	Emission factor (lignite for electricity production) (kg/MWh)	Emission factor (EL fuel oil for heat production) (kg/MWh)	Emission factor (natural gas for heat production) (kg/MWh)	Emission factor (fuel oil/ natural gas mix for heat production) (kg/MWh)	CO_2 equivalence factor (100-year lifetime) (kg CO_2/ kg emission)
CO_2	349	266.4	201.6	254	1
N_2O					310

Emissions from energy production using fossil fuels are compared below with emissions from energy production using MSW.

Calculation of climate-neutral CO_2 emissions

Starting with the given climate-relevant CO_2 emissions from MSW, the steps set out below serve to determine how much total energy existing MSW incineration plants must produce to substitute a corresponding amount of energy from fossil fuels and the resultant amount of CO_2 emissions. The energy produced from waste and used in excess of this amount can be said to be "climate-neutral" compared to energy from fossil energy, and for this energy we can speak of a "bonus for energy from waste incineration". Because of different emission factors of the relevant fossil fuels, separate computations were carried out for electricity and heat.

Substitution of electricity by MSW incineration: Formula (2) below shows that for CO_2, MSW incineration can be said to have an energy bonus compared to electricity production using fossil fuels (lignite) when electricity substitution exceeds 17.6×10^6 MWh electricity/year ($6,144 \times 10^3$ Mg CO_2/year divided by the emission factor for lignite of 349 kg CO_2). Actual electricity production by MSW incineration including own requirements is about 4.18×10^6 MWh/year, leaving a difference of 13.42×10^6 MWh/y which it would have to produce additionally to claim so-called climate-neutrality (energy bonus) for energy contributions equal to or greater than this amount.

Calculation example for CO_2 for substitution of electricity:
Formula (2)
Emission = $6,144 \times 10^3$ CO_2/y– **17.6 x 10⁶ MWh/y** x 0.349 Mg CO_2/MWh
Emission = 6,144 Gg CO_2/y – 6,144 Gg CO_2 (produced using fossil fuel)
Emission = 0 Gg CO_2/y

Substitution of heat by MSW incineration: Formula (2) shows that for CO_2, MSW incineration can be said to have an energy bonus compared to heat production using fossil fuels (80% EL fuel oil, 20% natural gas) when heat substitution exceeds 24.189×10^6 ($6,144 \times 10^3$ Mg CO_2/y divided by an emission factor of 0.254 Mg CO_2/MWh resulting from the emission factors for 80% EL fuel oil and 20% natural gas). Actual heat supply by MSW incineration is about 11.65×10^6 MWh/y, leaving a difference of 12.54×10^6 MWh heat/y which it would have to produce additionally to claim so-called climate-neutrality (an energy bonus) for energy contributions equal to or greater than this amount.

Calculation example for CO_2 for heat substitution
Formula (2)
Emission = $6,144 \times 10^3$ Mg CO_2/y – **24.189×10^6 MWh/y** x 0.254 Mg CO_2/ MWh
Emission = 6,144 Gg CO_2/y – 6,144 Gg CO_2/y (produced using fossil fuel)
Emission = 0 Gg CO_2/y

Discussion: The results of the two calculation exercises to determine how much energy has to be produced by MSW incineration and used as electricity or heat show that heat substitution is the more crucial factor. Their technical design makes it impossible for MSW incineration plants to generate as much electricity as would be needed for them to achieve CO_2 climate-neutrality compared to the fossil-fuel-produced energy to be substituted. For site-related reasons, an orientation to heat production cannot achieve this climate-neutrality for all plants to the same extent either. Consequently, the only possibility—on condition that the energy produced from waste is actually used to replace fossil fuel energy—is to make the contribution needed to achieve CO_2 climate-neutrality by improving energy supply from plants equipped with combined heat and power technology.

Calculation example for N_2O: Formula (2) shows that for N_2O, the energy figures calculated for CO_2, of about 24.2×10^6 MWh/y respectively 17.6×10^6 MWh electricity/y, are too high and would result in negative emissions of N_2O. In terms of this substance, one can speak of climate-neutrality and an energy bonus compared to energy from fossil fuels when the substituted amount of energy reaches 4.469×10^6 MWh/y (143 Mg N_2O/y divided by 32 kg N_2O/MWh). As can be seen from the calculation for N_2O, the main contribution to energy substitution comes from climate-relevant emissions of CO_2, and this CO_2 energy figure is therefore used in subsequent calculations relating to climate relevance.

> **Calculation example - N_2O:**
> *Formula (2)*
> Emission = 143 Mg N_2O/y – 4.469 x 10^6 MWh/y x 32 kg N_2O/MWh
> Emission = 0.143 Gg N_2O/y – 0.143 Gg N_2O/y (produced from fossil fuel)
> Emission = 0 Gg N_2O/y

Determination of total energy supply efficiencies based on gross fuel heat potential

Computation of total energy supply efficiencies for electricity or heat serves to determine what efficiency level—relative to the gross fuel potential—must be reached or exceeded to claim "climate-neutrality" compared to the fossil fuel energy actually replaced. In the following, this efficiency is referred to as "minimum supply efficiency" (Table 8.11).

For the heat produced in MSW incineration and the use of this heat it was assumed that it replaces mainly an energy mix from EL fuel oil and gas. Based on a supplied amount of heat of 24.189 x 10^6 MWh/a as determined above, a theoretical energy supply efficiency of 64.6% can be calculated for MSW incineration. In view of the current actual net heat supply of 11.65 x 10^6 MWh/a, this "minimum energy supply efficiency" required to obtain the CO_2 energy bonus can only be achieved by significantly increasing heat supply.

Table 8.11: Minimum heat energy supply efficiency to obtain a CO_2 energy bonus

Gross fuel heat potential	"Minimum energy supply" to obtain CO_2 energy bonus vs. fuel oil/gas	Minimum energy supply efficiency
37.448 x 10^6 MWh/a	24.189 x 10^6 MWh/a	64.6%

In computations carried out in 1999 (Johnke, 1999) for the waste incineration sector, it was found that for energy from MSW incineration CO_2 emissions can be assumed to be climate-neutral when the **total energy recovery efficiency** (i.e., including own requirements) is at least **32.4%**. This result was arrived at due to emission credits from the power plant mix. For reasons of simplifications, an emission factor of 690 kg CO_2 per MWh was used for comparison with fossil fuel energy production. This emission factor applies to the power plant mix and refers to net electricity production at the point of delivery (socket). This total energy recovery efficiency determined in (Johnke, 1999) was calculated based on a gross fuel heat potential of **29.7 x 10^6 MWh/a** and referred to an incinerated amount of MSW of 11.9 million x 10^6 Mg/a and an average calorific value of 9 x 10^6 kJ/Mg. The total energy supply by MSW incineration in excess of which CO_2 emissions can be said to be climate-neutral was calculated to be **9.62 x 10⁶ MWh/a**. However, the emission factor used in (Johnke, 1999) resulted in *too favorable an assumption*, which means that the percentage of 32.4 in excess of which CO_2 emission was assumed to be climate-neutral *can no longer stand*! (*Note: Owing to the new, very detailed ITAD data base for 2000 (individual data for 56 incineration plants, higher calorific value, higher waste quantities, use of new, differentiated emission factors for substitution of fossil energy) the percentages obtained in the calculations presented below differ significantly from those in Johnke, 1999.)*

For electricity produced in MSW incineration and its use, it is assumed that the substituted fuel is mainly lignite. Based on supplied electricity of 17.6 x 10⁶ MWh, as determined above, a theoretical energy supply efficiency of 46.9% can be calculated. In view of the current actual net electricity supply of 2.99 x 10⁶ MWh/a, this "minimum electricity supply" (Table 8.12) required to obtain the CO_2 energy bonus cannot be achieved even if electricity production and supply were significantly improved. Here, electricity production from waste comes up against technical limits.

Table 8.12: Minimum electricity supply efficiency to obtain a CO_2 energy bonus

Gross fuel heat potential	"Minimum electricity supply" to obtain the CO_2 energy bonus vs. lignite	Minimum energy supply efficiency
37.448 x 10⁶ MWh/a	17.6 x 10⁶ MWh/a	46.9%

Discussion

There are at present only a few sites which have the possibilities the plants need to exceed this minimum energy supply efficiency for heat or electricity and to thus generate less in climate-relevant CO_2 emissions— with regard to the additional energy they can produce and use from their waste—than what would be generated in electricity production from fossil fuels. Of the 56 plants which—from an energy perspective—have and could utilize such favorable site conditions, this is true (according to the ITAD survey) of only 5 plants with heat recovery only and 3 with cogeneration of electricity and heat.

Determination of additional energy potential

In order to determine the additional usable energy potential of the 56 existing plants, the optimal combination of cogeneration of heat and power was determined for each of the three types of energy use with the aid of available plant data. In so doing, it was assumed that the total amount of energy to be made available from the gross fuel heat potential per plant cannot reach more than 2.25 MWh per Mg waste. This assumption is based on best available energy production techniques and could be realized given best possible site conditions.

Scope for and limits to energy use in MSW incineration

In general, about 0.3 to 0.7 MWh of electricity can be produced in an MSW incineration plant from 1 Mg of MSW, depending on plant size, steam parameters and steam utilization efficiency. In the case of cogeneration of electricity and heat, about 1.25 to 1.5 MWh (full-load hours) of heat per Mg of waste can be used additionally, having regard for the incineration plant's site-dependent heat supply opportunities and depending on geographic location and normal (district) heat utilization periods (e.g., in Germany, 1,300-1,500 h/y out of a possible 8,760 h/y).

Given favorable site conditions, net efficiency can be increased to about 90% (not including own requirements) for an incineration plant operated at base load. In this energetically favorable case, up to 2.25 MWh per Mg of waste can be produced and supplied to external users as process steam or as energy mix (electricity and heat). However, where steam from a waste incineration plant is supplied to a power plant, that plant's efficiency in electricity production (maximum of 0.4) must be taken into account in the efficiency considerations.

Optimization of plants supplying electricity only

(Net) electricity production in the **7 plants** supplying electricity only was between 0.33 MWh/Mg waste and 0.7 MWh/Mg waste (Table 8.13).

Table 8.13: Status of plants supplying electricity only

Amount of waste incinerated (Mg/a)	Climate relevant CO_2 emission* (Mg)	Gross fuel heat potential (MWh/a)	Electricity supplied (MWh/a)	Electricity supplied (%)
1,050,033	496,666	2,860,711	480,136	16.8

* fossil fraction

In addition to increased electricity production, optimization of these 7 plants was considered to include an additional, simultaneous use of heat.

An increase in net electricity production to reach a higher average level of electricity supply per Mg of waste would involve the following potential for substitution of climate-relevant CO_2 from lignite (Table 8.14):

Table 8.14: Optimization potential of electricity supply

Net electricity production (MWh/Mg waste)	Total electricity supply (MWh/a)	Electricity supply (%)	Difference to actual supply (MWh/a)	Potential for substitution of CO$_2$ from lignite
0.375	393,762	13.8	-86,374	-
0.457	480,136	16.8	0	0
0.5	525,016	18.3	44,880	16 Gg CO$_2$ /a
0.625	**656,270**	**22.9**	**176,134**	**62 Gg CO$_2$ /a**
0.7	735,023	25.7	254,887	88 Gg CO$_2$ /a

If these plants additionally supplied heat to external users as district heat or process steam, heat supply could reach the following levels (Table 8.15).

Table 8.15: Optimization potential of heat supply

Net heat production (MWh/Mg waste)	Total heat supply (MWh/a)	Heat supply (%)	Potential for substitution of CO$_2$ from 80% oil/20% gas
0.3	315,010	11	80 Gg CO$_2$ /a
0.5	525,016	18.3	133 Gg CO$_2$ /a
1.0	1,050,033	36.7	267 Gg CO$_2$ /a
1.25	**1,312,541**	**45.8**	**333 Gg CO$_2$ /a**
1.5	1,575,050	55	400 Gg CO$_2$ /a

Optimization of plants supplying heat only

(Net) heat production in the **8 plants** supplying heat only was between 0.6 MWh/Mg waste and 2.25 MWh/Mg waste (Table 8.16).

Table 8.16: Status of plants supplying heat only

Amount of waste incinerated (Mg/a)	Climate-relevant CO$_2$ emission (Mg)	Net fuel heat potential (MWh/a)	Heat supply (MWh/a)	Heat supply (%)
2,451,301	1,159,465	5,541,618	4,849,162	87.5

In addition to increased heat use, optimization of these 8 plants was considered to also include an additional, simultaneous production of electricity.

If net heat production in these 8 plants were increased to reach a higher average level of heat supply per Mg of waste, the following heat supply levels and resultant potential for substitution of fossil CO$_2$ emissions could be realized (Table 8.17).

Table 8.17: Optimization potential of heat supply

Net heat production (MWh/Mg waste)	Total heat supply (MWh/a)	Heat supply (%)	Difference to actual heat supply (MWh/a)	Potential for substitution of climate-relevant CO_2 from 80% oil/20% gas
1.25	3,064,126	55.3	-	-
1.5	3,676,952	66.4	-	-
1.75	4,289,777	77.4	- 559,85	-
1.978	4,849,162	87.5	0	0
2.0	**4,902,602**	**88**	**53,40**	**13.6 Gg CO_2 /a**
2.25	5,515,427	99	666,65	169 Gg CO_2 /a

If these plants additionally produced electricity and delivered it to the grid, electricity supply could reach the following levels (Table 8.18):

Table 8.18: Optimization potential of electricity supply

Net electricity production (MWh/Mg waste)	Total electricity supply (MWh/a)	Electricity supply (%)	Potential for substitution of CO_2 from lignite
0.150	367,695	6.6	128 Gg CO_2 /a
0.200	490,260	8.8	171 Gg CO_2 /a
0.250	**612,825**	**11**	**214 Gg CO_2 /a**
0.275	674,108	12	235 Gg CO_2 /a
0.300	735,390	13	257 Gg CO_2 /a
0.325	796,673	14	278 Gg CO_2 /a
0.350	857,955	15.5	299 Gg CO_2 /a

Plants producing electricity and heat

In the 41 plants with cogeneration of heat and power, (net) electricity production was between 0.006 MWh/Mg waste and 0.47 MWh/Mg waste and (net) heat production was between 0.035 MWh/Mg waste and 1.78 MWh/Mg waste (Table 8.19).

Table 8.19: Status of plants supplying CHP

Amount of waste incinerated (Mg/a)	Climate relevant CO_2 emissions* (Mg)	Net fuel heat potential (MWh/a)	Electricity supply (MWh/a)	Heat supply (MWh/a)	Electricity supply (%)	Heat supply (%)
9,490,151	4,495,485	27,346,710	2,507,613	6,797,648	9.2	24.9

* fossil fraction

Optimization of these 41 plants was considered to include both improved heat use and improved electricity production. The computations presented below clearly show that a major contribution to CO_2 emissions reduction could be made by significantly increasing heat supply by plants with cogeneration of heat and power.

If net heat production by these 41 plants were increased to reach a higher average level of heat supply per Mg of waste, the following heat supply levels and the resultant potential for substitution of fossil CO$_2$ emissions could be realized (Table 8.20).

Table 8.20: Optimization potential of heat supply

Net heat production (MWh/Mg waste)	Total supply of heat (MWh/a)	Heat supply (%)	Difference to actual heat supply (MWh/a)	Potential for substitution of climate relevant CO$_2$ from 80% fuel 20% gas
0.5	4,745,075	17.4	- 2,052,573	-
0.716	6,797,648	24.9	0	0
0.75	7,117,613	26	319,965	81 Gg CO$_2$ /a
1.0	9,490,151	34.7	2,692,503	684 Gg CO$_2$ /a
1.25	11,862,689	43.4	5,065,041	1,287 Gg CO$_2$ /a
1.5	14,235,227	52	7,437,579	1,889 Gg CO$_2$ /a
1.75	16,607,764	60.7	9,810,116	2,492 Gg CO$_2$ /a
1.875	**17,794,033**	**65**	**10,996,385**	**2,793 Gg CO$_2$ /a**
2.0	18,980,302	69.4	12,182,654	3,094 Gg CO$_2$ /a

If net electricity production at these 41 plants were increased to reach a higher average level of electricity supply per Mg of waste, the following increases could be realized (Table 8.21).

Table 8.21: Optimization potential of electricity supply

Net electricity production (MWh/Mg waste)	Total electricity supply (MWh/a)	Electricity supply (%)	Difference to actual supply (MWh/a)	Potential for substitution of CO$_2$ from lignite
0.050	474,508	1.74	-	-
0.100	949,015	3.5	-	-
0.150	1,423,522	5.2	-	-
0.200	1,898,030	6.9	-	-
0.250	2,372,538	8.7	- 135,075	-
0.264	2,507,613	9.2	0	0
0.300	2,847,045	10.4	339,432	118 Gg CO$_2$ /a
0.350	3,321,553	12.1	813,940	284 Gg CO$_2$ /a

Potential for CO_2 saving through optimized energy use

Table 8.22: Plants supplying electricity only (and additionally heat)

Net electricity production (MWh/Mg waste)	Difference to actual supply (MWh/a)	Potential for substitution of climate-relevant CO_2 from lignite
0.625	176,134	62 Gg CO_2/a
Net heat production [MWh/Mg waste]	**Additional supply [MWh/a]**	**Potential for substitution of climate-relevant CO_2 from 80% oil/20% gas**
1.25	1,312,541	333 Gg CO_2/a
Total	**1,488,675 MWh/a**	**395 Gg CO_2/a**

Table 8.23: Plants supplying heat only (and additionally electricity)

Net heat production (MWh/Mg waste)	Difference to supply (MWh/a)	Potential for substitution of climate-relevant CO_2 from 80% oil/20% gas
2.0	53,440	13.6 Gg CO_2/a
Net electricity production [MWh/Mg waste]	**Additional supply [MWh/a]**	**Potential for substitution of climate-relevant CO_2 from lignite**
0.25	612,825	214 Gg CO_2/a
Total	**666,265 MWh/a**	**227.6 Gg CO_2/a**

Table 8.24: Plants with optimized supply of electricity and heat

Net heat production (MWh/Mg waste)	Difference to actual supply (MWh/a)	Potential for substitution of climate-relevant CO_2 from 80% oil/20% gas
1.875	10,996,385	2,793 Gg CO_2/a
Net electricity production (MWh/Mg waste)	**Difference to actual supply (MWh/a)**	**Potential for substitution of climate-relevant CO_2 from lignite**
0.200	-	-
Total	**10,995,385 MWh/a**	**2,793 Gg CO_2/a**

Additionally usable total energy potential and CO_2 emission savings

The sum of the efficiency increases in electricity production and a significantly improved use of heat energy would provide an additionally usable total energy potential of 13.15×10^6 MWh/a, and would allow a CO_2 saving of 3,416 Gg to be achieved in fossil fuel energy production (Table 8.25).

Table 8.25: Total potential saving of climate-relevant CO_2

Additionally usable total energy potential	Total potential saving of climate-relevant CO_2 from fossil energy sources
13.15 x 10⁶ MWh/a = 13.15 TWh/a	3,416 Gg CO_2/a

In specific terms, the additional supply of **1 x 10⁶ MWh** of energy (electricity and heat) from MSW incineration would make it possible to substitute **260 Gg of CO_2** from fossil fuel energy production (Table 8.26).

Table 8.26: Specific CO_2 saving potential

Specific amount of substituted energy	Specific CO_2 saving potential
1 x 10⁶ MWh/a	260 Gg CO_2/a

Determination of contribution additional usable energy potential could make to reduction of CO_2 emissions

Fulfillment of the waste management function means that a certain amount of climate relevant CO_2 emissions will always be generated in MSW incineration. It is therefore of special interest to make optimal use of MSW incineration plants' gross fuel heat potential. With a view to the amount of energy actually supplied by these plants, there is still additional energy recovery potential, which, insofar as this energy is actually used to substitute energy from fossil fuels, can contribute to a further reduction of climate-relevant emissions from energy production. This energy potential, still currently available at existing plants, can be exploited – at least in part—by improving both energy production efficiencies and heat supply opportunities. Based on a gross fuel heat potential of 37,448 x 10⁶ MWh/a contained in 12.99 x 10⁶ Mg of residual MSW, an additional usable energy potential of about 13.2 x 10⁶ MWh/a can be derived. This results in a theoretical potential reduction of **up to 3,416 Gg CO_2/a** in CO_2 emissions from energy production using fossil fuels (Table 8.27).

Table 8.27: Additional usable energy potential to reduce CO_2

Gross fuel heat potential	Actual energy supply	Additional usable energy potential	Potential reduction of CO_2 emissions from fossil fuel energy production
37.448 x 10⁶ MWh/a	14.6 x 10⁶ MWh/a	13.2 x 10⁶ MWh/a	3,416 Gg/a

Potential percentage saving of CO_2 emissions through use of energy from MSW incineration

By harnessing the additional energy potential available at existing MSW incineration plants, amounting to about 13.2 x 10⁶ MWh/a, a theoretical

CO_2 emission saving of **up to 0.5%** could be made relative to total CO_2 emissions from the energy sector (Table 8.28)

Table 8.28: Share of the CO_2 reduction potential MSWI/energy sector

Reduction in CO_2 emissions from fossil fuel energy production	Total CO_2 emissions from the energy sector (1999)	Potential CO_2 saving through use of additional energy potential in MSW incineration
3,416 Gg/a	641,000 Gg	0.5%

Table 8.29 below shows the effect of an increase in energy supply by MSW incineration plants in terms of CO_2 saving relative to total CO_2 emissions from the energy production sector. An increase of **1 x 10^6 MWh/a** in the amount of energy supplied by MSW incineration would result in a CO_2 saving of **260 Gg** relative to total CO_2 emissions from the energy sector.

Table 8.29: Effect of an increase in energy supply in terms of CO_2

Increase in energy supplied by MSW incineration	Resultant CO_2 saving	Potential CO_2 saving relative to total CO_2 emissions from energy sector
1×10^6 MWh/a	260 Gg CO_2	0.04%
2×10^6 MWh/a	520 Gg CO_2	0.08%
5×10^6 MWh/a	1,300 Gg CO_2	0.2%
10×10^6 MWh/a	2,600 Gg CO_2	0.4%
13.2×10^6 MWh/a	3,416 Gg CO_2	0.5%

Note: If an additional 13 million Mg of MWS per year currently deposited at landfill sites for domestic refuse were incinerated in MSW incineration plants with energy recovery, MSW incineration's contribution to CO_2 saving could be **doubled**, to reach **1%**.

Optimization measures to increase energy efficiencies

Possible measures to improve energy production and energy use could comprise:

- Improvement of the policy and/or economic framework for existing and new plants (creation of opportunities for increased supply of the heat generated at base load which cannot currently be supplied to external users (e.g. due to a lack of buyers), creation of more connections and development of the supply of district heat/process steam, measures to avoid competitive situations due to other plants using gas for the production of district heat and process steam, integration of MSW incinerators into regional and supraregional energy supply systems).

- Plant- and process-related optimization measures for existing and new plants (changing steam parameters (problem: cost-intensive due to the possible need to install a new boiler or turbine because of material-related questions, safety criteria); improving steam quality by the use of CCGT (external measure); heating (superheating) of steam (in-plant measure), e.g. of partially used (low-quality) steam, resulting in changed steam parameters (lower pressure, higher temperature); reduction of internal losses by making greater use of the energy contained in flue gases, e.g. when an economizer is installed upstream of the scrubber to achieve the necessary temperature decrease in flue gas treatment, use of the energy that becomes available during the temperature decrease from 250°C to 60°C).

Current contribution of MSW incineration to reduction of CO$_2$ emissions

By substituting for energy produced from fossil sources, the total energy supplied in 2000 by all 56 plants with grate-firing systems, amounting to **2.99 x 10^6 MWh of electricity** and **11.65 x 10^6 MWh of heat**, resulted in a CO$_2$ emission saving of about **4,000 Gg CO$_2$/a**. Of this total, electricity supply accounted for **1,040 Gg CO$_2$/a**, and heat supply for **2,960 Gg CO$_2$/a**.

CO$_2$ emission reduction targets under the national climate protection programme

The national climate protection program (BMU, 2000) lists measures and targets for CO$_2$ reduction until 2010. It includes a target of 25% for the reduction of CO$_2$ emissions by 2005, taking 1990 as the baseline, and expects that a reduction of the order of 18 – 20% (about 180,000 – 200,000 Gg of CO$_2$) will have been achieved by 2005 based on the CO$_2$ emission reduction measures taken since 1990. This means, according to BMU (2000), that from 1998 onwards, potential for an additional CO$_2$ reduction of 50,000 – 70,000 Gg will need to be tapped. An interministerial working group on CO$_2$ reduction has based its discussions on the reduction potentials within individual sectors:
- Private households and buildings 18,000 – 25,000 Gg of CO$_2$
- Energy sector and industry 20,000 – 25,000 Gg of CO$_2$
- Transport 15,000 – 20,000 Gg of CO$_2$

The following CO$_2$ reduction targets are to be achieved through additional reduction measures taken since 1998:
- **Renewable Energy Sources Act:** 10,000 Gg of CO$_2$ by 2005, an additional 5,000 Gg of CO$_2$ by 2010, total of 15,000 of CO$_2$
- **Market launch program for renewable energy sources:** 2,500 Gg

of CO_2 by 2005, 3,500 Gg of CO_2 by 2010, total of 6,000 Gg of CO_2

- **100,000-roofs photovoltaic program:** 200 Gg of CO_2 by 2005/2010
- **Waste management*, measures in the field of MSW:** 15,000 Gg of CO_2 by 2005*, 5,000 Gg of CO_2 by 2010, total of 20,000 Gg of CO_2

(*refers in part to CO_2 equivalents on the basis of avoided CH_4 emissions).

The climate protection program provides that should its implementation reveal that individual sectors are unable to meet their targets via certain measures, the initial priority should be to investigate other measures in that sector. Any outstanding deficit must then be compensated via greater efforts in other sectors.

Summary

At present, 56 out of a total of 60 thermal MSW treatment plants in Germany are incineration plants with a grate-firing system. Energy recovery is practiced at all plants with grate-firing. Depending on site conditions and season, the recovered energy is supplied to external users in the form of electricity, district heat, process steam and/or heating steam and used to meet own energy requirements.

In 2000, about 13 x 10^6 Mg of residual MSW were incinerated, and **2.99 x 10^6 MWh of electricity** and **11.65 x 10^6 MWh of heat,** i.e., a total amount of energy of about **14.6 x 10^6 MWh/a,** supplied to external users. The figure for electricity supply is equivalent to about 0.7% of gross electricity production in Germany. The actual energy supply efficiency relative to the plants' gross fuel heat potential is about **39%.** Inclusion of the plants' own electricity and heat requirements would bring the actual total energy recovery efficiency to about **45%.** This means that an additional energy potential remains untapped.

This available additional energy potential from MSW incineration could be used to further reduce climate-relevant emissions from energy produced using fossil fuels. As set out in the present status report, an emission saving of up to **3,416 Gg CO_2/a** could be achieved by **improving energy use,** taking into account feasible technical optimization measures in the plants concerned. This would correspond to a saving of **0.5%** relative to total CO_2 emission from energy production. However, in order to achieve this, an additional (hitherto unused) energy potential of **13.2 x 10^6 MWh/a** would need to be tapped in the form of electricity, district heat and/or process steam.

Note: If an additional 13 million Mg of MWS per year currently deposited at landfill sites for domestic refuse were incinerated in MSW incineration plants with energy recovery, MSW incineration's contribution to CO$_2$ reduction could be **doubled**, to reach **1%**. Therefore, the selection of new sites for incineration plants should be geared to energy supply aspects, and these plants should have a total energy recovery efficiency which affords climate-neutrality, or better yet, an energy credit!

As regards the CO$_2$ emission reduction potential to be additionally activated by 2005 under the national climate protection program, amounting to 50,000 – 70,000 Gg of CO$_2$, improved energy use in MSW incineration could make a contribution of **about 5%** to that reduction.

References

BMU Umwelt 11/2000. Sonderteil "Nationales Klimaschutzprogramm".

IPPC Guidline 2000. Good Practice Guidance and Uncertainty Management in National Greenhouse Gas Inventories. IPCC/OECD/IEA Program for National Greenhouse Gas Inventories. WMO Intergovernmental Panel on Climate Change, UNEP.

ITAD, 2001. Energieerzeugung durch Müllverbrennungsanlagen im Jahre 2000 in Deutschland. Unpubl. internal survey by M. Treder (waste incineration plant in Hamm).

Johnke, B. 1999. German Proposal for a Draft Meeting Report "Good Practice in Inventory Preparation for Emission from Waste Incineration". IPCC/OECD/IEA Program for National Greenhouse Gas Inventories, WMO Intergovernmental Panel on Climate Change, UNEP, Sao Paulo, Brazil, 27-29 July 1999.

Johnke, B. 2003. Abfallverbrennung - Ein Beitrag tum Klimaschutz in Deutschland, TK Verlag, Neuruppin, ISBN 3-935317-13-1.

Johnke, B. 2003. Klimaschutz und Energieeffizient, Verlag Saxonia Freiberg, Abfall kolloguium 2003, ISBN 3-934409-21-0.

Reimann, D.O., and Hämmerli, H. 1996. Verbrennungstechnik für Abfalle in Theorie und Praxis, Schriftenreihe Umweltschutz, Bamberg.

Section III
HUMAN RIGHTS

Petitioning for Adverse Impacts of Global Warming in the Inter-American Human Rights System

Donald M. Goldberg[a] and **Martin Wagner**[b]

[a]Senior Attorney, Center for International Environmental Law
1367, Connecticut Ave. New Washington, DC 20036-1860, USA
email : dgoldberg@ciel.org
[b]Director of International Programs for Earthjustice
Earthjustice, 426 17th Street, 6th floor Oakland, CA 94612-2820
email : mwagner@earthjustice.org

Greenhouse gases are accumulating in Earth's Atmosphere as a result of human activities, causing global mean surface air temperatures to rise.

U.S. Climate Action Report 2002

Introduction

It is beyond dispute that human activities are causing global warming, as even the U.S. government now admits. Although the world's least developed communities have contributed little to the problem of global warming, they likely will bear its heaviest impacts, and certainly will be least able to adapt to them.[1] The direct impacts of global warming include higher temperatures, sea-level rise, melting of sea ice and glaciers, increased precipitation in some areas and drought in others. Indirect social, environmental, economic and health impacts will follow, including increased death and serious illness in poor communities, decreased crop yields, heat stress in livestock and wildlife, and damage to coastal ecosystems, forests, drinking water, fisheries, buildings and other resources needed for subsistence.[2]

For the indigenous inhabitants of the Arctic—the Inuit, the Yupik and the Inupiat, to name but a few—these impacts threaten fundamental rights protected by regional and universal human rights systems. These include the rights to life and personal security; to use and enjoyment of property; to residence and movement; to inviolability of the home; to preservation of health; to the benefits of culture; to work and fair remuneration; to means of subsistence; and to free disposition of natural resources. Many of the impacts of global warming will be especially problematic for indigenous communities, who are recognized to have special status under international law. Citizens of wealthier countries may be able to insulate themselves from the impacts of global warming, at least in the near term, and perhaps it is partly for this reason that the United States, the world's largest contributor to global warming, has chosen to reject international response measures, such as the Kyoto Protocol to the United Nations Framework Convention on Global Warming.

The options for bringing a human rights complaint against the United States are examined here. Because international human rights law gives states primary responsibility for ensuring the protection of human rights, we confine this analysis to a claim against the United States. However, because some human rights institutions have in some instances recognized the responsibility of corporations for human rights violations,[3] global warming may be a basis for claims against US corporations as well.

Only two international human rights regimes are available to bring a claim against the United States: the UN human rights system and the Inter-American system established under the Organization of American States (OAS). While the United States participates in both regimes, it is not a party to several key agreements of each. In particular, it has not signed the UN Optional Protocol to the International Covenant on Civil and Political Rights, by which states accept the jurisdiction of the Human Rights Committee to consider the human rights claims of private individuals. Nor has it ratified the American Convention on Human Rights, which subjects consenting member states to the jurisdiction of the Inter-American Court of Human Rights.

On balance, we believe the Inter-American system would be the more responsive forum for several reasons. First, the Inter-American Commission on Human Rights ("Commission") has the authority to receive petitions by private citizens directed against any OAS member state.[4] Second, the Commission has recognized the relationship between human rights and the environmental impacts of development activities, and its interpretation of this relationship suggests that it would recognize the human rights implications of the effects of global warming.[5] Third, the Commission has wide-ranging power to look at new developments in human rights law, even if they arise in other systems.

The severe impacts of global warming already being felt by many indigenous Arctic communities that could give rise to a human rights claim in the Inter-American system are discussed first. Next, some of the procedural elements of such a claim, with particular emphasis on issues of jurisdiction are considered. Then some of the rights protected by the Inter-American human rights system that might be implicated in a human rights claim and the remedies that might be available to claimants are examined.

Plight of the Indigenous Peoples of the Arctic[6]

The Arctic, the area above 66º, 30' North Latitude, is the aboriginal homeland of the Inuit, Inupiat, Yupik, and several other native groups. Most indigenous Arctic inhabitants reside along coastlines and in river valleys, living off the land in the traditions of their ancestors, using knowledge passed down for hundreds of generations. Their subsistence livelihoods depend on fish, marine mammals, and other wildlife. The activities associated with the harvest of these resources also make important contributions to the health, culture, and identity of native Arctic peoples. During the past several decades, the Arctic has warmed at an alarming rate and it is projected that it will continue to warm by as much as 18º F by 2100.[7] This warming trend has had a devastating impact on Arctic ecosystems, including sea ice, permafrost, forests and tundra.[8]

Melting Sea Ice

Melting sea ice affects populations of marine mammals, caribou, and polar bears and the subsistence livelihoods of people that depend on them. Because sea ice forms a natural breakwater against storm wave action, ice melting allows larger storm surges to develop and causes erosion, sedimentation, and coastal inundation.[9] For centuries, native Arctic peoples like the Inupiat and Yupik have based their hunting seasons on the yearly freezing and thawing cycles of Arctic ice. Rising temperatures have disrupted these cycles, and Yupik hunters have noticed that seals have moved farther north and that walruses are becoming thinner.[10] According to scientists, the retreat of sea ice has reduced the platform that seals and walruses traditionally use to rest between searches for fish and mussels; weakened and less productive, they provide less sustenance for the Yupiks.[11] Inuit hunters find it increasingly hard to hunt caribou, long a staple of Inuit diet, because the caribou are falling through once solid sea ice. Yupik and Inupiat hunters themselves are increasingly at risk of falling through thinning ice, making hunting more difficult. This, combined with a shorter hunting season due to a shorter freezing period also makes hunting less productive. In the remote towns inhabited by native Arctic peoples, where store-bought meat can cost up to $22 a pound, the negative effects of an unproductive hunting season also pose economic threats. Ironically, climate change is also likely to impair transport by shortening the seasonal use of ice roads, making it harder for native Arctic peoples to purchase food they might not be able to afford in the first place.

Thawing Permafrost

Thawing permafrost in the Arctic has damaged houses, roads, airports and pipelines, and caused landscape erosion, slope instability, and landslides. Local coastal losses to erosion of up to 100 feet per year have been observed in some locations in the Siberian, Alaskan and Canadian Arctic.[13] Several villages in this region are sufficiently threatened by increased erosion and inundation that they must be protected or relocated.

In Shishmaref, Alaska, a small Inuit village in the Chukchi Sea, seven houses have had to be relocated, three have fallen into the sea, and engineers predict that the entire village of 600 houses could disappear into the sea within the next few decades. Shishmaref's airport runway has almost been met by rising seawater, and its fuel tank farm, which seven years ago was 300 feet from the edge of a seaside bluff, is now only 35 feet from the bluff. The town dump, which has sea water within 8 feet of it, could pollute the nearby marine environment for years if inundated. Advancing sea water has contaminated Shishmaref's drinking water supply.[13]

Attempts to mitigate the impending disaster in Shishmaref have been futile. The US Army Corps of Engineers built a succession of sea walls that were demolished by the sea, and town leaders say there is no longer any safe place to relocate threatened houses. Relocation of the entire town seems inevitable, but will cost tens of millions of dollars.[14] Even if such a project is feasible, the Inuit in Shishmaref will more than likely have to move to the outskirts of a large town, such as Nome or Kotzebue, a step that many feel would extinguish their subsistence lifestyle culture. Shishmaref's situation is not unique. Coastal villages along the Beaufort and Chukchi seas such as Barrow, Kivalina, and Point Hope are experiencing similar fates.[15]

Damage to Forest and Tundra Ecosystems

Forest and tundra ecosystems are important features of the Arctic and subarctic environment that native peoples rely on to practice their subsistence life style. In Alaska, substantial changes in patterns of forest disturbance, including insect outbreaks, blowdown, and fire, have been observed in both the boreal and southeast coastal forest.[16] Rising temperatures have allowed spruce bark beetles to reproduce at twice their normal rate.[17] A sustained outbreak of the beetles on the Kenai Peninsula has caused over 2.3 million acres of tree mortality, the largest loss from a single outbreak recorded in North America.[18] Outbreaks of other defoliating insects in the boreal forest, such as spruce budworm, coneworm, and larch sawfly, also have increased sharply in the past decade.

Climate warming and insect infestations make forests more susceptible to forest fire. Since 1970, the acreage subjected to fire has increased steadily from 2.5 million to more than 7 million acres per year. A single fire in 1996 burned 37,000 acres of forest and peat, causing $80 million in direct losses and destroying 450 structures, including 200 homes. As many as 200,000 Alaskan residents may now be at risk from such fires, with the number increasing as outlying suburban development continues to expand. The increase in forest fires also harms local wildlife, such as caribou,[19] that native Arctic peoples depend on for subsistence.

Seeking Redress in the Inter-American Human Rights System

The impacts described in the preceding section, severe as they are, may be only a foretaste of impacts Arctic peoples will face in the future. Unchecked, global warming threatens to destroy their culture, render their land uninhabitable, and rob them of their means of subsistence. At present, it may not be possible to fully redress these harms through the Inter-American system. The Commission does not have the authority to force countries to curtail their emissions, nor can it compel states to compensate individuals for human rights violations. The Inter-American Court does have such power, at least in theory, but two barriers bar access to the Court by Arctic inhabitants seeking to sue the United States. First, the Convention does not permit a private citizen to submit a case directly to the Court. Second, because the United States has not ratified the Convention, it is not subject to the jurisdiction of the Court.

Nevertheless, a report by the Commission examining the connection between global warming and human rights could have a powerful impact on worldwide efforts to address global warming. It would demonstrate that the issue is not merely an abstract problem for the future, but is instead a problem of immediate concern to all people everywhere. Recognition by the Commission of a link between global warming and human rights may establish a legal basis for holding responsible countries that have profited from inadequate greenhouse gas regulation and could provide a strong incentive to all countries to participate in effective international response efforts.

Procedural Issues Arising Under the Rules of the Inter-American Commission

General Requirements

Anyone alleging a human rights violation by the government of a nation that is a member of the OAS may submit a petition to the Inter-American Commission. If the accused state is party to the American Convention, that document, the Commission's Statute and its Rules of Procedure establish jurisdiction and procedure.[20] If, like the United States, the accused state is not a party to the American Convention, but is a member state of the OAS, the Commission's Rules of Procedure and past practice recognize that the obligations of the Declaration apply and the Commission may hear claims asserting violations by that state: "Pursuant to the [OAS] Charter, all member states undertake to uphold the fundamental rights of the individual, which, in the case of non-parties to the Convention, are those set forth in the American Declaration, which constitutes a source of international obligation."[21]

Petitioners may be the victims themselves, third parties, or any "non-governmental entity legally recognized in one or more of the member states of the OAS."[22] The petition must identify the state responsible, "by act or omission," for the violations of any of the applicable human rights, and describe the acts or situation leading to the violations.

The petition must also identify the steps taken to exhaust domestic remedies, or the impossibility of doing so, and indicate whether the complaint has been submitted to another international settlement proceeding.[23] Regarding exhaustion, the Commission's Statute requires it to "verify, as a prior condition to the exercise of [its authority to accept a petition], whether the domestic legal procedures and remedies of each member state not a Party to the Convention have been duly applied and exhausted."[24] The Commission has recognized a number of exceptions to the exhaustion requirement, including the absence of effective remedies and, in certain circumstances, the inability of the petitioner to exhaust remedies for lack of resources.[25]

While the respondent government has the burden of showing that effective domestic remedies exist, once it has done so, the petitioner has the burden of proving that those remedies were exhausted or that the case falls into one of the exceptions to the exhaustion requirement.[26] In the case of a claim for violations resulting from the effects of global warming, the primary question would be whether a tort action against the United States could provide a viable remedy. The US government is immune to suit for tort except "under circumstances where the United States, if a private person, would be liable to the claimant in accordance with the law of the place where the act or omission occurred."[27] The government is not subject to suit, however, for acts or omissions that are the result of discretionary functions.[28] Because US courts would probably consider most actions that enhance and diminish global warming effects to be discretionary, there would probably not be a domestic remedy available that could force the government to take such actions. The government is also not subject to suit for conduct that violates the US Constitution, such as failure to compensate for the loss of property.[29]

The Arctic region includes parts of the United States, Canada, Greenland/Denmark, Iceland, Norway, Sweden, Finland, and Russia. An important question is whether all inhabitants of the Arctic have rights to bring claims against the United States in the Inter-American system, no matter which country they reside in. Article 1 of the American Convention suggests that a nation has a human rights obligation only with respect to individuals subject to its jurisdiction: "The States Parties...undertake to respect the rights and freedoms recognized herein and to ensure to all persons subject to their jurisdiction the free exercise of those rights and freedoms, without any discrimination...."[30] The American Declaration of the Rights and Duties of Man contains no similar limitation, but the Commission has implied one (*see* the *Coard* case, described below).

Arctic peoples residing in Alaska clearly are subject to the jurisdiction of the United States and the obligations of the United States under the Convention and Declaration apply to them without question. There are important reasons to extend these protections to inhabitants of other regions of the Arctic as well. Unlike the types of violations anticipated when the basic international human rights agreements were drafted, violations arising out of the effects of global warming are clearly not limited to the territory of the nations responsible for those effects. The impacts of US greenhouse gas emissions on Siberian, Canadian, and Scandinavian Arctic communities are fundamentally the same as those on Alaskan Arctic communities. Because the primary contributors to anthropogenic global warming are few (the United States being the primary culprit), and the victims of global warming effects are residents of nations all around the globe, it is important that those nations responsible for global warming not be shielded from responsibility for impacts outside their territory by outmoded limitations on human rights.

Fortunately, the Inter-American Commission is one of several international institutions that have recognized that responsibility for human rights is not circumscribed by national borders. The Commission has interpreted the notion of jurisdiction broadly, citing the Declaration's Preamble:

> The American States have on repeated occasions recognized that the essential rights of man are not derived from the fact that he is a national of a certain state, but are based upon attributes of his human personality.

The Commission took up the question of the extraterritorial application of the Declaration *sua sponte* in *Coard v. United States*, in which citizens of Grenada alleged violations of the Declaration by the United States arising out of the U.S. military invasion of Grenada.[31] The Commission determined that the fact that the alleged violations occurred outside the United States did not bar it from considering the petition:

> [U]nder certain circumstances, the exercise of [the Commission's] jurisdiction over acts with an extraterritorial locus will not only be consistent with but required by the norms which pertain. The fundamental rights of the individual are proclaimed in the Americas on the basis of the principles of equality and non-discrimination – "without distinction as to race, nationality, creed or sex." Given that individual rights inhere simply by virtue of a person's humanity,

each American State is obliged to uphold the protected rights of any person subject to its jurisdiction. While this most commonly refers to persons within a state's territory, it may, under given circumstances, refer to conduct with an extraterritorial locus where the person concerned is present in the territory of one state, but subject to the control of another state – usually through the acts of the latter's agents abroad.[32]

In another case, *Saldaño v. Argentina*, the Commission stated:

> The Commission does not believe...that the term "jurisdiction" in the sense of Article 1(1) is limited to or merely coextensive with national territory. Rather, the Commission is of the view that a state party to the American Convention may be responsible under certain circumstances for the acts and omissions of its agents which *produce effects* or are undertaken outside that state's own territory.[33]

Thus, the Commission recognizes that in certain circumstances states must protect the rights of people outside their territory from the effects of acts or ommissions by their agents. While these effects will usually result from actions taken abroad, they include effects of actions taken domestically. The Commission cites a decision by the European Court of Human Rights that clearly articulates this view:

> [A]lthough Article 1 (art. 1) sets limits on the reach of the Convention, the concept of "jurisdiction" under this provision is not restricted to the national territory of the High Contracting Parties.... [T]he responsibility of Contracting Parties can be involved because of acts of their authorities, whether performed *within or outside* national boundaries, which produce effects outside their own territory.[34]

The Commission has further noted that the respondent government has the burden "to prove the existence of a provision or permissible reservation explicitly limiting or excluding the application of some or all of the provisions of the instrument to a particular class of individuals."[35]

The Commission in *Saldaño* did not describe all the circumstances under which it is appropriate to extend jurisdiction beyond national boundaries. The negotiating history of the International Covenant on Civil and Political Rights provides some insight into the matter, however. In considering whether to base a similar limitation strictly on national boundaries, "it was contended that it was not possible for a State to protect the rights of persons subject to its jurisdiction when they were outside its territory; in such cases, action would be possible only through diplomatic channels."[36] It thus appears that the intent of the jurisdictional limitation was not to prevent states from being held responsible for any violations they caused, but rather to ensure that states not be held responsible for violations they could do nothing about. In the case of transboundary environmental harm, obviously, the state causing or permitting the harm can prevent the violation.[37]

Relevant Rights Protected by the Declaration

The rights that apply to OAS member states, such as the United States, that are not parties to the Convention, are those that are contained in the American Declaration. The Commission and the Inter-American Court on Human Rights have recognized, however, that related rights in the Convention or other human rights documents, even those in other systems, may be used to elaborate the rights in the Declaration, as well as to understand the human rights obligations of OAS member states. In the *Coard* case, for example, the Commission used international humanitarian law to interpret the obligations of the Declaration, stating that "it would be inconsistent with general principles of law for the Commission to construe and exercise its Charter-based mandate without taking into account other international obligations of member states which may be relevant."[38] The Court has even recognized that "a treaty can *concern* the protection of human rights, regardless of what the principal purpose of the treaty might be."[39] Many rights have given rise to a body of law that, to some extent, transcends the document or regime in which they are enumerated.

Many rights contained in the Declaration may give rise to complaints based on the adverse impacts of global warming. These include the right to life (Art. I), the right to residence and movement (Art. VIII), the right to inviolability of the home (Art. IX), the right to preservation of health and to well-being (Art. XI), the right to benefits of culture (Art. XIII), and the right to work and to fair remuneration (Art. XIV). The Commission has applied several other rights that would be relevant to our petition, including the nonderogable right of all peoples to their own means of subsistence[40] and the right to freely dispose of natural resources.[41] In addition, the Commission has recognized that indigenous peoples are entitled to special protections, especially in the case of threats to the environment on which their physical and cultural lives depend.[42] While a full treatment of all of these rights is beyond the scope of this paper, we shall briefly discuss a few that may be of particular interest.

Rights to Life and Preservation of Health and Well-Being

By placing their means of subsistence, infrastructure, and environment at risk, global warming threatens the life, health, and well-being of native Arctic peoples. Melting sea ice, in addition to posing a direct threat to hunters who must traverse the ice in search of game, limits the availability of seals, walruses, polar bears, caribou, and other wildlife that are the staples of the Arctic diet. Thawing of the permafrost destroys buildings, roads, and other critical infrastructure. Erosion has allowed drinking water supplies to become contaminated with advancing sea water, and insect infestations have greatly increased the vulnerability of native peoples and their settlements to forest fires.

The right to life is the most fundamental human rights doctrine and is recognized in every basic international human rights instrument, including the American Declaration.[43] This right is "unanimously considered to be . . . enforceable in respect of all persons, even where there is no treaty obligation. The right to life is included among the peremptory norms (*jus cogens*) from which 'no derogation is permitted.'"[44] The right to health is also fundamental.[45]

The Inter-American Commission has recognized the connection between environmental health and the rights to life, health and personal security. The Commission has found that oil exploitation activities that resulted in severe air and water pollution violated the rights of local residents to life and health. In the Commission's words, "[t]he right to have one's life respected is not . . . limited to protection against arbitrary killing. States Parties are required to take certain positive measures to safeguard life and physical integrity. Severe environmental pollution may pose a threat to human life and health."[46] Likewise, the Commission has held that a government's failure to prevent mining and other activities from degrading the environment of traditional indigenous lands violated the rights to life, health and personal security.[47] Other international human rights tribunals have reached similar conclusions.[48] In the words of one respected jurist, "States . . . may be criminally or civilly responsible under international law for causing serious environmental hazards posing grave risks to life."[49]

Rights to Privacy, Residence and Protection of the Home

The effects of global warming have already begun to impact the homes and communities of many individuals and groups in the Arctic. Subsidence due to permafrost melting is destroying homes, roads and other vital structures in the Arctic. Effects such as these violate the rights of each individual to protection of "private and family life,"[50] "the inviolability of his home"[51] and, in some situations, "not to leave [the territory of the state of which he is a national] except by his own will."[52]

International tribunals have recognized that harm to the environment that affects one's home can violate these rights. For example, in *Lopez Ostra v. Spain*, the European Court of Human Rights held that Spain's failure to prevent a waste treatment plant from polluting nearby homes violated the petitioner's "right to respect for her home and her private and family life," and held the State liable for damages.[53]

Right to Property

Article XXIII of the American Declaration provides: "Every person has a right to own such private property as meets the essential needs of decent living and helps to maintain the dignity of the individual and of the home." When entire communities are forced to abandon their homes and ancestral lands, the violations of this right could not be less ambiguous. The Commission has noted that "various international human rights instruments, both universal and regional in nature, have recognized the right to property as featuring among the *fundamental rights of man*" [emphasis added].[54] The Commission adds that the Declaration, while not legally binding, establishes "rules which have become rules of international customary law and, as such, are considered obligatory in the doctrine and practice of international law."[55]

Conceptually, the right to property falls somewhere between civil/ political rights and economic/social rights. It has received its strongest support from Western countries, led by the United States.[56] It is not limited to any particular type of property and may include both movable and immovable property.[57] The property right may be particularly relevant when the petitioner is a member of an indigenous community because of the strong link between property and culture in those communities and their inherent ownership rights in the land they traditionally occupy. These rights are well established in the Inter-American human rights system.[58]

A notable feature of the right to property is the obligation to pay "just compensation" when the state deprives a person of property.[59] This obligation is not spelled out in the Declaration, but it may be considered a natural consequence of the right to property. Indeed, in some instances it is only through compensation that the property right can be given effect, as human rights law permits the state, in certain circumstances, to deprive persons of their property.[60]

Right to One's Own Means of Subsistence

Global warming is rapidly making unsustainable the subsistence life style of indigenous Arctic communities. Customary international law provides special protection for the right of all people to their own means of subsistence. Two major international human rights agreements, provide "In no case may a people be deprived of its own means of subsistence."[61] Global warming has already begun to deprive people and communities their means of subsistence. For example, the movement of seal populations threatens a critical Inuit food source. There can be no doubt that this effect of global warming constitutes a human rights violation.

Right to Culture, Especially for Indigenous Peoples

The ancient culture of the indigenous Arctic peoples cannot be separated from the land and the subsistence life style it supports. As the land is altered in ways that make subsistence living difficult, if not impossible, their culture is put at grave risk. The Inter-American Commission has recognized that "certain indigenous peoples maintain special ties with their traditional lands, and a close dependence upon the natural resources provided therein – respect for which is essential to their physical and cultural survival."[62] As a result of the intimate connection between indigenous peoples and their lands, "displacement [from indigenous] lands or damage to these lands invariably leads to serious loss of life and health and damage to the cultural integrity of indigenous peoples."[63] As one expert international jurist has written :

> "[Cultural] disintegration is compounded by destruction of the ecology and habitat upon which indigenous groups depend for their physical and cultural survival. Deforestation, particularly of rain forests, and pollution introduced by outsiders jeopardize the *modus vivendi* of indigenous groups. The social nexus binding members of the group to the environment is thus annihilated."[64]

The Commission has further noted that, "for historical reasons and because of moral and humanitarian principles, special protection for indigenous populations constitutes a sacred commitment of the states."[65] For these reasons, the Commission has agreed that "[i]ndigenous peoples have the right to a safe and healthy environment, which is an essential condition for the enjoyment of the right to life and collective well-being."[66]

The Commission has applied these principles to find human rights violations where environmental degradation affects indigenous peoples. For example, the Commission has found that involuntary relocation of indigenous peoples due to development activities constituted a human rights violation that could be justified only in time of war or national emergency.[67] Other international tribunals have agreed. In *Lubicon Lake Band v. Canada*, the UN Human Rights Committee found that expropriation and destruction of indigenous band's land for oil and gas exploitation threatened the way of life and culture of the band, and therefore violated their right to enjoy their culture, as guaranteed by article 27 of the International Covenant on Civil and Political Rights.[68]

As noted above, global warming has already begun degrading lands traditionally occupied by indigenous peoples. This degradation and other effects of global warming inevitably affect other aspects of indigenous culture as well. Indigenous communities risk being forced from their traditional lands, losing traditional food sources, and the destruction of religious or culturally important sites. Global warming has particular human rights implications for these indigenous communities.

Remedies

The Commission cannot force the United States to take any particular action, whether to reduce the causes of global warming or compensate for the effects. However, a favorable outcome to a claim based on those effects could contribute significantly to global efforts to address global warming. In the best case, the Commission would accept the claim and encourage the parties to negotiate a solution. Assuming the United States and the petitioners would not agree to a mutually satisfactory remedy (a safe assumption in this case), the Commission likely would undertake an independent investigation of the facts underlying the claim, probably including site visits to affected regions, and would then issue a report on the petition. The report would set out the Commission's conclusions concerning the relationship between global warming and human rights.

Each significant phase of the Commission's consideration of the claim would provide an opportunity to raise public awareness concerning the human rights implications of global warming. A Commission report finding that global warming results in human rights violations would be important. As an authoritative interpretation of international law, such a finding would help bring a rights-based approach to global warming discussions. Governments or private individuals wishing to pressure the United States and other governments to take meaningful action to address the causes of global warming would welcome the ability to cite the Commission's findings. Plaintiffs in domestic judicial proceedings could use the findings to supplement their claims (or, in some judicial systems, as an independent basis for a claim), and domestic tribunals could use them to justify favorable decisions.

Conclusion

The jurisprudence of human rights law has grown in recent years and continues to develop rapidly. This brief article has touched on a few of the issues relevant to establishing accountability for global warming and its impacts. We conclude that a petition to the Inter-American Commission has significant potential benefits and merits further consideration.

Many rights contained in the American Declaration, including the rights to life and personal security; to residence and movement; to inviolability of the home; to the benefits of culture; and to work and to fair remuneration, could serve as the basis of a complaint. Other rights – such as the rights to means of subsistence, to freely dispose of natural resources, and special protection for indigenous communities – have been recognized by the Commission, even though they are not explicitly mentioned in the Declaration, because they are relevant to understanding rights in the Declaration as they relate to the effects of global warming.

Formal recognition by an international authority such as the Inter-American Commission of the connection between global warming and human rights would have a powerful impact on worldwide efforts to address global warming. Such recognition would demonstrate that the issue is not an abstract problem of degrees per decade and statistical probabilities, but is instead a vital concern of all people everywhere. It would bring to the global warming discourse a basis for holding responsible those who have profited from poorly regulated greenhouse gas emissions, and for placing limits on such emissions in future. And it would be consistent with the growing international recognition that a healthy environment is fundamental to the enjoyment of nearly all of the most fundamental human rights.

Footnotes

1. Intergovernmental Panel on Global Warming, Global warming 2001: Impacts, Adaptation and Vulnerability, Summary for Policy-makers and Technical Summary of the Working Group Two Report (2001), at 7.
2. *Idem.*, at 26.
3. *See, e.g.*, Report on the Situation of Human Rights in Ecuador, OEA/Ser.L/V/ II.96, Doc. 10 rev. 1, 24 Apr. 1997.
4. Statute of the Inter-American Commission on Human Rights, Art. 20; Rules of Procedure of the Inter-American Commission on Human Rights (*hereinafter* Rules of Procedure), adopted at its 109th special session, Dec. 4-8, 2001, Art. 49. The Commission has the authority to receive communications from states complaining of human rights violations, but only if the alleged violator is a party to the American Convention on Human Rights, which the United States is not. American Convention on Human Rights, Art. 45; Rules of Procedure, Art. 48.
5. *Supra*, n.4.
6. The authors thank Andrés Pérez for his contribution to this section.
7. Climate Change Impacts on the United States: The Potential Consequences of Climate Variability and Change—Overview: Alaska. US Global Change Research Program, 2000, pp. 74-75.
8. Potential Consequences of Climate Variability and Change for Alaska. USGCRP Foundation Report (1998), Chapter 10, p. 289.
9. *Idem.*, at 292.
10. Infoterra: testament of Caleb Pungowiyi, Yupik elder from Nome, Alaska.
11. *Supra* n. 9, at 366.
12. *Idem.*, at 293.
13. *Close-up Amid Global Warming Debate, Alaska's Landscape Shifts*, Kim Murphy, Los Angeles Times, Jul. 8, 2002.
14. *Idem.* According to the US Army Corps of Engineers (in 1998) plans to protect and relocate these villages included constructing a $4-6 million sea wall in Shishmaref (a 10-15 year *interim* solution), and relocating Kivalina on higher ground at an estimated cost of $54 million.
15. *Idem.*
16. *Supra* n.9, at 296.
17. *In Alaska, Hotter Weather Provokes Startling Changes*, Timothy, Egan, <u>The New York Times</u>, June 18, 2002
18. This figure constitutes 70-80% of the trees in the entire area. The federal government has given the Kenai Borough $10 million for forest regeneration and to protect communities from beetle-induced fires. The spruce bark beetle has destroyed more than 2 billion board feet of timber in Alaska in the last 25 years. *Battling the Bark Beetle: As Global Warming Rises, so do Tree-Killing Infestations*. Emagazine.com November/December 2001

19. *Idem.*
20. Rules of Procedure, Art.27.
21. Report No. 109/99, Case 10.951, *Coard v. United States*, Sept. 29, 1999, at para. 36. *See also* Articles 49 and 50 of the Commission's Rules of Procedure, which state that procedures for receiving petitions alleging violation of the Convention also apply to the receipt of petitions alleging violations of the Declaration.

    David Padilla, Assistant Executive Secretary of the Commission, has described the authority of the body over nonparties to the Convention:

    [F]or those OAS member states that have yet to ratify the American Convention . . . the Commission's Statute, a binding instrument adopted by a unanimous General Assembly vote, provides that:

    In relation to those member states of the Organization that are not parties to the American Convention on Human Rights, the Commission shall have the following powers, in addition to those designated in Article 18:

    b) to examine communications submitted to it and any other available information, to address the government of any member state not a Party to the Convention for information deemed pertinent by this Commission, and to make recommendations to it, when it finds this appropriate, in order to bring about more effective observance of fundamental human rights. [Statute, article 20]

    This provision obligates the Commission to act where it finds that a complaint from any source meets the admissibility requirements contained in its rules and determines that a Convention-protected right, or a right set forth in the American Declaration of the Rights and Duties of Man, in the case of nonstate parties, has *prima facie* been violated. This is a truly liberal provision compared to those governing other international human rights systems. Private actors become active participants in an international forum. Also, member states *ipso facto* voluntarily commit themselves to participate in the Commission's quasi-judicial process aimed at clarifying and, when it so determines, remedying violations of internationally recognized human rights.

    David J. Padilla. 1993. *The Inter-American Commission on Human Rights of the Organization of American States: A Case Study*, 9 Am. U.J. Int'l L. & Pol'y 95: 99-100.
22. Rules of Procedure, Art. 23. Petitioners may designate an attorney or other person to represent them before the Commission. *Idem.*
23. Rules of Procedure, Art. 28.
24. Statute of the Inter-American Commission on Human Rights, Art. 20(c).
25. *See, e.g.,* Rules of Procedure of the Inter-American Commission on Human Rights, Art. 31.
26. *See* Inter-American Court of Human Rights, *Velásquez Rodriguez* case ("The Court now affirms that if a State which alleges non-exhaustion proves the existence of specific domestic remedies that should have been utilized, the opposing party has the burden of showing that those remedies were exhausted or that the case comes within the exceptions of Article 46(2)."). Inter-American Commission's decision in the Cherokee Nation case. It must be noted, however, that the Inter-American Court of Human Rights has stated: "It must not be rashly presumed that a State party to the Convention has failed to comply with its obligation to provide effective domestic remedies." *Velásquez Rodriguez* case.
27. Federal Tort Claims Act, 28 U.S.C. § 1346(b).
28. *Idem.* § 2680(a).

29. *See FDIC v. Meyer*, 114 S. Ct. 996, 1001 (1994).
30. The drafters of the Inter-American Convention affirmatively rejected the idea of an explicit territorial limitation on responsibility for human rights violations. An early draft of Article 1 of the Convention stated: "The States Parties undertake to respect the rights and freedoms recognized herein and to ensure to all persons *within their territory and subject to their jurisdiction* the free and full exercise of those rights and freedoms." (Emphasis added.) Subsequently, the negotiators unanimously agreed to omit the territorial limitation. *See. Human Rights: the Inter-American System* (Thomas Buergenthal and Robert Norris, (eds.), 1982-1993), Part II, Booklet 13, p. 2; *and Meetings of the Second Session of Committee I*, Doc. 36 (Nov. 11, 1969) in *Human Rights*, Part II, Booklet 12, p. 28.
31. Report No. 109/99, Case 10.951, Sept. 29, 1999, <http://www1.umn.edu/humanrts/cases/us109-99.html>.
32. *Idem.*, at para. 37.
33. Report N° 38/99, Inter-Am. C.H.R., OEA/Ser.L/V/II.95 Doc. 7 rev. at 289 (1998), at para. 17. The petitioner alleged that Argentina had violated the rights of an Argentine citizen being held on death row in the United States by not presenting an inter-state complaint to the United States on the citizen's behalf. The Commission found that it did not have jurisdiction to consider the petition because the petitioner failed to show that Argentina exercised any authority or control over Mr. Saldaño in relation to his arrest and conviction, and did not show "any act or omission by Argentine authorities that implicate that state in the alleged violations arising out of Mr. Saldaño's prosecution in the United States so as to subject him to Argentina's jurisdiction within the meaning of Article 1(1) of the American Convention." *Idem.*, paras. 21, 22.
34. *Loizidou v. Turkey*, 310 Eur. Ct. H.R. (Ser. A) (1995), para. 62 (cited in *Saldaño* at para. 19).
35. *Rafael Ferrer-Mazorra v. United States*, Report No. 51/01, Case 9903, April 4, 2001. In *Rafael Ferrer-Mazorra,* the Commision considered the rights of Mariel Cubans detained in the United States. In response to the US argument that the American Declaration did not apply because, under US immigration law, the detainees were not officially within US territory, the Commission responded that :

    > *OAS member states are obliged to guarantee the rights under the Declaration to all individuals falling within their authority and control*, with the onus falling upon the State to prove the existence of a provision or permissible reservation explicitly limiting or excluding the application of some or all of the provisions of the instrument to a particular class of individuals, such as excludable aliens.
    > *Idem.*, para. 180.

36. Marc J. Bossuyt. 1989 *Guide to the 'Travaux Preparatoires' of the International Covenant on Civil and Political Rights,* 53.
37. The UN Committee on Civil and Political Rights has also discussed this issue. In considering a communication arising out of the abduction from Argentina of Uruguayan citizen by agents of the Uruguayan government, the Committee stated that "it would be unconscionable to so interpret the responsibility under article 2 of the Covenant as to permit a State party to perpetrate violations of the Covenant on the territory of another State, which violations it could not perpetrate on its own territory." *Saldias de Lopez v. Uruguay*, UN GAOR, 36th Sess., Supp. No. 40, UN Doc. A/36/40 (1981), at 183.
38. *Coard v. United States, supra*, para. 40.

39. *The Right to Information on Consular Assistance in the Framework of the Guarantees of Due Process of Law.* Advisory Opinion OC-16/99, Oct. 1, 1999, Inter-Am. Ct. H.R. (Ser. A) No. 16 (1999) (citing *"Other Treaties" Subject to the Advisory Jurisdiction of the Court (Art. 64 American Convention on Human Rights),* Advisory Opinion OC-1/82, Sept. 24, 1982. Series A No. 1; opinion, para. 1) [emphasis in original].

40. International Covenant on Civil and Political Rights (*hereinafter* ICCPR), Art. 1; International Covenant on Economic, Social and Cultural Rights (*hereinafter* ICESCR), Art. 1.

41. ICCPR, Art. 1; ICESCR, Art. 1.

42. In the Report on the Situation of Human Rights in Ecuador, OEA/Ser.L/V/ II.96, Doc. 10 rev. 1, 24 Apr. 1997, the Commission noted that:"[C]ertain indigenous peoples maintain special ties with their traditional lands, and a close dependence upon the natural resources provided therein – respect for which is essential to their physical and cultural survival." The Commission further noted that; "[W]ithin international law generally, and inter-American law specifically, special protections for indigenous peoples may be required for them to exercise their rights fully and equally with the rest of the population. Additionally, special protections for indigenous peoples may be required to ensure their physical and cultural survival – a right protected in a range of international instruments and conventions."

43. American Declaration, Art. 1. *See also:* American Convention, Art. 4; Universal Declaration of Human Rights, Art. 3; International Covenant on Civil and Political Rights (ICCPR), art. 3, 999 U.N.T.S. 171 (entered into force Mar. 23, 1976).

44. *Review of Further Developments in Fields with which the Sub-Commission has been Concerned, Human Rights and the Environment: Final Report by Mrs. Fatma Zohra Ksentini, Special Rapporteur,* p. 44. UN ESCOR, Hum. Rts. Comm., UN Doc. E/ CN.4/Sub.2/1994/9 (1994).

45. *See: American Declaration,* Art. XI; Universal Declaration of Human Rights, Art. 25.1; International Covenant on Economic, Social and Cultural Rights, Art. 12; UN Convention on the Rights of the Child, Art. 24.

46. *IACHR Ecuador Report* at 88.

47. IACHR Res. No. 12/85, Case 7615 (*Yanomami Indians v. Brazil*), Mar. 5, 1985, reprinted in Annual Report of the IACHR 1984-85, OEA/Ser.L/V/II.62, Doc. 10 rev. 1, Oct. 1, 1985.

48. *See: e.g., EHP v. Canada,* in which the UN Human Rights Committee determined that the dumping of nuclear wastes near a town posed a threat to the right to life. Communication No. 67/1980, 4.1. *In:* United Nations, 2 Selected Decisions of the Human Rights Committee under the Optional Protocol 20, UN Doc. CCPR/C/OP/2 (1990).
National courts have recognized the connection as well. For example, the Supreme Court of Costa Rica has stated that life "is only possible when it exists in solidarity with nature, which nourishes and sustains us — not only with regard to physical food, but also with physical well-being. It constitutes a right which all citizens possess to live in an environment free from contamination. This is the basis of a just and productive society." Constitutional Chamber of the Supreme Court, Vote No. 3705, July 30, 1993.

49. B.G. Ramcharan. 1983. *The Right to Life.* Int'l L. Rev. 30: 310-311.

50. American Declaration, Art. V. *See also:* Universal Declaration, Art. 12; UN Convention of the Rights of the Child, Art. 16; European Convention, Art. 8.

51. American Declaration, Art. IX. *See also:* Universal Declaration, Art. 12; ICCPR Art. 17; UN Convention of the Rights of the Child, Art. 16; European Convention, Art. 8.
52. American Declaration, Art. VIII.
53. Series A, No. 303-C, Application No. 16798/90, 1994. *See also: Hatton and Others v. United Kingdom,* European Court of Human Rights, Decision 36022/97, decided 10/2/01 (noise pollution from Heathrow Airport's overnight flights violated local residents' rights to privacy and inviolability of the home and family).
54. IACHR. 1994. "Nicaragua": The Right to Property. *Annual Report of the Inter-American Commission on Human Rights* 1993, OEA/Ser.L/V/II.85, doc.9, rev., Feb. 11, 1994, at 464-77.
55. *Idem.*
56. Asbjorn Eide, Catarina Krause and Allan Rosas, 1995. Economic, *In*: Social and Cultural Rights: A Textbook. Martinua Nijhoff (ed.), at 144. Conversely, the United States has generally opposed the inclusion of economic and social rights in the Inter-American human rights system. *See:* Matthew Craven, *The Protection of Economic, Social and Cultural Rights under the Inter-American System of Human Rights.*
57. Asbjorn Eide, Catarina Krause and Allan Rosas. 2001. Economic, Social and Cultural Rights: A Textbook. Martinua Nijhoff (ed.), at 150.
58. *See, e.g.:* Inter-American Court of Human Rights. 2001. The Case of the Mayagna (Sumo) Awas Tingni Community v. Nicaragua, Judgment of August 31, 2001.
59. *See, e.g.:* the American Convention on Human Rights, Art. 21.
60. *Idem.*
61. ICCPR, Art. 1.2; ICESCR, Art. 1.2.
62. *IACHR Ecuador Report* at 106; *and* 114.
63. *IACHR Ecuador Report, supra,* at 114 (quotation omitted).
64. *Ksentini Final Report, supra,* para 79 (quoting Prof. V. Muntarbhorn, background paper, 1989 UN Seminar on the Effects of Racism and Racial Discrimination on the Social and Economic Relations Between Indigenous Peoples and States. E/CN.4/1989/22, annex III A, 27-28).
65. Resolution of the IACHR, OEA/Ser.L/V/II.29, Doc. 38 rev. (1972). *See also: Report on the Situation of Human Rights of a Segment of the Nicaraguan Population of Miskito Origin* 76. IACHR, OEA/Ser.L/V/II.62, Doc. 10 rev. 3, at 81, 29 Nov. 1983 (hereafter *Miskito Report*) ("[F]or an ethnic group to be able to preserve its cultural values, it is fundamental that its members be allowed to enjoy all of the rights set forth by the American Convention on Human Rights, since this guarantees their effective functioning as a group, which includes preservation of their own cultural identity.").
66. Proposed American Declaration of Indigenous Rights, Art. XIII.1 (approved by the Inter-American Commission on Human Rights February 26, 1997). *See also: Yanomami Case, supra* (holding that destruction of Brazilian rainforests violated the human rights of indigenous forest dwellers).
67. *Miskito Case, supra,* at 84-85.
68. Decision of Mar. 26, 1990 (Lubicon Lake Band v. Canada). UN Hum. Rts. Comm., UN Doc. CCPR/C/38/D/167/1984 (1990). *See also: Ilmari Lansman v. Finland,* Case No. 511/1992 UN Hum. Rts. Comm., Oct. 26, 1994, Reprinted in Official Records of the General Assembly, 50th Sess., Supp. No. 40, A/50/40, vol. I (approval of mining activities in areas essential to Finnish indigenous peoples' culture and spiritual practices threatened their cultural integrity and thereby violated Article 27).

Section IV
COUNTRY CASE STUDIES

CHAPTER 10

Climate Change and Water Resources Management in Semiarid Southern Africa

Umoh T. Umoh[a], Santosh Kumar[b] and **A. A. Oladimeji[c]**
[a]Dept. of Environmental Science, Private Bag 0022, Gaborone,
University of Botswana, Botswana
[b]Department of Mathematics and Statistics, University of Melbourne, Victoria
3010, Australia and School of Computer Science & Mathematics,
Victoria University, PO. Box 14428, Melbourne City, MC 8001, Australia
[c]Federal University of Technology, Dept. of Biological Sciences,
PMB 65, Minna, Nigeria.

Introduction

Trends in the climate of the world these days point to an extreme type of variation that appears to be undirectional. Thus, one of the problems of global concern today relates to climate change and its effects on environmental variables. For example, it is now agreed that there is global warming as a result of a number of factors, the chief being accumulation of greenhouse gases in the troposphere (Boyd, 1989, Desanker and Magadza, 2001). Global warming is also linked with the extreme weather conditions set in motion by the El-Nino Southern Oscillation (ENSO) in many parts of the world. The trends in temperature and sea-level anomalies, particularly since the 1990s, have led to growing concern that the frequency and intensity of the ENSO may be responsible for global warming. Other factors, usually linked with global warming are changes in solar irradiance and major volcanic eruptions. From the beginning of the last century up to date, increase in atmospheric greenhouse gases emerges as the dominant forcing mechanism for global warming (Oldfield, 1998, Mann et al., 1998).

Most increase in greenhouse gases is human induced. Developing countries that occupy the southern African subregion contribute a substantial amount of greenhouse gases to the atmosphere. This they do through bush burning, natural gas flaring and use of very old vehicles and machinery. The response of environment to climate change varies from one place to another. In the Polar Regions ice melting occasioned by flooding can occur, leading to a general sea-level rise. In the tropics, climate change is causing an increase in incidents of drought and aridity, decline in agricultural productivity and incursion of desertification. Effects of global warming in some temperate areas may be flooding and humid conditions while in arid and semiarid areas, further drying up is experienced. Thus one of the expected results of global warming in semiarid southern Africa will be a further drying up of the region through increasing occurrence of drought and desertification.

The negative effects of climate change on water resources are internationally recognized. One of the commitments of the Kyoto Protocol is for all the parties to *'cooperate in preparing for adaptation to impacts of climate change, develop and elaborate appropriate and integrated plans for coastal zone management, water resources and agriculture and for the protection and rehabilitation of areas, particularly in Africa affected by drought and desertification as well as flood"* (Kyoto Protocol, 1997, Article 4(e)). The southern African climate system is prone to flood, drought, and extreme weather events, and displays an inherent degree of seasonal and interannual variability.

Compounding the vagaries of climate is the high degree of dependence by society on climate system, especially in key sectors of water resources, agriculture and health. This vulnerability poses a significant impediment to socioeconomic development and in recent years has caused huge financial losses totaling billions of Pula, severely impacting society at all levels from rural individual to the national government. Mitigation of such impacts can only be undertaken with adequate forward planning, which in turn is fully dependent on understanding and acceptance of the nature of climate change.

The eleven countries that comprise the southern African subregion are shown in Figure 10.1.

Much of southern Africa below latitude 18°S is basically semiarid. Rainfall amount is low and highly variable over space from one year to another. Drought is a recurrent element in southern African climate. Rains fall mostly during summer months between November and March. Rates of evaporation are very high, thus limiting the effectiveness of low rainfall. Water is one of the major keys of the development of the subregion. It is a basic need for human beings as well as for economic development. This very precious resource is scarce and determines to a large extent the density and spatial distribution of the population. This chapter examines water resources management under the increasing effects of climate change in the semiarid countries of southern Africa. Thereafter climatic variability and current trend of drought and desertification in the subregion are examined in the light of the water resources of the region and the management strategies adopted to control the situation. Lastly, some suggestions are proffered for effective water resource management.

Overview of Regional Water Resources

Water resources are inextricably linked with climate, so the prospect of global climate change has serious implications for water resources and regional development (Riebsane et al., 1995). Efforts to provide adequate water resources for southern Africa confront several challenges, including population pressure, problems associated with land-use, such as erosion/siltation and ecological consequences on the hydrological cycle. Climate change—especially change in climate variability through droughts and desertification—complicates these problems, the impact is greater on those countries with limited access to water resources.

Hydrological performance of southern Africa in terms of runoff yield is less than that of subregions of other countries. Apart from the Zambezi, the major southern African rivers traverse semiarid to arid lands on their way to the coast. The Zambezi passes through eight southern African

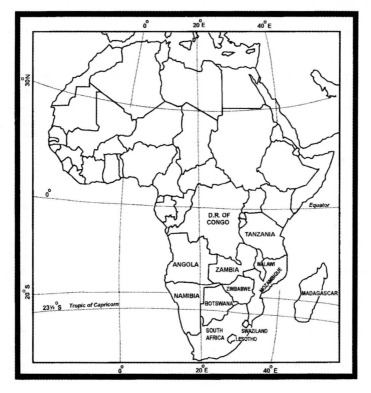

Fig. 10.1: Africa, showing the countries of the Southern African sub region.

countries namely: Angola, Namibia, Botswana, Zimbabwe, Zambia, Malawi, Tanzania, and Mozambique. Furthermore, because these rivers originate within the tropics, where temperatures are high, evaporative losses also are higher than rivers in temperate regions. This situation is further exacerbated by less precipitation which in turn leads to reduced runoff.

Most lakes have a delicate balance between precipitation and runoff. All large lakes show less than 10 % runoff to precipitation ratio (Talling and Lemoalle, 1998). An important water basin such as the Okavango Delta (Angola, Botswana, Namibia, and Zimbabwe) has no outflow because of evaporation and precipitation balance runoff. In semiarid southern Africa, the incidence of seasonal flow cessation is on the increase, as shown by some streams in Zimbabwe (Magadza, 2000). Droughts periods now translate into critical water shortages for industrial and urban domestic supplies (Magadza, 1996).

Table 10.1 shows estimates of ranges of percentage changes in precipitation, potential evaporation and runoff in African river basins. A change in hydrographs of large basins has been observed. Between the mean annual discharge of humid and drought periods, the percentage reduction varies from 40 to 60 % (Olivry et al., 1993). In recent years there have been significant interruptions of hydropower generation as a result of severe droughts.

Table 10.1. Estimates of ranges of percentage changes in precipitation, potential evaporation and runoff in African river basins.

Basin	Change in precipitation (%)	Change in potential evaporation (%)	Change in runoff (%)
Nile	10	10	0
Niger	10	10	10
Volta	0	4 to –5	0 to –15
Schebeli	–5 to 18	10 to 15	–10 to 40
Zaire	10	10 to 18	10 to 15
Ogooue	–2 to 20	10	–20 to 25
Rufifi	–10 to 10	20	–10 to 10
Zambezi	–10 to –20	10 to 25	–26 to –40
Ruvuma	–10 to 5	25	–30 to –40
Limpopo	–5 to –15	5 to 20	–25 to –35
Orange	–5 to 5	4 to 10	–10 to 10

Source: Desanker and Magadza, 2001

Arnell (1999) indicated that the greatest reduction in runoff by the year 2050 will be in the southern African region, also indicating that as the water use-to-resource ratio changes, some countries, e.g. Zimbabwe, will shift into the high water-stress category. The Zambezi River has the worst scenario for decreased precipitation (about 15%), increased potential evaporative losses (about 15-25%), and diminished runoff (about 30-40%). Cambula (1999) has shown a decrease in surface and subsurface runoff of five streams in Mozambique, including the Zambezi under various climate change scenarios. For the Zambezi basin, simulated runoff under climate change is projected to decrease by about 40% or more.

Climate change exerts remarkable effects on river flows, groundwater recharge and other biophysical components of the water resource base, and demands for that resource. The consequences, or impacts, of such changes on risk or resource reliability depend not only on the biophysical changes in stream flow, recharge, sea-level rise and water quality, but also on the characteristics of the water management system. Possible changes in water resources and demand will impact on water supply, flood risk, power generation, navigation, pollution control, recreation, habitats and ecosystems services in the absence of planned adaptation to climate change. In practice, of course, the *actual* impacts of climate change will be rather different because water managers will make incremental or autonomous adaptations to change—generally based on imperfect knowledge – and the impact of change will be a function of adaptation costs and residual impacts. It is important to assess the effect of climate change by, say, the 2050s in the context of the water management system in existence then.

The sensitivity of a water resource system to climate change is a function of several physical features and, importantly, societal characteristics. Physical features that are associated with maximum sensitivity include :

- Current hydrological and climatic regime that is marginal for agriculture and livestock.
- Highly seasonal hydrology as a result of seasonal precipitation.
- High rates of sedimentation of reservoir storage.
- Topography and land-use patterns that promote soil erosion and flash-flood conditions.
- Lack of variety in climatic conditions across the territory of the national state, leading to inability to relocate activities in response to climate change.

Societal characteristics that maximize susceptibility to climate change include:

- Poverty and low income levels, which prevent long-term planning and provisioning at the household level.
- Lack of water control infrastructures.
- Lack of maintenance and deterioration of existing infrastructure.
- Lack of human capital skills for system planning and management.
- Lack of appropriate, empowered institutions.
- Absence of appropriate land-use planning.
- High population densities and other factors that inhibit population mobility.

- Increased demand for water because of rapid population growth.
- Conservative attitudes toward risk [unwillingness to live with some risks as a trade-off against more goods and services (risk aversion)].
- Lack of formal links among the various parties involved in water management.

Table 10.2. Yields of existing and potential reservoirs in Botswana

River	Storage site	Catchment Area (km³)	Storage volume (10^6m³)	Mean annual inflow (10^6m³)	Annual yields (10^6m³)
Notwane	Gaborone	3,982	144	29.7	9.4
Shashe	Shashe	3,650	88	87.3	25.3
Boteti	Mopii	—	95	—	—
Nywane	Nywane	238	2.3	1.7	—
Metsemolthaba	Bokaa	3,570	35	12.2	4.9
Mahalapswe	Mahalapye	754	13	10.4	2.7
Motlouse	Letsibogo	5,480	125	64.9	30.8
Shashe	Lower Shashe	7,800	408	146	73
Limpopo	Cumberland	42,000	1,000	179	50
Limpopo	Martins Drift	97,000	1,000	471	120
Limpopo	Point Drift	1,58,000	1,000	620	175

Source: Botswana National Water Master Plan 1991.

Surface water resources in the subregion consist of rivers, most of which are ephemeral and wetlands notable the Okavango Delta and various salt pans. Most of the rivers have low flows that are highly variable and erratic. Suitable dam sites in regions with relatively flat topography are lacking. These factors combine with high rates of evaporation to account for significant reduction in sustainable yields from reservoirs as indicated in Table 10.2 for Botswana. The extractable volume of groundwater in Botswana is estimated to be about 1,00,000 million m³ (Khupe, 1994). But only 1% of this amount is rechargeable by rainfall given the semiarid climate characterized by low rainfall and high rates of evaporation as well as the nature of the geology and the aquifers.

Climatic Variability

The Earth's climate reflects, in part, the presence of so-called greenhouse gases in the atmosphere. These gases (including carbon dioxide, methane, nitrous oxide, and others) serve to trap energy reflected by the Earth's surface and cause the planet to be much warmer than it would otherwise be. The magnitude of greenhouse effect is driven by the atmospheric concentrations of these greenhouse gases. In future, the ongoing increase in greenhouse gas concentrations is expected to increase the average global temperature and cause other changes in global climate. The average global temperature increased between 0.4°C and 0.8°C over the past century, with most of the warming occurring prior to 1940 and over the past 25 years (Aldy et al., 2001).

The historical climate record for Africa shows warming of approximately 0.7⁰C over the continent during the 20th century, a decrease in rainfall over large portions of the Sahel, and an increase in rainfall in east central Africa (Desanker and Magadza, 2001). Climate change scenarios for Africa, based on results from several general circulation models using data collated by the Intergovernmental Panel on Climate Change (IPCC) Data Distribution Center (DDC), indicate future warming across Africa ranging from 0.2⁰C per decade to more than 0.5⁰C per decade. This warming is greatest over the interior of semi-arid margins of the Sahara and central southern Africa.

The two most recent expert assessments of the science of climate change – the Intergovernmental panel on Climate Change (IPCC) Third Assessment Report and the National Research Council response to the White House – confirm that human activity has influenced global climate change over the past century and is projected to have potentially significant impacts on global climate over this century. For example, the opening of recent National Research Council report noted:

Greenhouse gases are accumulating in the Earth's atmosphere as a result of human activities, causing surface air temperatures and subsurface ocean temperatures to rise. Temperatures are, in fact, rising. The changes observed

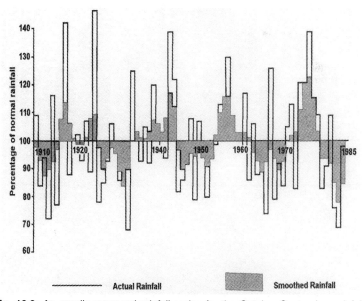

Fig. 10.2: An areally averaged rainfall series for the October-September rainfall in the summer rainfall region for the period 1910/11 to 1983/84. (after Tyson, 1987).

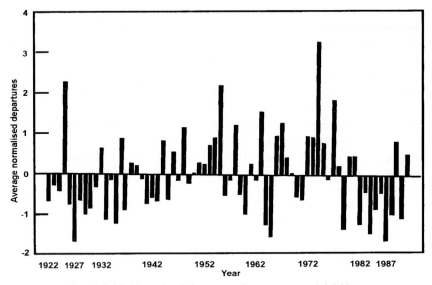

Fig. 10.3 (a): Normalized departures from average rainfall-Maun

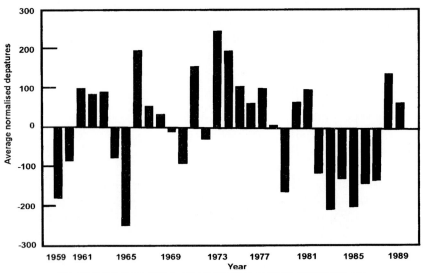

Fig. 10.3 (b): Normalized departures from average rainfall-Rakops

over the last several decades are likely mostly due to human activities, but we cannot rule out that some significant part of these changes are also a reflection of natural variability. Human-induced warming and associated sea-level rise are expected to continue through the 21st century (NRC,2001).

Higher greenhouse gas concentrations would increase average global temperatures over the period 1990 to 2100 by 1.4⁰C to 5.8⁰C, according to climate models. The rate of warming over the next century would very likely be greater than any temperature change experienced over the past 10,000 years. This warming would increase summer mid-latitude continental drying and the associated drought risk.

Many studies have explored linkages between recognizable patterns of climatic variability, in particular El Niño and North Atlantic Oscillation- and hydrological behavior, in an attempt to explain variations in hydrological characteristics over time. The study in southern Africa by Shulze (1997) emphasized variability not just from year to year, but also from decade to decade, although patterns of variability vary considerably from region to region. In general, runoff tends to increase where precipitation has increased and decrease where it has declined over the past years. Variations in flow from year to year are related to precipitation changes rather than to temperature (e.g. Krasovskaia, 1995: Risbey and Entekhabi, 1996).

The climate of southern Africa is strongly influenced by the position of the sub continent in relation to the major circulation features of the Southern Hemisphere. Rainfall is highly seasonal. Precipitation over the interior northern regions of southern Africa follows an annual cycle and is almost entirely a summer phenomenon. More than 80% of annual rainfall in these regions occurs between October and March (Tyson, 1987). The manner in which rainfall has varied over southern Africa has received considerable attention (Tyson, 1971; 1978, 1980; 1983; Tyson and Dyer, 1975; 1978; Tyson et al., 1975). A remarkable regular series of alternating wet and dry spells has been observed over the years in the region. During dry spells, years of below-normal rainfall have predominated. An areally averaged series for summer rainfall region is shown in Figure 10.2. Contained within the generally random year-to-year rainfall variability is an underlying nonrandom component, which has varied systematically over the years. The decade 1950-1959 was characterized by above-normal precipitation over southern Africa. Later, during the period 1960-1969, this rainfall anomaly pattern dramatically reversed in sign, with rainfall deficits observed. More recently, the pattern has been one of increased aridity throughout the subregion. From 1931 to date mean rainfall has decreased by more than 15%.

Studies have shown that the Sahel zone has been experiencing declining rainfall since the turn of the last century and more particularly since the 1950s (Lamb, 1985; Lamb and Peppler, 1991; Smith, 1992; Evans, 1993). In contrast, a study of north-central Botswana (Arntzen et al., 1994) indicated that although general geological chronology points to an aridification trend, the long-term trend is not reflected by short-term data. Normalized average departures from mean rainfall for both Rakops and Maun (Botswana) do not indicate a declining trend (Figure 10.3). Instead, analysis confirmed the existence of periodic cycles of below average rainfall in line with the observations of Tyson et al. (1975).

It is useful to view the rainfall conditions for each country in the region, and to compare these with evaporation figures and total surface runoff, to obtain an overall idea of the water situation. These figures are shown in Table 10.3. In all the countries runoff is generally less than 20% of rainfall, i.e., most of the rain is returned to the atmosphere as evaporation and transpiration. Angola, for example, with 130 km³ runoff and 997 km³ total rainfall has 13% while Botswana has 0.15%.

Table 10.3. Rainfall, evaporation and estimated total surface runoff statistics for southern African countries.

Country	Rainfall range (mm)	Average rainfall (mm)	(km³)	Potential evaporation range (mm)	Total surface runoff (mm)	(km³)
Angola	25-1,600	800	997	1,300-2,600	104	130.0
Botswana	250-650	400	233	2,600-3,700	0.6	.35
Lesotho	500-2,000	700	21	1,800-2,100	136	4.13
Malawi	700-2,800	1000	119	1,800-2,000	60	7.06
Mozambique	350-2,000	1100	879	1,100-2,000	275	220.0
Namibia	10-700	250	206	2,600-3,700	1.5	1.24
South Africa	50-3,000	500	612	1,100-3,000	39	47.45
Swaziland	500-1,500	800	14	2,000-2,200	111	1.94
Tanzania	300-1,600	750	709	1,100-2,000	78	74.0
Zambia	700-1,200	800	602	2,000-2,500	133	100.0
Zimbabwe	350-1,000	700	273	2,000-2,600	34	13.1
Total			4,665			**599.27**

In his study of the impact of the 1991-92 drought on the environment and people of Zambia, Tiffen (1995) noted that rainfall in Zambia reduced in quantity and reliability from north to south. During the three heaviest rainfall months (December, January, and February) the north normally received 250-300 mm per month, the middle section of the country 225-250 mm per month, and the southern province 140-160 mm per month. The southern province and parts of the central, eastern and western areas of the country suffered the most from the drought which hit the greater part of southern Africa. The 1991/92 drought had a serious impact on the water resources of Zambia. Major rivers such as the Zambezi and Kafue had very little flow. The water level in Lake Kariba was drastically reduced. Due to low inflows from the rivers the Kariba reservoir hit a dangerously low level, leading to power rationing and load shedding. Smaller streams, small dams and wetland areas in the valleys dried up early, reducing water for human consumption, livestock use, and vegetable production.

Drought and Desertification in Southern Africa

There is no doubt that southern Africa has been experiencing increasing incidence of drought and desertification since the beginning of the 20th century. Periodic dry spells occurred in the 1920s, 1940s, 1960s, 1980s (Tyson, 1987) and 1990s. The early 1980's drought was one of the most severe on record (Dent et al., 1987). Between 1982 and 1993 two rainy seasons recorded less than 75% of normal rainfall, in 1982/83 (average rainfall, 408 mm) and 1992/93 (average rainfall, 484 mm) (Laing, 1992). In Botswana, the annual total precipitation fell short of the long-term mean by as much as 61% during the 1983/84 drought. Aridity is on the increase and is now a continuous trend in the southern African subregion. Drought and desertification are endemic features in southern Africa. The mean total annual rainfall and mean annual temperature recorded by all stations for different years in different parts of the region show a trend toward decreasing amounts of rainfall. Yet the annual mean temperature has shown no decline. Indeed, the mean annual temperature values tend to be highest in years of lowest rainfall. The causes of these climatic phenomena have been well documented in the literature (e.g. Tyson, 1981, 1987; Harrison, 1984; Lindesay, 1988; Schulze, 1988; Mason, 1990).

Desertification and the attendant problems of declining biological productivity, deterioration of the physical environment, and increasing hazards for human settlement and life affect southern Africa. The process of desertification is defined as "the impoverishment of arid, semiarid and some humid ecosystems by the combined impact of man's activities and drought" (Dregne, 1977). Figure 10.4 shows the latest assessment of the desertification potential of southern Africa by UNCOD (1977). The whole of Namibia and Botswana and more than half of South Africa are rated as potential desert, with large areas of Cape Province and part of the northern Transvaal at very high risk.

Increasing aridity and increasing pressure of population on the drought-prone environment readily lead to desertification. This is because the land is not allowed to recover from the stress imposed by drought before cattle are let loose for grazing or people begin to cultivate it. One major occurrence in desertified areas is a general reduction in quantity and quality of available water resources. Decline in water resources may take the form of lack of rain for prolonged periods, exhaustion of groundwater sources, lack of river flow and depletion of soil moisture. In the study of desertification in north-central Botswana (Artnzen et al., 1994), it was found that below-average flows have persisted in the Boteti river since the mid -1980's drought. This in turn has precluded recharge to aquifers in the area while concomitantly increasing dependence on groundwater sources, resulting in a drop in water table level both locally and regionally. This has necessitated deepening of hand-dug wells in the Boteti river channel many times in the past few years.

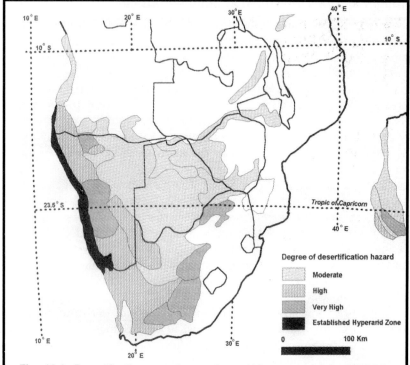

Fig. 10.4. Desertification map for southern Africa prepared for UNCOD. Desertification is here defined as "the intensification or extension of desert conditions; it is a process leading to reduced biological activity, with consequent reduction in plant biomass, land-carrying capacity for livestock, crop yields, and human well-being."

Over exploitation of groundwater resources often leads to deterioration in water quality due to higher levels of salinity which occur as fresh water is pumped out and replaced by saline water.

Reduction in quantity of water is also caused by silting up of reservoirs by eroded soils. In a study of desertification in Zimbabwe (Darkoh, 1986), many dams were found to be completely silted up, i.e., storage capacity had been reduced to zero. Several dams constructed relatively recently are silting up very rapidly; for example the Gizwe dam built in 1970 was 66% silted by the mid-1980s. Siltation of water reservoirs has also been reported from Tanzania (Payton and Shishira, 1991), Namibia, Lesotho and Swaziland. In a study of runoff generation in a small catchment (10.1 km²) near Mochudi (Botswana), it was found that a 15-minute rainfall intensity accounted for more than 80% variation in peak flows. Furthermore the time between the rainfall peak and runoff peak (basin lag) for the first rain of the season was only 2 hours; a second storm a few hours later had a lag of 1.75 hours. This is a strong indication of lack of appreciable infiltration; the entire catchment behaved as an impervious surface due to the presence of erosion crusts which had developed from compaction of silty-loam soils.

Implications of Drought and Desertification for Hydrology and Water Resources in the Subregion

The only source of recurrent water in the region is rainfall. It supplies water for both surface (overland flow and stream flow) and groundwater runoff. Rainfall is seasonal in the region and most rivers dry up during the dry season. A close relationship between rainfall, overland flow, and stream discharge has been established. Drought and desertification disrupt this relationship since they decrease both surface flow and groundwater storage. In a desertified area, much of the rainfall, usually heavy downpours, runs off rapidly. Such conditions impede groundwater recharge and create flash flows on the surface. When drought occurs in a desertified area, far less water is available for these two processes because the rainfall is less and a high proportion of the water is evapotranspired.

One of the expected effects of climate change in the region, if not the only one, is increasing drought incidence occasioned by intensification of the process of desertification. The implications of the increased occurrence of drought and expanding desert-like terrain on water resources are demonstrated by the coincidence of low discharge, recharge and runoff with periods of drought. Low discharge and recharge affect the surface

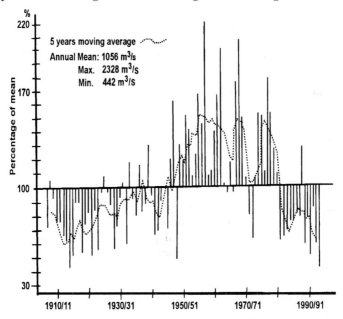

Fig. 10.5. Graph showing annual flow of water in the Zambezi River at Victoria Falls from 1907 to 1995 (after Skoftland, 1995).

and groundwater of the region adversely. Under such conditions, there is very little water available on the surface and much less available as groundwater. An indication of long term climate changes in the subcontinent is depicted by a 90-year record of flow in the Zambezi River at Victoria Falls shown in Figure 10.5. Average flow over the study period was 1056 m^{-3}/s^{-1}. Annual flows were mostly below the average for 38 years prior to 1945. Lower than average flows were recorded from 1910 to 1940 and 1980 to 1995. The lowest ever recorded flow was in 1995.

One method that could be adapted for combating drought and checking flash flows in the region is the construction of dams and the impoundment of reservoirs behind them. There are numerous dams and reservoirs in operation in this region today. However, it is now known that this management option encourages drying out of downstream reaches where regulated water is not released into a natural channel.

The most impacted land is fadama land. Both channel and valley side erosion also occur in such reaches in perpetuity (Olofin, 1994). Secondly, the reservoirs make water available for evaporation throughout the year, and in higher proportions during the periods of no input. Indeed, impounded and natural lake waters are severely reduced during drought periods.

Of course, compounding the effects of drought and desertification on water resources in southern Africa, is the explosive urban growth, placed at 4% per year for developing countries (Uitto, et al., 1998). These researchers contend that the provision of clean water in big cities of the developing world will be one of the challenging tasks of development strategists in the 21st century. It is already known that urban water has to compete with agricultural water requirements. Groundwater levels are known to be dropping rapidly in many places in the Third World as a result of demands of nearby urban centers. When such urban growth with its concomitant water demand is set against declining water availability due to climate change, one will agree with Nakayama (1998) that water may be the 21st century's oil. It is easy to expect that conflict over an equitable distribution and use of water will increase in the face of crippling scarcity caused by growing population, urbanization and environmental degradation occasioned by climate change. Nakayama (1998) is of the view that such conflicts will escalate substantially in arid and semiarid regions of the world.

On a global scale, the number of people living in water-stressed countries (i.e. countries with more than 600 people per flow unit) was estimated as 300 million in 1990. This number is expected to increase to over 3,000 million by the year 2025, most living in Africa and South Asia. By 2025, most countries in the Nile, Zambezi and Limpopo River basins will be water stressed, or even suffer chronic water scarcity.

At present, over 130 million people live in southern Africa. The region has one of the fastest growing populations in the world, with growth rates between 2.2 and 3.6% per year. At such rates, population doubles or even triples every 20 to 30 years. In such a situation, water supply is bound to be a major problem. A flow unit is equal to 1 million m^3 of water. More than 600 people per flow unit creates conditions of water stress, above 1,000 of absolute water scarcity. The two driest countries in the region, namely Botswana and Namibia, are already in this situation, as is Malawi. Furthermore, South Africa and Zimbabwe are expected to reach these critical levels within the next 25 years. The high population growth rate of southern African countries indicates that massive efforts will be needed to secure safe water supplies for all.

The foregoing discussion shows that the water resources of southern Africa are very vulnerable to climate change through occurrence of droughts and desertification. The most important input that can be made to cushion the adverse effect of climate change on water resources in the region is to adopt a water management option that would stand up to the extremes of climatic variations that is currently plaguing the region.

Suggestions for Water Resources Management

Water conservation is vital for sustainable use of the region's meagre water resources. Components of such a water conservation policy would include: effective water pollution control, water reuse, water recycling, water demand management involving restriction/reduction in water use, control of distribution losses and the exploitation of new sources of water such as rain-water harvesting and desalination. The use of economic instruments would play a pivotal role in water demand management in southern Africa. However, care must be taken not to price potable water out of reach of the poor, especially in rural areas as water quality has a vital role in enhancing the health of the people.

The type of water resources management needed in southern Africa therefore should go beyond the construction of hydraulic structures such as dams, reservoirs and boreholes. This must be done so that the effects of climate change on water resources are contained in the region. The design of the structures themselves should undergo some modification. To this end, dams should be the temperature-exporting types that will retain cool water in the reservoirs and discharge warm water, usually from the top to minimize the rate of evaporation from the reservoirs. Contingency measures must be taken to ameliorate channel erosion that may result from top-dam release water. Also, since evaporation is related to surface area, the designs of reservoirs should minimize surface area and focus on depths. To this end, reservoir sites could be dredged before impoundment.

Holistic, basin-wide water budgeting and implementation should be embarked upon to achieve equity and compensating allocation of water to avoid unnecessary conflicts. The basin-wide approach should include reservoir catchment protection to enhance sustainable water availability and environmental quality. The observation in other regions such as the Sudan-Sahelian belt of West Africa is that once a reservoir is impounded, agricultural and settlement activities are increased in the catchment area (Olofin, 1996). Such increase leads to escalated environmental degradation, sediment generation through erosion, and reservoir siltation.

There is a need to focus on the recycling and reuse of water. This is because of the fact that climate change will bring about drastic reductions in the amount of available water in the region. For example, recycled wastewater can be used for flushing toilets, washing cars, household gardening and small-holder irrigation. Similarly, areas of wasteful application of water should be identified and corrected.

There is need for more comprehensive legislation on water than currently available to address all issues pertaining to development, users and uses of water with clear enforceable sanctions for offenders. Indeed, what one is advocating, in line with Issar (1998), is to simulate non-conventional but appropriate, interdisciplinary innovations and technological applications in all that involves water development, management and method of use. This approach is essential in containing the anticipated impacts of climate change on the water resources of southern Africa.

Conclusion

The most likely manifestation of climate change in southern Africa would be global warming. The clearest impact of global warming to expect in the subregion is a further drying-up of the arid and semiarid conditions through increasing incidence of droughts and desertification. The subregion is already showing signs of drying-up as various countries within the region are facing serious water crisis of dimensions not experienced before. Most rivers are ephemeral or seasonal with highly variable flow rates. Existing groundwater accumulated during earlier geological periods when the climate was wetter than now. Availability of this water is controlled by the geology of the region so that borehole yields vary widely. The groundwater is saline in arid zones and subjected to pollution especially by nitrates in densely populated areas. Despite the limited quantities of available water resources, the demand of water in the region has been on the increase as a result of expanding urbanization, increasing population, rising standard of living of this population and expansion in mining, commercial, and industrial activities. These water problems are expected to escalate as drought and desertification intensify. Such a situation would create substantial conflicts over the use of water and lead to human tragedy if timely solutions are not found. Fortunately, some suggestions have been proffered for effective water resource management in the face of increasing effects of climate change through correctly focused investment, appropriately applied technology, strong political will, and efficient management strategies.

References

Aldy, J.E., Orszag, P.R., and Stiglitz, J.E. 2001. Climate Change: An agenda for global collective action, Paper presented at Conference on "The Timing of Climate Change Policies", Pew Centre on Global Climate Change, October 2001.

Arnell, N.W. 1999. Climate change and global water resources. Global Environmental Change 9: S31-S50.

Arntzen, J., Chanda, R., Musisi-Nkambwe, S., Ringrose, S., Sefe, F. and Vanderpost, C. 1994. Desertification and possible solutions in the Mid-Boteti River Area: A Botswana Case Study for the Intergovernmental Convention to Combat Desertification (INCD). Consultancy report for Ministry of Agriculture, Government of Botswana, Gaborone.

Botswana National Water Master Plan Study. (1991). Final Reports, vols. 1, 5 and 6, Gaborone, Botawana.

Boyd, D. 1989. "State of the Environment". Our Planet. UNDP 1(14): 4-7.

Cambula, P. 1999. Impacts of climate change on water resources of Mozambique, Republic of Mozambique. Final Report Mozambique/U.S. Country Study Programme Project on Assessment of Vulnerability of Economy of Mozambique to Projected Climate Change, Maputo, Mozambique.

Darkoh, M. B. K. 1986. Combating desertification in Zimbabwe. Desertification Control Bull. 72/73:17-28.

Dent, M.C., Schulze, R. E., Wills, H.N.M., and Lynch, S.D. 1987. Spatial and temporal analysis of drought in summer rainfall region of southern Africa. Water South Africa 13: 37-42.

Desanker, P. and Magadza, C. 2001. Africa, In: Climate Change 2001: Impacts, Adaptation and Vulnerability, J.J. Mc Carthy, O.F. Canziani N.A., Leary, D. J. Dokken, and K. S. White, (Eds.). Cambridge Univ. Press, Cambridge, UK.

Dregne, H.E. 1977. Status of desertification in the hot arid lands. United Nations Conf. Desertification, Nairobi, Kenya, 74/31: 4-6.

Evans, T. 1993. Sea-level rise in the Caspian Sea: Is this the Greenhouse effect?. AGID News 72/73: 7-12.

Harrison, M.S.J. 1984. A generalized classification of South African summer rain-bearing synoptic systems. J. Climatology 4: 547-560.

Issar, A.S. 1998. Climate, history and water. Work in Progress, United Nations University 15(2): 16-17.

Khupe, B. B. 1994. 'Integrated water resource management in Botswana' In: Proc. Integrated Water Resources Management Workshop. J. Gould (eds.) A Giesk and Kanye, Botswana.

Krasovskaia, I. 1995. Quantification of the stability of river flow regimes. Hydrological Sci. J. 40: 587-598.

Kyoto Protocol. 1997. Kyoto Protocol to the United Nations Framework Convention on Climate Change. FCCC/CP/1997/L.7/Add.1,10 December.

Laing, M. 1992. Drought update 1991-1992, South Africa. Drought Network News, (Nebraska) 4: 15-17.

Lamb, P. J. (1985) 'Rainfall in Sub-Saharan West Africa during 1941-1983. Zeitschrift für Gletscherkunde und Glazialgeologie 212: 131-139.

Lamb, P. J. and Peppler R. A. 1991. West Africa In: Teleconnections Linking Worldwide Climate Anomalies. R.W., Glantz, R.W. Katz, and N. Nicholls, (eds) Cambridge University Press, UK, pp. 121-191.

Lamb, P. J. and Peppler R. A. 1992. 'Further case studies of tropical Atlantic surface atmospheric and oceanic patterns associated with sub-Saharan drought, J. Climate, 5: 476-488.

Lindesay, J. A. 1988 Southern African rainfall, the Southern Hemisphere semi-annual cycle, J. Climatology 8: 7-30.

Magadza, C. H. D. 1996. Climate change: some likely multiple impacts in southern Africa In: Climate Change and Food Security. T.E. Downing (ed.). Springer-Verlag, Dordrecht, The Netherlands, pp. 449-483.

Magadza, C. H. D. 2000. Climate change impact and human settlements in Africa: prospects for adaptation. Environmental Monitoring, 61,193-205.

Mann, M., Bradley and Hughes, M. K. 1998. Global scale temperature patterns and climate forcing over the past six centuries. Nature 382: 779-788.

Mason, S. J. 1990. Temporal variability of sea surface temperatures around southern Africa: a possible forcing mechanism for the eighteen-year rainfall oscillation. South African J. of Sci. 86: 243-252.

Nakayama, M. 1998. Water: The 21st century's oil? Work in Progress, United Nations University 15(2): 18-19.

National Research Council 2001. Assessment of Science of Climate Change, National Acad. Press, Washington, DC, USA.

Oldfield, F. 1998. For research on climate change, past is the key to the future. EOS, Transactions, American Geophysical Union 80(3): 493-494.

Olivry, J. C., Briquet, J. P., and Mahe, G. 1993. 'Vers un Apprauvrissment Durable des Resources en Eau de l'Afrique Humide? IAHS Publ. No. 261, pp. 67-78.

Olofin, E. A. 1994. Dam construction and fadama development. In: Strategies for Sustainable Use of Fadama Lands in Northern Nigeria. A. Kolawale et al. (eds.). CSER/ABU (Zaria) and IIED (London), pp. 69-75.

Olofin, E. A. 1996. Dam-induced drying out of the Hadejia-Nguru Wetlands, Northern Nigeria, and its implications for the fauna. In: Proc. 1993 African Crane and Wetland Training Workshop. R. D. Beilfuss, et al. (eds.). Int. Crane Foundation, Baraboo, USA, pp. 141-146.

Payton, R. and Shishira, E. K. 1991. The effects of soil erosion and sedimentation on land quality: defining the baselines in the Haubi Basin of Central Tanzania. Working Paper No. 17, Environment and Development Studies Unit, School of Geography, Stockholm University.

Riebsame, W. E., Strzepek, K. M., Wescoat, J. L., Perrit, Jr. R., Graile, G. L., Jacobs, J. and Yates, D. 1995. Complex river basins, In: As Climate Changes, International Impacts and Implications. K.M. Strezelek and J.B. Smith (eds.). Cambridge Univ. Press, Cambridge, UK, pp. 57-91.

Risby, J. S. and Entekhabi, D. 1996. Observed Sacramento Basin streamflow response to precipitation and temperature changes and its relevance to climate impact studies. J. Hydrology. 184: 209-223.

Schulze, G. C. 1988. El Niño en La Niña-die seun en die dogter (the son and daughter), South African Weather Bureau Newsletters 475: 7-8.

Schulze, R. E. 1997. Impacts of global climate change in a hydrologically vulnerable region: challenges to South African hydrologists, Prog. Phy. Geog. 21: 113-136.

Skoftland, E. 1995. Background of water resources in the SADC region. Water resource management in southern Africa—a vision for the future. Conf. SADC Ministers Responsible for Water Resources Management: Pretoria 23-24 November, 1995. Dept. Water Affairs and Forestry, South Africa.

Smith, K. 1992. Environmental Hazards: Assessing Risk and Reducing Disaster, Routledge, London, UK.

Talling, J.F. and Lemoalle, J. 1998. Ecological Dynamics of Tropical Inland Waters, Cambridge Univ. Press, Cambridge, UK.

Tiffen, M. 1995. The Impact of the 1991-92 drought on environment and people in Zambia. In: People and Environment in Africa, T. Binns (ed.). John Wiley and Sons, Chichester, UK.

Tyson, P. D. 1971. Spectral variation of rainfall spectra in South Africa. Annals, Assoc.of Amer. Geog., 61: 711-720.

Tyson, P. D. 1978. 'Rainfall changes over South Africa during the period of meteorological record. In: Biogeography and Ecology of Southern Africa, M. J. A., Werger and W. Junk, (Eds.). The Hague, pp. 53-69.

Tyson, P. D. 1980. Temporal and spatial variation of rainfall anomalies in Africa south of latitude 22° during the period of meteorological record. Climate Change 2: 363-371.

Tyson, P. D. 1981. Atmospheric circulation variations and the occurrence of extended wet and dry spells over southern Africa. J. Climatology 1: 115-130.

Tyson, P. D. 1983. The great drought, Leadership SA, 2(3): 49-57.

Tyson, P. D. 1987. Climate Change and Variability in Southern Africa. Oxford Univ. Press, Cape Town, Republic South Africa.

Tyson, P. D. and Dyer, T. G. J. 1975 'Mean annual fluctuations of precipitation in the summer rainfall region of South Africa. South African Geog. J. 57: 104-110.

Tyson, P. D. and Dyer, T. G. J., 1978. 'The predicted above-normal rainfall of the seventies and the likelihood of droughts in the eighties in South Africa. South African J. Sci., 74: 372-377

Tyson, P. D., Dyer, T. G. J., and Mametse, M. N. 1975. Secular changes in South African rainfall: 1880 to 1972. Quart. J. Roy. Meteor. Soc. 101: 817-833.

Uitto, J. I., Biswas, A. K. and Tortajada-Quiroz, C. 1998. City Water: 21st century challenge. Work in Progress, United Nations University 15(12): 8-9.

UNCOD. 1977. Report of the United Nations Conference on Desertification, Nairobi, Kenya. A Conference 74/31, 4-6.

CHAPTER 11

Adaptation to Effects of Climate Change in Southern Africa

Umoh T. Umoh[a,] A A. Oladimeji[b] and S. Kumar[c]

a. University of Botswana, Dept. Environmental Science, Private Bag 0022, Gaborone, Botswana.
b. Federal University of Technology, Dept. Biological Sciences, PMB 65, Minna, Nigeria.
c. Dept. of Mathematics and Statistics, University of Melbourn, Victoria 3010, Australia, and Victoria University, School of Computer Science and Mathematics PO Box 14428, Melbourne City, MC 8001, Australia.

Introduction

Climate means "average weather" and its long–term variability over a particular period of a month, season, year or several months. Over the past few years, climate change and its effects on environmental variables has emerged as one of the most important issues facing the international community. The simple reason is that all countries share a common planet and lack a viable alternative living space, should we destroy the atmosphere of this planet. Trends in climate these days, of the world in general and southern Africa in particular, point to an extreme type of variation that appears to be unidirectional. Southern Africa is highly vulnerable to various manifestations of climate change.

Responses to climatic variations in the subregion include: increasing incidence of droughts, desertification, and aridity resulting from changes in rainfall and intensified land use; food security at risk from declines in agricultural production and uncertain climate; water shortages for plants, animals, and man; natural resources productivity at risk and biodiversity that might be irreversibly lost; vector- and water-borne diseases, especially in areas with inadequate health infrastructure; coastal zones vulnerable to sea-level rise, particularly roads, bridges, buildings, and other infrastructure that are exposed to flooding and other extreme events. The database is poor, despite attempts in the 1970s to improve climatological information, development of drought warning systems, and increasing use of remote-sensing techniques. Increasing aridity and increasing pressure of population on drought-prone environment, easily leading to desertification have impoverished many nations. Major droughts in southern Africa in the 20th century occurred in 1913-15, 1940-41, 1948-49, 1972-73, 1983-85, 1987, and 1992-94 (Collins, 1993; Wilhite, 1993, 1997; Tyson, 1986; Dent et al., 1987). The causes of these climatic phenomena and the impact of such events on the rural poor have been documented in the literature (e.g. Tyson, 1981, 1986: Harrison, 1984, 1986; Lindesay, 1988; Mason, 1990; Abrams et al., 1992; AFRA, 1993; Nicholson and Palao, 1993; Vogel, 1994).

The ability of human systems to adapt to and cope with climate change depends on such factors as wealth, technology, education, information, skills, infrastructure, access to resources, and potential management capabilities. There is potential for developed and developing countries to enhance and/or acquire adaptive capabilities. Populations and communities vary markedly in endowment with these attributes. The adaptive capacity of human systems in southern Africa is low due to lack of economic resources and technology, and vulnerability high as a result of heavy reliance on rainfed agriculture, frequent droughts and floods, and poverty.

Timely and effective adaptation strategies to consequences of climate change are no doubt necessary, although there are constraints and problems related to implementation feasibility. Using examples from different countries in southern Africa, this chapter discusses different adaptation strategies to contain effects of the impacts of climate change in agriculture; wildlife resources; rangeland and livestock production; human health; hydrology and water resources; drought and desertification; implications of adaptive strategies; mechanism and priorities for adaptive responses.

Concept of Adaptation

Adaptation is a very broad concept and, as applied to climate change, can be defined in three different ways. Firstly, adaptation refers to all those responses to climate change that may be used to reduce vulnerability or any action designed to take advantage of new opportunities that may arise as a result of climate change. Secondly, adaptation to climate change can be viewed as a process through which people reduce adverse effects of climate on their health and well-being, and take advantage of the opportunities that their climatic environment provides (Burton, 1997). Thirdly, adaptation may be defined as any adjustment, whether passive, reactive or anticipatory, proposed as a means for ameliorating the anticipated adverse consequences associated with climate change.

According to the IPCC (1996), adaptability refers to the degree to which adjustments are possible in practice, processes, or structures of systems to projected or actual changes of climate. Also, according to the IPCC, adaptation can be spontaneous or planned and can be carried out in response to, or in anticipation of changes in conditions (IPCC, 1996).

No matter what definition is employed it is important to note that there are many adaptation measures that may be adopted in response to climate change. The Second Assessment Report of IPCC working group II describes 228 different adaptation measures (IPCC, 1996). Seven of these different measures, commonly used, have been applied in southern Africa : (i) hear losses, (ii) share losses, (iii) modify the threat, (iv) prevent effects, (v) change use, (vi) change location, and (vii) educate, inform and encourage behavioral change. Another category of adaptation included by the IPCC Technical Guidelines (Carter et al., 1994), is called "restoration". This category aims to restore a system to its original condition following damage or modification due to climate. From the perspective of adaptation as a continuous process, and as a learning process, the notion of restoration might even be considered maladaptive, if by restoration is meant return to a preexisting state.

Climate Charge

Weather changes all the time the average pattern of weather, termed climate, also varies as a result of actions of man. The single human activity that is most likely to have a large impact on climate is the burning of "fossil fuel" such as coal, oil, and gas. These fuels contain carbon and burning them releases carbon dioxide gas. Climate change is caused by carbon dioxide and other gases released by human activities. El Niño also contributes to climate change in the region. This phenomenon has become a descriptive term of the extensive warming of the upper ocean in the tropical eastern Pacific lasting three or more months. El Niño events are linked with a change in atmospheric pressure between the western and central regions of the Pacific Ocean (known as the Southern Oscillation), a shifting of focus of tropical rainfall from the western to the central Pacific Ocean, a weakening of the Pacific trade winds, and sea-level changes. Because the ocean and atmospheric changes are so closely linked with each other, they are collectively known as the El Niño/ Southern Oscillation, or ENSO. This is recognized as a major driving force behind changes in the weather from year to year. For example, droughts are more frequent during and immediately following El Niño in Brazil, Australia and southern Africa. El Niño has been linked to world food crises because it affects many countries simultaneously (WMO, 1999).

The historical climate record for Africa shows warming of approximately 0.7°C over the continent during the 20th century, a decrease in rainfall over large portions of the Sahel, and an increase in rainfall in east central Africa (Desanker and Magadza, 2001). Climate change scenarios for Africa, based on results from several general circulation models using data collated by the Intergovernmental Panel on Climate Change (IPCC) Data Distribution Center (DDC), indicate future warming across Africa ranging from 0.2ºC to 0.5ºC per decade. This warming is greatest over the interior of semiarid margins of the Sahara and central southern Africa.

Contribution to climate change by southern African countries is presented in Table 11.1, highlighting information about the emissions of greenhouse gases in each of the countries of the region.

Table 11.1: Estimated emissions of greenhouse gases in southern African countries (World Resources Institute, 1996).

Country	CO_2 emissions from energy-related activities (million tons), 1992	CO_2 emissions from land-use change (million tons), 1991	Methane from anthropogenic sources (thousand tons), 1991
Angola	4.5	16.0	340.0
Botswana	2.2	3.2	110.0
DRC	4.2	280.0	380.0
Lesotho	Na	Na	44.0
Malawi	0.7	11.0	72.0
Mozambique	1.0	15.0	98.0
Namibia	Na	1.8	96.0
South Africa	290.3	14.0	2,400.0
Swaziland	0.3	0.4	25.0
Tanzania	2.1	22.0	760.0
Zambia	2.5	34.0	150.0
Zimbabwe	18.7	5.3	230.0
Southern Africa	326.3	402.7	4,705.0
World	22,339.4	4,100.0	270,000.0
Southern Africa as percentage of world total	1.5	9.8	1.7

The contributions these countries make to global totals are also shown in Table 11.1. From the data, two things are apparent: first, South Africa is the region's main contributor (for example, 89% of southern Africa's carbon dioxide from energy-related activities comes from South Africa); and second, even with South Africa, the region's total contribution to global emissions is fairly low, less than 3% of total carbon dioxide emissions (Rowlands, 1998).

Hulme (1996) enumerated potential climate change impacts in southern Africa over the next 60 years to include:

(i) temperature rise of 1.5ºC;

(ii) modest drying over large parts of the region;

(iii) decline in grasslands, replaced by thorn scrub savannah; expansion of dry forest biomes; expansion of desert areas;

(iv) 15 to 20% of large nature reserves and national parks experiencing a change in biome, which has consequences for biological diversity;

(v) variations in runoff, and greater annual variation in the same;

(vi) crop yield increasing generally, though with some adverse impacts in semiarid regions;

(vii) changes in distribution of disease-bearing insects;

(viii) decline in distribution of ungulate species richness;

(ix) other impacts – for example, sea-level rises.

It is believed that climate effects could be largely negative for Africa. For example, a report by the IPCC concluded that:

The African continent is particularly vulnerable to the impacts of climate change because of factors such as widespread poverty, recurrent droughts, inequitable land distribution and over dependence on rain-fed agriculture. Although adaptation options, including traditional coping strategies, theoretically are available, in practice the human, infrastructural and economic response capacity to effect timely response actions may well be beyond the economic means of some countries. (Watson et al., 1997).

The vulnerability of human populations and natural systems to climate change differs substantially across regions and across populations within regions. Some adaptive capacity, vulnerability, and key concerns for Southern Africa according to the Report of the IPCC Working Group II (2001) are:

- Adaptive capacity of the human system is low due to lack of economic resources and technology, and vulnerability high as a result of reliance on rain-fed agriculture, frequent droughts and floods, and poverty.
- Major rivers are highly sensitive to climate variation; average runoff and water availability would decrease.
- Extension of ranges of infectious disease vectors would adversely affect human health in the subregion.
- Desertification would be exacerbated by reduction in average annual rainfall, runoff, and soil moisture.
- Increases in droughts, floods and extreme events would add to stress on water resources, food security, human health and infrastructures, and would constrain development in the region.
- Significant extinctions of plant and animal species are projected and would impact on rural livelihood, tourism, and genetic resources.

- Grain yields are projected to decrease for many scenarios, diminishing food security, particularly in small food-importing countries.
- Coastal settlements along the East-Southern African coast, for example, would be adversely impacted by sea-level rise through inundation and coastal erosion.

Economic Impact of Climate Change

Scientists believe that significant climate change will take place gradually over a period of many decades. If the change is gradual, the overall economic impact on wealthy countries such as the United States and the United Kingdom will be modest. Whether it is fast or slow, climate change has greater impact on African countries (developing countries) than on the rich countries. Two factors lead to this conclusion. First, southern African countries are forced to live "closer to the edge" and have less capacity to adapt to changes. Compare the flooding of the Mississippi River in 1993 with various major floods in southern Africa, such as that of Mozambique in 2000. While the Mississippi floods were serious, the US was able to adjust to them remarkably smoothly. Very few people died, aid was supplied by other parts of the country, food prices were hardly affected, and people got on with their lives. A similar flood in any southern African country would kill tens of thousands of people and cause massive disruption in food supply, widespread diseases, and economic dislocation for many years. The second reason is that people in some southern African countries live traditional lives in cultures that depend much more directly on a specific climate. Their agricultural practices, their housing and many aspects of their way of life, are adapted to local climate conditions. These traditional ways have been passed down for countless generations. Because of low education levels and strong cultural traditions, changing these ways in response to climate change may be very difficult.

Climate has a big influence on plants and animals in the natural environment, on oceans, and on human activities such as agriculture, water supplies, and heating and cooling. The effects of climate depend upon how much change there is, how fast it occurs, and how easily the community can adapt to the new conditions. The effects of climate on people would change from place to place. Economically developed societies, such as those in North America, Europe, and Japan, could use technology to reduce direct impacts. For example, they might develop new crop varieties, construct new water systems, and limit coastal development.

In contrast, economically less developed societies, such as those in southern Africa, depend much more directly on climate and could be hit much harder by sudden or large changes. Coastal areas and low-lying islands are usually flooded by storms or rising sea level. Droughts are becoming more serious. Southern African countries have far fewer resources for adapting to such changes. They may not be able to afford large projects such as sea walls or aqueducts. Peasant farmers may have difficulty adopting new agricultural practices. The resultant social tensions could lead to more political unrest, large-scale migration, and serious international problems such as terrorism and wars.

Responses to Climate Change

On understanding that climate change is real, the international community as part of a comprehensive strategy, has thought of appropriate methods for adapting to global climate change, in particular through the UNFCC. Some work has been done in this direction. For instance, much of the 1996 report of the IPCC's Second Working Group was concerned with adaptability: "the degree to which adjustments are possible in practice, the processes or structures of systems to projected or actual changes in climate" (IPCC, 1996). Moreover, the importance of adaptation has also been recognized in the international negotiations to date: the international treaty guiding action, the United Nations Framework Convention on Climate Change (UNFCCC), commits all parties to "formulate, implement, publish and regularly update ... programmes containing ... measures to facilitate adequate adaptation to climate change" (UNFCC, 1992, Article 4.1 (b). It also pledges them to cooperate "to prepare for adaptation to the impact of climate change" (UNFCC,1992, Article 4.1 (e)), and commits North-South assistance in "meeting costs of adaptation to the adverse effects of climate change for the developing country parties that are particularly vulnerable" (UNFCC, 1992, Article 4.4).

At a meeting of the conference of parties to the UNFCCC in 1995, governments reviewed the adequacy of existing international commitments to achieve this goal and decided that additional commitments were required. Thus, the Ad Hoc Group on the Berlin Mandate (AGBM,1995) was established to identify appropriate actions for the period beyond 2000, including strengthening of commitments through adoption of a protocol or another legal instrument. The AGBM process culminated in adoption of the Kyoto Protocol in December 1997 (United Nations, 1997). In the Kyoto protocol, industrialized countries (Annex 1 Parties to the UNFCCC) agreed to reduce their overall emissions of six GHGs by an average of 5% below 1990 levels between 2008 and 2012. Also at Kyoto, a "clean development mechanism" (CDM) was introduced into the discussions (Kyoto Protocol, 1997, Article 12).

The purpose of the clean development shall be to assist Parties not included in Annex 1 in achieving sustainable development and in contributing to the ultimate objective of the Convention, and to assist Parties included in Annex 1 in achieving compliance with their quantified emission limitation and reduction commitments. (Kyoto Protocol, 1997, Article 12.2)

The developing countries do not have specific mitigation obligations under the terms of the UNFCCC, nor under the Kyoto Protocol to the convention (United Nations, 1997). This differentiation of commitments between developed and developing countries is accepted by all parties, given that "the largest share of historical and current global emission of greenhouse gases has originated in developed countries, that per capita emissions in developing countries are still relatively low and that the share of global emission originating in developing countries will grow to meet their social and development needs, [it is accepted that] the developed country parties should take the lead in combating climate change and adverse effects thereof" (UNFCCC, 1992, Preamble and Article 3.2).

Other international bodies are also taking up the challenge of climate change. These organizations include the World Bank, the United Nations Environment Program (UNEP), the UN Development Program (UNDP), and the Global Environment Facility (GEF), as well as a variety of regional institutions.

Many countries have developed national climate strategies based on a diverse range of policy instruments, such as economic instruments, regulation, research and development, and public awareness and information. Energy efficiency, fuel switching, public transportation and renewable energies are promoted. At the local level, many cities—mainly in industrialized countries—have adopted GHG emission reduction targets and have taken measures to implement them, mostly in the energy and transport sector. Large multinational corporations such as Shell International and BP Amoco have declared that they will voluntarily observe elements of the Kyoto Protocol (van der Veer, 1999: Browne, 2000). Similarly, environmental nongovernmental organizations (NGOs) around the world have initiated climate campaigns with the aim of convincing citizens and governments to strengthen the Kyoto Protocol.

Adaptation Measures

Agriculture

There are many possible adaptation options for responding to climate change and sea-level rise in agriculture. Among these adaptation options are altered planting dates, change to a crop more adaptable to the new climate, application of irrigation, changes in levels of fertilization, and changes in agricultural systems. In general, some details of on-farm possible adaptation choices, already being applied or which can be applied

Table 11.2: On-farm adaptation choices

Adaptation by crop choice	Action	Impact
All season crops	Plant quicker (or slower) maturing varieties	Ensure maturation in growing season shortened by reduced moisture or thermal resources or maximize yield under longer growing season
All crops	Plant drought- or heat resistant crops	Reduce crop loss or yield reductions under reduced moisture conditions or reduced irrigation requirements
All crops	Plant pest-resistant crops	Reduce yield reduction where altered climatic conditions have encouraged increases in wields or insect pests
All seasonal crops	Use altered mix of crops	Reduce overall yield variability due to climate change

Adaptation by altered tillage and husbandry	Action	Impact
Altered tillage	Use minimum or reduced tillage	Reduce loss of soil organic matter, reduce soil erosion, reduce nutrient loss
	Use terracing, ridging	Increase soil moisture availability to plants
	Level land	Spread water and increase infiltration
	Use deep ploughing	Break up impervious layers or hard-pan to increase infiltration
	Change fallow and mulching practices	Retain moisture or organic matter
	Alter cultivation	Reduce weed infestation
	Switch seasons for cropping	In particular, change timing of cropping to avoid increased drought or to match altered precipitation patterns.
Alter crop husbandry	Alter times of sowing etc. Alter row and plant spacing	Increase root extension to soil water
	Intercropping	Reduce yield variability, maximize use of moisture

Adaptation by Alteration of inputs	Action	Impact
Altered irrigation	Introduce new schemes to dry land areas	Avoid losses due to drought
	Improve irrigation efficiency (e.g. Drip irrigation)	Avoid moisture stress
	Use water harvesting	Increase water availability
	Vary amounts of application	For example, increase NO_2 to take full advantage of CO_2 effects, or decrease to minimize input costs
Alter use of fertilizer	Alter timing of application	Match application to altered patterns of precipitation
Alter use of chemical control.	Vary timing and amounts of application	Avoid pest and weed damage

to southern Africa are shown in Table 11.2. As can be seen from the Table, three categories of choices are possible. These include adaptation by crop choice, adaptation by altered tillage and husbandry, and adaptation by alteration of inputs.

Coping with agricultural crises during drought periods in southern Africa, is generally characterized by a mix of technology, economic and social responses, which can operate best for a short period. Mechanisms used for coping with drought and famine crises locally may be categorized into four groups, namely, agricultural adaptations, finding alternative ways of making money in order to buy food and other essential needs, searching for alternative sources of food, and depending on other people or relief aids.

The possible activities used in obtaining money for food and other essential needs during periods of crisis are very varied in different countries of the subregion. A number of these avenues exist even without the occurrence of climatic crisis, but during a crisis, they become intensified. For example, many of the farmers in drought-prone areas and many members of their families already have secondary occupations that give them money. Wherever possible, whatever the natural resource base could offer (e.g. hunting, fishing, firewood cutting or using grass or wood to make craft products) has been exploited or intensified.

Range land and livestock production

There are some specific adaptation measures for range land and livestock production. For example, in terms of human management decisions, livestock farmers could decide to change (a) the timing, location and duration of management; (b) the mix of grazers and browsers; (c) supplemental feeding; (d) the location of watering points; (e) breeding management; (f) operation production strategies; and (g) market strategies. Selection of the adaptation measures is determined by the impacts of the change in climate.

Planned adaptation in South African countries will be mainly those adaptations which result in government or public policy actions. Such adaptations would include (a) reduction in stock rates; (b) improvement in nutritional plane by using protein, vitamin and mineral supplements; (c) change in mix of grazing and browsing animals; (d) alteration of animal distribution, for example, through watering points; (e) restoration of graded areas; (f) increase in range-land vegetation and/or adapted species; (g) modification of price supports and other governmental program to encourage cattle farmers to respond quickly to climate change; (h) development of breeding program; (i) development of agroforestry systems; and (j) provision of education. Most of these are already practiced in south African countries.

Hydrology and water resources

In southern Africa, as in many other parts of the world, there are only two sources of water—surface and groundwater. In contrast to the two sources of water supplies, there are many avenues of water demand in the domestic sector, the commercial sector, the industrial (primary and secondary) sector, etc. Thus, when addressing adaptation measures in the water resources sector, it is important to recognize that water adaptation can be divided into two major classes, namely, supply and demand.

(a) Supply adaptation measures can take any of the following three forms in Southern Africa:

- Modification of existing physical structures,
- Construction of new structures,
- Alternative management of existing water supply systems.

(b) Demand adaptation measures, which can also take three forms, namely:

- Conservation and improved efficiency,
- Technological change,
- Market/price-driven transfers to other activities.

All the above measures have been applied in different countries of southern Africa in one form or the other and can be more effectively applied in the event of climate change.

Table 11.3 shows adaptation measures to promote conservation and improve water-use efficiency for domestic, agricultural, industrial, and energy uses while Table 11.4. highlights adaptation measures to promote technology change.

Table 11.3: Adaptation Measures to Promote Conservation and Improve water Use Efficiency

Type of water use	Measure
Domestic	Low-flow toilets; low-flow shower; reuse of cooking water; more efficient appliance use; leak repair; water recycling for domestic or other uses; rain-water collection for domestic uses
Agricultural	Night time irrigation, lining of canals, introduction of closed conduits; improvement in measurements to find losses and apply water more efficiently, drainage reuse; use of wastewater effluent; better control and management of supply network
Industrial	Reuse of acceptable quality water, recycling
Energy	Keeping reservoirs at lower head to reduce evaporation; changing releases to match other water uses; taking plants off-line in lowflow times; cogeneration (beneficial use of waste heat)

Table 11.4: Adaptation measures to promote technology change

Type of water use	Measure
Domestic	Water-efficient toilets; water-efficient appliances; water recycling
Agricultural	Low water use crops; high value per water use crops; drip; microspray; low energy; precision application irrigation systems; salt tolerant crops that can use drain water; drainage water mixing stations
Industrial	"Dry" cleaning technologies; closed cycle and/ or air cooling
Energy	Additional run of the river hydropower; more efficient hydropower turbines; alternative thermal cooling systems; cooling ponds; wet tower and dry towers

Alternatives for balancing supply and demand of water can be achieved by any of the following approaches: the do-nothing approach, increasing water supply, and decreasing water demand (Table 11.5).

Table 11.5: Alternatives for balancing supply and demand of water

Do-nothing approach	Increasing water supply	Decreasing water demand
(a) Accept Shortage (unplanned rationing)	(a) Increase system capacity (divert surface water, develop new groundwater supplies, conjunctive use of groundwater and surface water)	(a) Metering
	(b) Improve efficiency (reduce evaporation, detect leaks)	(b) Pricing elasticity
	(c) Modification of weather	(c) Restrictions
	(d) Desalination of water	(d) Educational campaign emphasizing water-use conservation
	(e) Renovation of water	

The "do-nothing" approach implies accepting the problems of water deficits or shortages. Usually in contemporary times this approach is not common in South African countries. Consequently, efforts are always made to adapt by either increasing supply or decreasing demand of water. In many cases, water management adaptations choose a combination of both strategies.

Basically, there are at least five major alternatives of increasing water supply. These include increasing capacity (including interbasin water transfer), improving efficiency, modifying weather, desalination of water, or renovating waste water.

Among adaptation measures that can be used to decrease water demand are metering, pricing policy, restrictions on use and legal measures (Table 11.5). Such measures may be used not only in reducing overall consumption, but also in redistributing patterns of peak system load. Introduction of metering and pricing policy into a water system would no doubt have major effects on the consumption of, and consequently demand for water, as consumers would respond to increase in prices by decreasing or increasing water consumption. However, rate structures inevitably favor, explicitly or implicitly, certain uses or users and water rates involve issues of social equity as well as system efficiency. However in South African countries, any proposals to decrease water demand through metering and pricing policy will always be met with considerable suspicion and probably, resistance. This is understandable because conscious decisions about controlling water demand have important consequences for individuals and for the growth and development patterns of whole sectors of the economy. There is also the possibility of political implications. In spite of this, pricing policy and metering could become important water management tools as they are unique and adjustable in meeting the goals of adaptation measures, and can be implemented without great capital outlay or significantly increased operating costs.

Restrictions and rationing of water, especially in cities, have never been popular in Africa. As a policy tool for adapting to the consequences of climate change, they could be effectively implemented since such measures are quite common locally during serious problems of water shortages. In Botswana, as in other countries of the south African subregion, restrictions have been voluntarily or legally imposed; they may be based on hours of use or types of activity, or confined to peak periods of water demand. Usually in south African countries, communities faced with potential water shortages are quick to formulate and impose program of water-use restrictions. Thus with public education and awareness of the need for restrictions, it may not be too difficult to apply the measures necessary to solve some aspects of the consequences of climate change. In general, restrictions and rationing of water can result in conservation and improved efficiency.

Human Health

Adaptation strategies concerned with human health include administrative or legislative, engineering-based/technological or personal (behavioral). For example, legislation could potentially affect a very large population of the region. Also, technological advances, independent of legislative or administrative mandates, might well bring some substantial benefit to the country. For example, advances in sanitary treatment facilities would prevent an enormous burden of illness in the country. Of course, legislative and administrative mandates are required for the effective dissemination and population-wide adaptation of such technology. Individual preventive responses to health threats are however very unreliable, however, because of widespread misperception among the population of what constitutes relative health risks.

Population level public policy adaptation measures include:

- Reduction of heat-related mortality and morbidity,
- Reduction of transmission of vector-borne diseases,
- Reduction of agricultural stresses,
- Reduction of impacts of extreme weather events and sea-level rise,
- Reduction in general population vulnerability,
- Personal adaptive steps.

Drought and Desertification

Drought is a periodic reduction in moisture availability below average conditions. Simply, drought may be defined as the nonavailability of adequate water for man, animals, and plants. Three types of drought are recognized: meteorological, agricultural and hydrological. This classification is based on the different uses made of precipitating water by man. Drought results in the depletion or exhaustion of soil and shallow groundwater and administers shocks to the ecological system. Droughts have been recurrent phenomena in southern Africa. Countries within the subregion experience considerable distress during drought occurrence. Mass starvation, famine and cessation of economic activities are some of the adverse impacts of draught. Disasters caused by drought are also exacerbated by such diverse factors as poor agricultural practice, increase in population density, and the country's inability to provide alternative supplies of food, water, and employment.

Desertification is a general name given to the processes whereby ecosystems lose the capacity to revive or to repair themselves. This results from interactive processes of drought and man. Climate change might exacerbate desertification through alteration of spatial and temporal patterns in temperature, rainfall, solar insulation, and wind. Arid, semiarid, and dry subhumid areas of southern Africa experience declines in rainfall, resulting in decrease in soil fertility and agricultural, livestock, forest, and range-land production. Ultimately, these adverse impacts lead to socioeconomic and political instability.

In understanding the type of adaptation strategies to use for combating droughts and desertification, it is also significant to note the major processes which result in drought and desertification, which include (a) the destruction or removal of vegetation cover and its nonreplacement, (b) accelerated erosion of soil by wind or water and the consequent exposure of subsoil or rock, (c) soil salinization associated with vegetation loss, (d) waterlogging and salinization associated with irrigation, (e) declining soil fertility, (f) compaction and crusting of soils leading to decreased infiltration and increased runoff, (g) alkalinization of soils, and (h) alteration of microclimates. Most of these processes occur as a result of many causes, both natural and human-induced, some of which are interactive. Thus, apart from adaptation strategies needed for the consequences of climatic variations and variability processes in general, the main adaptation strategies for combating drought and desertification processes in particular, must include (a) strategies for restoration or rehabilitation of overgrazed and degraded dry-land pastures and range-land degradation, (b) prevention of degradation on rain-fed and irrigated agricultural land, (c) minimization and management of biomass burning, (d) minimization of forests and woodland destruction, (e) restoration of degraded forests and woodlands, and (f) social, economic, and political aspects of sustainable management of the dry zones.

Table 11.6: Techniques for controlling erosion by water (from Mannion and Bowlby, 1992)

Mechanical Measures	Biological measures
Bench terraces	Cover cropping
Contour bunds	Mulching
Tie ridging	Contour cultivation
Strip cropping	Minimum and no-till cultivation

Table 11.7: Techniques for controlling erosion by wind (from Mannion and Bowlby, 1992)

Reduction of wind velocity	Reduction of soil erodibility
Vegetation measures	*Manure Conservation*
Cover cropping	Mulching
Close growing crops	Tillage
Sand-dune stabilization (grass and	Timing seedbed preparation
afforestation)	Irrigation
	Terracing
Cultivation measures	Contour cultivation
Mulching	Strip cropping
Rotation grazing	
Crop rotation	*Topsoil conditioning*
Planting crops normal to prevailing winds	
Field and strip cropping	Correct timing of tillage
Primary and secondary tillage	Minimum tillage
	Crop rotation
Mechanical measures	Manuring
Windbreaks	Chemical stabilizers
Shelterbelts	
Dune stabilization by bush matting or	
stones	

A number of adaptation and control measures, including preventive and curative, have been used against wind and water erosion and some have been applied in southern Africa. In general, most of these measures are geared toward soil protection. The categories of adaptation strategies include mechanical and biological techniques. Specifically, some of the strategies include those involving blockage of sand at source and transfer areas and stabilization of sand, sand sheets and dunes. In this regard, conservation techniques such as windbreaks, shelterbelts, fallow, area protection and sand-dune fixation are used. Tables 11.6 and 11.7 show some of the techniques for controlling erosion by water and wind. Some of these measures are already being applied.

Windbreaks consist of linear planting of trees in alternate rows around valleys, fields, orchards etc. Shelterbelts are linear plantations around orchard, valleys and cultivated fields, etc. Windbreaks and shelterbelt measures have been successfully applied in Botswana and can be used in the event of expected climate change. Other techniques, significant for wind erosion include area protection and sand-dune fixation. Table 11.8 shows actions required for prevention and reversal of desertification.

Table 11.8: Action required for prevention and reversal of desertification (adapted from Kemp, 1990)

Prevention		Reversal	
(a)	Good land-use planning and management: (e.g. cultivation only where and when precipitation is adequate; animal population based on carrying capacity of land in very dry years; maintenance of woodland where possible	(a)	Prevention of further soil erosion: e.g. by contour ploughing, gully in-filling, planting or constructing windbreaks
(b)	Irrigation appropriately managed to minimize sedimentation, salinization, and waterlogging	(b)	Reforestation
(c)	Plant breeding for increased drought resistance	(c)	Improve water use: e.g. storage of runoff, well-managed irrigation
(d)	Improved long-range drought forecasting, coupled with social and economic infrastructure to use the forecasts	(d)	Stabilization of moving sand: e.g. using matting, by reestablishment of plant cover; using oil waste mulches and polymer coating,
(e)	Social, cultural and economic controls: e.g. population planning; planned regional economic development education	(e)	Social, cultural, and economic controls: e.g. reduction of grazing animal herd size; population resettlement

Other aspects of strategies for combating drought and desertification include adaptation measures for biomass burning, minimization of forests and woodland destruction, restoration of degraded forests and woodlands, and social, economic and political aspects of sustainable management resources in the dry-land areas.

Implications of Adaptive Strategies

A number of implications will arise from applications of adaptation measures to the impacts of climate change. Such implications will include those related to several aspects of environmental processes and characteristics, including physical, ecological, socioeconomic, financial, and legal aspects.

In order to implement adaptation measures for impacts of climate change, there is urgent need to address a number of mechanisms. These "implementation mechanisms" represent the primary vehicles through which national, regional and international responses to climate and climate systems can be brought into force. These mechanisms include (a) scientific and technological means, (b) data monitoring and collection, (c) information and public awareness, as well as education and training, and (d) financing and other mechanisms.

Conclusion

The impacts of climate change in southern Africa are expected to be both diverse and extensive, and will include alterations of physical, chemical, biological, and socioeconomic elements. Consequently, the environmental and social problems associated with the potential impacts of climate change in the countries within southern Africa may prove to be among the major problems facing the subregion. The consequences may exacerbate the already critical situation in many parts of the region. Timely and effective adaptive strategies to the consequences of climate change are no doubt necessary although there are constraints and problems related to implementation feasibility. These constraints and problems call for priority setting of the adaptation measures and monitoring systems and data collection and some other issues for finding solutions to the various constraints and problems. Following IPCC (1992), these problems and constraints may be categorized into (a) legislative/institutional/organizational, (b) economical/financial, (c) technical, and (d) cultural/social. Within these categories, different problem levels may be distinguished. At present, legislative/ institutional/organizational setups are generally ineffective. The economic and financial resources are inadequate to support production costs. The technologies are also inadequate, while there are many problems with the cultural and social institutions. Clearly the challenges facing the south African subregion on the implications of climate change and the adaptation measures for the impacts of climate change are enormous. There is need therefore to apply effective adaptation measures and specifically set out priorities on adaptation measures, which will prevent or at least reduce the adverse consequences of climate variation and change. No doubt, good policies based on good information lead to sustainable development. It is now urgent for countries in southern Africa to get started and save the future generation from the adverse consequences of the potential climate change.

References

Abrams, L., Short, R. and Evans, J. 1992. Root Cause and Relief Constraint Report. National Consultative Forum on Drought, Johannesburg, Transvaal, South Africa.

AFRA (Association for Rural Advancement) 1993. Drought Relief and Rural Communities, Special Report, No. 9, Pietermaritzburg, Notal, South Afirca.

AGBM 1995. Report of Ad Hoc Group on the Berlin Mandate on the Work of Its First Session. Geneva 21-25 August 1995. [On-line: http://www.unfcc.int/resource/docs/1995/agbm/02.htm]

Agnew, C. T. 1995. Desertification, Drought and Development in the Sahel In :

Binns T. Ed. People and Environment in Africa. John Wiley & Sons, New York, NY.

Browne, J. 2000. Rethinking corporate responsibility: Reflections. SoL Journal 1 (4): 48-53.

Burton, I. 1997. Vulnerability and adaptive response in context of climate change. Climate Change 36: 185-196.

Carter, T. R., Parry, M. L. Harasawa, H. and Nishioka, S. 1994. IPCC Technical Guidelines for Assessing Climate Change Impacts and Adaptations. University College, London, UK.

Collins, C. 1993. Famine defeated, African Recovery 9: 1-12.

Dent, M. C., Schulze, R. E., Wills, H. N. M. and Lynch, S. D. 1987. Spatial and temporal analysis of drought in the summer rainfall region of southern Africa, Water South Africa 13: 37-42.

Desanker P. and Magadza C. 2001. Africa In: Climate Change 2001: Impacts, Adaptation, and Vulnerability. J. J. Mc Carthy, O. F. Canziani, et al. Eds. Cambridge Univ. Press, Cambridge, UK.

Grove A. T. 1990. The Changing Geography of Africa. Oxford Univ. Press, Oxford, UK.

Harrison, M. S. J. 1984. A generalized classification of South African summer rain-bearing synoptic systems. J. Climatology. 4: 547-560.

Harrison, M. S. J. 1986. A synoptic climatology of South African rainfall variations. Unpubl. Ph.D. thesis, University of Witwatersrand, Johannesburg, Transvaal, South Africa.

Hulme, M. (ed.). 1996. Climate Change and Southern Africa: An Exploration of Some Potential Impact and Implications in the SADC Region, Climate Research Unit and WWF, Norwich, Norfolk, England.

IPCC, 1996. Climate Change: Impacts, Adaptation and Mitigation of Climate Change: Scientific-Technical Analyses. Contribution of Working Group II to the Second Assessment Report of the Intergovernmental Panel on Climate Change R.T. Watson, M.C. Zinyowera, M.C. and R. S. Moss (eds.) Cambridge Univ. Press, Cambridge, UK.

IPCC, 2001. Climate Change 2001: Impact, Adaptation, and Vulnerability. Contribution of Working Group II to the Third Assessment Report of the Intergovernmental Panel on Climate Change. J.J., McCarthy O. F., Canziani, N. A., Leary, D. J., Dokken, K. S. White, (eds.). Cambridge Univ. Press, Cambridge, UK.

Kemp, 1990. Global Environmental Issues: A Climatological Approach. Routledge, London.

Lindesay, J. A. 1988. Southern African rainfall, the Southern Hemisphere semi-annual cycle. J. Climatology 8: 17-30.

Mason, S. J. 1990. Temporal variability of sea surface temperatures around southern Africa: a possible forcing mechanism for eighteen-year rainfall oscillation. South African J. Sci. 86: 243-252.

Nicholson, S. E. and Palao, I. M. 1993. A re-evaluation of rainfall variability in the Sahel, Int. J. Climatology, 13: 371-389.

Nwarie, J. 1998. Africa and the reality of capitalism. The Gambia [On-line: http://www.angelfire.com/pq/hippy/africa.html]

Rowlands, I. H. (ed.) 1998. Climate Change Cooperation in Southern Africa. Earthscan Publ. Ltd., London, UK.

Tyson P. D. 1981. Atmospheric circulation variations and the occurrence of extended wet and dry spells over southern Africa. J. Climatology, 1: 115-130.

Tyson P. D. 1986. Climate Change and Variability in Southern Africa. Oxford Univ. Press, Cape Town, South Africa.

United Nations, 1997. Kyoto Protocol to the United Nations Framework Convention on Climate Change. Kyoto, Japan [On-line: http://www.unfcc.int/text/resource/docs/cop3/07a01.pdf.]

van der Veer, J. 1999. Profits and principles, the experiences of an industry leader. In: Proc. Greenport'99 Conference, April 1999. World Business Council for Sustainable Development, Geneva, Switzerland [Online http://www.wbcsd.ch/Speech/s73.htm.]

Vogel, C. H. 1994. Mismanagement of droughts in South Africa, past, present and future. South African J. Sci. 90, 4-5.

Warren, A and Khogali, M. 1992. Assessment of Desertification and Drought in Sudano-Sahelian Region 1985-1991. UNSO, New York, NY.

Watson, R. T., Dixon, J. A., Hamburg, S. P., Janetos, A. C., and Moss R. H. 1998. Protecting Our Planet, Securing Our Future. UN Environment Programme, US National Aeronautics and Space Administration, and The World Bank, Washington, DC, USA.

Watson, R. T., Zinyowera, M. C. and Moss, R. H. (eds.) 1997. Summary for policymakers, The Regional Impacts of Climate Change: An Assessment of Vulnerability. Special Report IPCC Working Group II, published for the Intergovernmental Panel on Climate Change. [On-line: http://www.ipcc.ch]

UNDP, 1999. Human Development Report, 1999. Oxford Univ. Press, New York, NY.

UNFCCC, 1992. United Nations Framework Convention on Climate Change, Reprinted in International Legal Materials, vol 31, pp. 849-71.

Wilhite, D.A. 1993. Editorial. Drought Network News 6 (1): 1-4.

Wilhite, D. A. 1997. Responding to drought: Common threads from the past, visions for the future: J. Amer. Water Resources Assoc., 33(5): 951-959.

World Bank. 1998. World Development Indicators. World Bank, Washington, DC.

World Meteorological Organization. 1999. Weather, Climate and Health. WMO Special Publication, No. 892.

World Resource Institute. 1996. World Resources 1996-97. Oxford Univ. Press, London, UK.

CHAPTER 12

Climate Change, Vulnerability and Adaptation in Bangladesh

Saleemul Huq[a] and Khondkar Moinuddin[b]
[a] Chairman, Bangladesh Centre for Advanced Studies, International Institute for Environment and Development, London, UK.
[b] Bangladesh Centre for Advanced Studies, Dhaka, Bangladesh

Bangladesh lies in the eastern part of South Asia latitudinally between 20°34′ and 26°33′ N and longitudinally between 88°01′ and 92°41′ E (Fig. 12.1). The geographic area of the country covers 1,47,570 km². It is surrounded by India on three sides—west, north and east and has a common border with Myanmar in the southeast. The Bay of Bengal lies south of Bangladesh.

Three of the world's largest rivers (Ganges, Brahmaputra and Meghna) flow through the country and form the largest delta in the world. As part of the Bengal basin, the territory of Bangladesh has been filled by sediments washed down from the highlands, especially from the Himalayas where the slopes are steeper and the rocks less consolidated (Haroun, 1984). Floodplains of the major rivers, occupying 80% of the country's geographic area, are generally smooth relief comprising broad and narrow ridges (former river levels) and depressions.

Population

According to the latest census (2001), the population of Bangladesh is about 129 million. It is the most densely populated country in the world with 965 people per km². The present population growth rate (2001) is 1.42% compared to 2.17% a decade earlier (1991). About 42% of the population is below 15 years of age meaning that it has a high growth potential. The crude birth rate (per thousand population) declined from 32 in 1990 to 21 in 1997 and crude death rate (CDR) reduced from 11.3. to 5.5 during the same period. Total fertility rate (TFR) has reduced substantially, from 4.33 (1990) to 3.10 (1997). An intensive family planning programme over the years has contributed to the reduction in population growth.

Infant mortality rate has substantially reduced from 94 (per 1000 live births) in 1990 to 57 in 1997. The nationwide immunization program implemented with external assistance over the years is aimed at protecting children from common diseases.

Social Development

Life expectancy (at birth) in Bangladesh increased from 48 years in 1980 to 59 years in 1998 and to 61 years in 2000. Although there has been significant progress in providing safe water to the people, the majority of the people in urban areas do not have access to safe water. The percentage of urban population having access to safe water increased to 43% during 1990-96 compared to 29% during 1980-82. The massive influx of rural poor people to the cities has led to the rapid growth of urban slums having little or no water supply and sanitation facilities. In rural areas 85% of the population had access to clean water in 1990-96 compared to 43% in 1980-82 (World Indicators). According to recent government statistics, 97.90% of the population had access to clean water (1999). Access to sanitation recorded a rise from only 4% (1982-85) to 35% of the total population in 1990-96. Recent government statistics show a further improvement in access to sanitation by 40% of the population in 1999.

Literacy

In recognition of the importance of human resources in economic growth and development, greater emphasis has been given to education and skill development under the development plans and programs. Public sector development budgets have gradually allocated a higher share for the education sector, especially for primary education. As a result, school enrollment has significantly increased. Primary school enrollment ratio increased from 74% in 1995 to 82% in 1997. The enrollment ratio for males increased from 81% to 84% and for female from 73% to 79% between 1995 and 1997. The introduction of nonformal education and adult literacy program has also been a major success. Apart from the government, NGOs are playing an important role implementing nonformal and adult literacy programs.

Natural Resources

(i) Land

Formed by three mighty rivers – the Ganges, the Brahmaputra and the Meghna—the Bengal (Bangladesh) basin is one of the youngest and most active deltas in the world. The annual sediment load of the Ganges, Brahmaputra, Meghna river systems is estimated to be 2.4 billion tons of cobbles, sand, and silt (Rashid and Paul, 1987). The deltaic floodplain of Bangladesh has a wide range of soil types. The land system in Bangladesh generally consists of three land types: (a) floodplain, (b) terrace and (c) hills. This floodplain, covering 80% of the land area, is crisscrossed by innumerable tributaries and distributaries of the three major river systems. The terrace area consists of just 8% of the total land surface at an elevation of 10 to 20 feet above the surrounding floodplain. Hill area consists of northern and eastern hills which together cover about 12% of the total land area (Fig. 12.1).

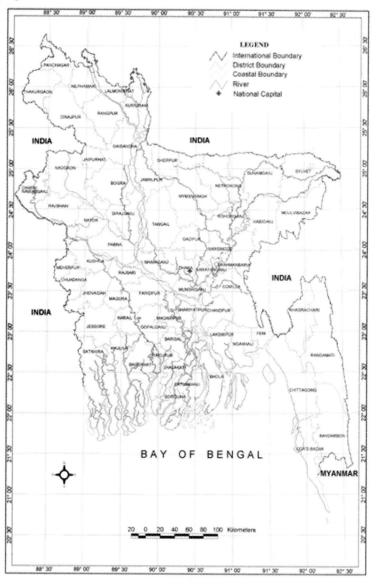

FIG. 12.1: Map of Bangladesh

Water Regime

A huge amount of water equivalent to 953 million acre feet flows through Bangladesh. About 90% of the water flows from the upstream rivers originating from other countries of the region. This vast outflow is second only to that of the Amazon system (Haroun, 1984). The country also has a reasonably good groundwater resource. Heavy rainfall and annual inundation contribute to substantial recharging of groundwater annually.

Forests

The major forest areas of Bangladesh comprise the Sundarbans, the Chittagong Hill Tracts, and deciduous forests. The Sundarban mangrove forest occupying 5,500 km^2 is the largest patch of productive mangrove forest in the world. A part of the Sundarban forest has been declared a world heritage site by the UNESCO in view of its rich biodiversity. The Hill Tracts house long-rotation high-value timber species. These are now being rapidly depleted due to increasing demand for timber woods. The deciduous forest in central Bangladesh is an important source of commercial timber and fuel wood. It is also under heavy pressure due to expansion of agriculture and industrial areas.

Biodiversity

Although Bangladesh is geographically a small country, it is rich in biodiversity. Its terrestrial and aquatic areas support a good variety of biological population in terms of both flora and fauna.

Climate

Like all tropical countries, Bangladesh has a hot and humid climate during most of the year and a mild winter stretching from late November to early March. The climate is fairly uniform throughout the country. There are basically three seasons: summer, rainy season and winter. Summer stretches from March to June with hot and high humid weather. The monsoon season lasts from June to October with heavy rainfall. The country experiences relatively cooler and drier weather from November to March. Annual rainfall ranges from 1,200 mm to 2,550 mm, on average. There are significant spatial and temporal variations in rainfall—the eastern zone having higher rainfall than the western and 80% of the rainfall occurring in five months, between June and October. Winter evaporation appears to be higher than available moisture, resulting in an acute shortage of topsoil moisture.

System of Governance

The constitution of Bangladesh provides for a unitary and parliamentary system of government. The fundamental rights of the citizens and the role of the government have been enshrined in the constitution, which acknowledges the supremacy of the people in all affairs of the state.

Although due emphasis on democracy and social justice has been given in the constitution, there have been aberrations from the stated principles in the real world. The introduction of a one-party rule, abolishing the multiparty democratic system, the change-over of power through violent means in 1975, and the subsequent military rules especially during the 1980s, caused major obstacles to democratic governance and people participation in the affairs of state and government. The constitution was also changed during this period a number of times to suit the needs and aspirations of the ruling class, disregarding the will of the people. The general masses, intelligentsia, and civil society, however, have continued to demand the return to a multiparty democratic system and pro-people reforms leading to good governance. Moreover, the development partners and donors have emphasised the need for political reforms for the sake of socioeconomic development of the country which has helped to build a general consensus for return to a parliamentary form of democratic government.

Since the early 1990s the country has been holding general elections with broad participation of all its major political parties. Already three parliamentary elections have been held, including the most recent in 2001, in a free and fair manner under a neutral and caretaker government.

In spite of restoration of democracy, changeover of government through free, fair and democratic elections, the parliament has not been fully effective. The long boycott by the opposition from the parliament and the lack of cooperation and consensus between the opposition and the ruling parties over trifle matters appear as serious problems. Many important decisions are delayed and reforms not implemented. The people, including civil society and conscientious citizens as well as the development partners, continue to voice their concern arising out of the non-cooperation and negative attitudes of the political parties.

For administrative purposes, apart from the Ministries and Departments, at the national level the country is divided into six divisions and 64 districts, each district divided on average into 5 to 6 subdistricts or upazilas, each comprising 5 to 10 unions which are the lowest level local bodies in the rural areas. Municipalities or pourashovas are urban counterparts of Union Parishads. The local government bodies at the union and pourashavas levels are directly elected by the people. There are no representative local bodies at the upazila, district, and divisional levels; these are administered by trained officials appointed by the central government. Although attempts have been made in the past to elect local bodies at the upazila level, this could not be given an institutional shape on a sustainable basis due to lack of strong political commitment from the central government. There is, however, a strong likelihood of representative local government bodies at the upazila and district levels in the near future, given the growing demand for people participation and decentralization of the political and administrative system. Representative local government bodies, however, operate in the municipalities or pourashavas as well as the city corporations including the capital city of Dhaka.

Size of Economy and Growth Trend

Bangladesh with a per capita GNP of $ 350 in 1998, ranked 173 among the 206 economics included in the World Development Indicators (World Bank, 2000). The country, however, ranked 53 in terms of GNP at $ 44.2 billion in the same year.

Annual growth of GDP and per capita GDP over the long term (1965–1998) recorded a growth of 3.9% and 1.4% respectively. Private consumption increased by 3.7% per annum and gross domestic fixed investment also increased at an equal rate (3.7%) during the same period (1965 – 98).

The recent growth performance of the economy is marked by some improvement during the 1990s compared to the preceding decade and the long-term trends.

Table 12.1 shows the trend of growth rate in GDP.

Table 12.1: Growth rate of GDP in Bangladesh

Period	Annual GDP growth rate (%)
1981-90	3.84
1991-95	4.06
1996-2000	5.76

Source: Economic Survey 2001, Ministry of Finance, Government of Bangladesh.

The growth rate in GDP was 6.7% in 2000. According to the recently published document of the Ministry of Finance, the per capita GDP increased to $ 369 in 2000. The per capita Gross National Income (GNI) increased to $ 377. The reason for GNI or GNP being greater than GDP is the net factor income from abroad, especially through remittance of wage income.

Structural Change in GDP

The composition of GDP has undergone significant changes over the past two decades. The share of agriculture in CDP, although it remains the single largest sector, has steadily declined over the years. Agriculture now contributes 19.49% to GDP (1999/2000) compared to 24.66% a decade earlier (1990/91). According to the World Development Indicators (World Bank), the share of agriculture in GDP declined from 42% in 1970 to 22% in 1998. Agriculture absorbed 81% of the total labor force in 1970, which declined to 65% in 1998.

Value additions from industrial output and services sector accounted for 24.0% and 38% of GDP respectively in 1980 but their shares increased to 28.0% and 50.0% respectively in 1998. In other words, the declining share of agriculture has been matched by rising shares of industrial and services sector.

Key Sectors

Agriculture

Agriculture, despite its reduced share in GDP, is one of the key sectors of the economy. Rice is the principal crop and occupies more than 80% of the total cultivable land. Production of rice has increased significantly over the years due to the efforts under the national development plans and of programmers to attain self-sufficiency and food security for the vast population, especially the poor. Rice production increased from 17.86 Mt in 1990 to 24.90 Mt in 2000. Although wheat contributes a small share in total food grain production, there has been rapid growth in its production relative to rice—more than twofold from 0.89 Mt to 1.82 Mt between 1990 to 1998. This growth in food grain production has been accompanied by a rise in use of chemical fertilizers and expansion of irrigation. Use of chemical fertilizer per hectare increased from 495 kg in 1981 to 1,453 kg in 1997. Intensive cultivation using chemical fertilizers is believed to have a negative impact on the natural productivity of soils, howerer. There has also been a substantial rise in the production of potatoes, from 1.06 Mt in 1990 to 1.55 Mt in 1998. While the production of rice, wheat, and potatoes has gone up, production of other crops has either stagnated or declined over the years. Production of pulse, jute, sugarcane and some vegetable has declined due to reduction in acreage under these crops. Cultivable land as a percentage of total land reduced from 63.3% in 1980 to 60.8% in 1997. The scarcity of land and increasing demand for housing and industrial sectors are the principal factors for reduction in cultivable land.

Industrial Production

The quantum index of industrial production shows that there has been rapid growth in the production of garments, textile dyeing and printing, books and periodicals, pharmaceuticals etc. Production of garments increased more than sevenfold during the decade 1988 to 1998. The quantum index of pharmaceuticals increased to 312.59 (1990 – 99 = 100).

A marked decline in the manufacture of machinery and equipment has occurred over the past decade, with a quantum index of 50 in 1998 compared to the base year 1988. There has also been a reduction in the production of cotton textile and jute textile, the quantum index of which declined to 72.19 in 1998/99 compared to the base year (1980-81 = 100).

Overall industrial production has, however, steadily increased over the years. The quantum of industrial index went up to 204 in 1998 compared to the base year (1988 = 100) (Table 12.2).

Table 12.2: Growth of industrial production during the 1990s

(Base: 1988 = 100)

Year	Quantum index of industrial production
1992	141.81
1993	153.89
1994	163.33
1995	173.50
1996	179.30
1997	195.95
1998	204.17

Source: Bangladesh Bureau of Statistics, Statistical Yearbook 1999.

Foreign Trade

Liberalization of the economy and the impact of globalization has led to significant rise in foreign trade over the years. Foreign trade (including both exports and imports) as percent of GDP increased from 17% in 1980 to 33% in 1998 (World Bank). The country has faced an unfavorable trade balance over the decades owing to its limited export base and heavy dependence on imports of both consumer and capital goods as well as raw materials. In recent years the trade deficit as percentage of GDP has reduced from 5.2 in 1995/96 to 2.0 in 1999/2000 due to faster growth of export than import. In absolute terms, the trade gap, however, increased from US $ 1.5 billion in 1991 to $ 2.7 billion in 1998 (Export Promotion Bureau, 1999). The terms of trade have also been unfavorable for Bangladesh as the index of export price has lagged behind the index of import price. The index of export price increased to 141.3 while the index of import price increased to 158.0 in 1999/2000 compared to the base year (1981/82 = 100).

Export

Earnings from export totaled $ 5.8 billion in 1999-2000 compared to $ 4.42 billion in 1998/99. The export of ready-made garments comprising both oven apparels and knitwears stood at $ 4 billion which accounted for about 76% of the total export in 1999-2000. The value of garment exports increased from less than a billion dollars in 1990/91 to more than four billion dollars in 1999/2000. Apart from ready-made garments, jute and frozen food (shrimp) are two important exportables of Bangladesh. In 1999-2000, jute and frozen food accounted for about 5.9% and 6% of the total export respectively. The share of jute in total export has declined over the years. Jute and jute goods, which contributed more than 50% to the country's export earning until 1985/86, now account for less than 6% in total export. Leather and tea, two other traditional exportables of Bangladesh, now contribute to export earnings some 3.4 % and 0.5% respectively (Economic Survey 2001).

Remittance

It may be noted that a large contingent of Bangladeshis, most unskilled laborers, are working abroad and regularly remit their hard-earned cash back to Bangladesh. Export of manpower is one of the principal sources of foreign exchange earnings. In 1990-2000 fiscal year the total remittance stood at US $1949 million, some 14.2% higher than in the previous fiscal year.

Inflation Rate

In spite of moderate annual fluctuations in the price level of various commodities and consumer items, the inflation rate has been contained well below 10% over the decade. It fell from 8.9% in 1998/99 to 3.4% in 1999/2000. Both food and nonfood price indices have followed similar trends over the past decade. The consecutive good harvests of rice and other crops have been the primary factor for the stable price of food grain and low inflation rate.

Foreign Development Aid and Debt

The experience of the country with foreign aid spans over three decades. Although the fact that the country is rated as a moderately indebted one by the international community (WB/ADB, 1998), by now the debt accumulated by successive governments, has snowballed to a colossal sum. External debt increased from $ 4.23 billion in 1980 to $ 16.38 billion in 1998. It is about time to start repaying these debts with interests. Unfortunately, the debt burden is getting to a point where its servicing will seriously undermine the country's ability to repay the debts (Ali, 2001).

Poverty

In spite of progress achieved in macroeconomic stability and social conditions, reduction of poverty incidence and improvement of living conditions continue to be a primary challenge for Bangladesh. The incidence of national poverty (head-count index based on consumption expenditure data) has, of course, declined by 5.4 percentage points (from 58.50 to 53.08) in a twelve-year period between 1983/84 and 1995/96 (World Bank, 1998). This implies an annual rate of reduction in the order of about 0.5%. However, compared with other countries, the pace of progress in reducing poverty has remained very slow. Even this slow pace may not be sustainable as the possibility of reversal remains high. Any bad harvest or sudden flooding can subvert the trend. The absolute number of poor, in fact, increased from 49 million in 1983/84 to 54 million in 1995/96. Slow economic growth and rising income disparity are the primary reasons for the slow progress in poverty reduction. Slow poverty reduction has also been the result of the low level of human development and the poor environment.

Contribution to Climate Change

As a least-developed country Bangladesh makes minimal contribution to global emission of greenhouse gases (GHGs) through various anthropogenic activities. The sectoral economic activities which contribute to GHGs include energy, agriculture, forestry, transportation, and households. Although the global energy system is responsible for about 80% of the global CO_2 emissions and 65% of all greenhouse gases (Berk et al., 2001), the energy sector of Bangladesh contributes only 30% of its total emissions of GHGs.

Energy Sector of Bangladesh

Bangladesh has one of the lowest per-capita energy consumptions in the world. Per-capita energy consumption in 1997 was 197 kg of oil equivalent in Bangladesh compared to 500 kg in the low-income countries of the world. Total energy production and use were 21,894 and 24,327 thousand tons of oil equivalent respectively in 1997. The country imported 10% of the total energy consumption in 1997. Total energy consumption increased at an annual average rate of 3.1% and per-capita energy consumption increased by 0.9% per annum during 1980-97 period (World Development Indicators, World Bank). Total energy and per-capita energy consumption for the low-income countries of the world increased by 3.9% and 2.0% during the same period. Per-capita annual growth in Bangladesh was only half of the growth in low-income countries.

Traditional energy comprising fuel wood, crop residue, bagasse, animal and vegetable waste accounted for 43% of the total energy in 1997, down from 81% in 1998. The share of commercial energy comprising natural gas, oil, coal and hydel is on the rise as the supply of traditional energy sources shrinks. Natural gas is the principal indigenous energy resource accounting for 62% of the total commercial energy in 1990 (National Energy Policy, 1990).

The share of natural gas in commercial energy is on the rise and estimated at 75% in 2001. Petroleum oil is the other main energy which is almost entirely imported into the country. Hydro electricity and coal constitute less than 3% of the total commercial energy. The lion's share (50%) of indigenous natural gas is used for power generation followed by industrial fertilizer production (25%).

Electricity in Bangladesh is mainly produced from natural gases. About 85% of the total electricity was produced from natural gas in 1997 compared to 47% in 1980. The share of oil in electricity production declined from [26.6% in 1980 to only 9.4% in 1980] whereas the share of hydropower declined from 24% to 6% during the same period. Despite 10% increase in electricity production per annum, per-capita consumption is only 76 KWh in Bangladesh compared to 448 KWh in the low-income countries and 2,053 KWh in the world.

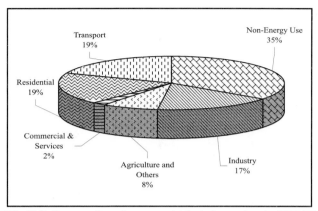

FIG. 13.2: Consumption of commercial fuels by sectors in 1994-95

The consumption pattern of energy shows that nonenergy use (e.g. fertilizer production) accounts for a significant share of the total commercial energy. In 1994-95, 35% of the total commercial energy was used for non-energy purposes, followed by residential and transport sectors, which accounted for 19% and industry 17% respectively (Fig. 12.2). Agriculture used 8% while the commercial and service sectors consumed only 2% of the total commercial energy (Alam, 2000). Transport and household sectors have experienced a rapid growth in energy consumption compared to other sectors over the years.

Emission of Greenhouse Gases (GHGs)

According to the World Development Indicators, the total emission of CO_2 in Bangladesh was 23 Mt in 1997, up from 7.6 Mt in 1980. Per-capita emission of CO_2 was 0.2 ton in Bangladesh, compared to 1.6 ton in the low-income countries and 4.0 tons in the world in 1996. Energy Efficiency (GDP per Kgoe) improved in Bangladesh from 2.9 in 1980 to 6.8 in 1997.

According to the ALGAS (Asian Least-cost Greenhouse Gas Abatement Strategy) study per-capita net emission of GHGs for Bangladesh in terms of CO_2 equivalent was 0.67 ton in 1990 and a total net emission of 7,200 kilo tons excluding traditional biomass burning (Table 12.3).

The energy sector (total combustion plus fugitive) contributed 29.43% to total GHGs for the year 1990. Emissions due to fuel combustion are attributable to energy transformation (6.10%), household (7.67%), and transport (2.60%) (Fig. 12.3).

The agriculture and livestock sector contributed 39.62% GHGs to total emissions. The relatively high emission of the agricultural sector is due to the emission of methane from rice fields which occupy 80% of the cultivable area of Bangladesh. Of the 37% emission from the agriculture sector, 22% is due to rice cultivation under rain-fed conditions and use of fertilizers and

Table 12.3: Bangladesh national GHGs inventory in 1990 (kt)

Source and Sink	CO$_2$ Emission	CO$_2$ Removal	Net CO$_2$	CH$_4$ Gg	N$_2$O Gg	NO$_1$ Gg	CO Gg	CO$_2$-Equipment (excluding CO$_2$ emissions from (TBB))	Percent of total CO$_2$ equivalent
Total (Net) National Emissions	39,900	5,809	34,092	1,739	4.51	203	4,309	72,000	100.00%
1. All Energy (Fuel Combustion + Fugitive)	12,863		12,863	331	4.40	200	4,205	21,186	29.43%
A. Fuel Combustion									
1. Energy and Transformation Industries	4,392		4,392			40	5.00	4,392	6.10%
2. Industry	2,420		2,420	24	0.44	16	394	3,050	4.24%
3. Transport	1,875		1,875			15	7	1,875	2.60%
4. Commercial Institutional	239		239	1	0.01	0.2		259	0.36%
5. Agriculture	680		680					680	0.94%
6. Residential	2,082		2,082	138	1.75	63	1,699	5,523	7.67%
7. Others (please specify)	400		400					400	0.56%
8. Traditional Biomass Burned for Energy	62,084	62,084		162	2.20	79	2,100	4,084	5.67%
B. Fugitive Fuel Emissions				7					
1. Oil and Natural Gas Systems								149	0.21%
2. Coal Mining (NA)									
Statistical Difference	775		775						
2. Industrial Processes	1,491		1,491				6.49	1,491	2.07%
A. Comment Production	153		153					153	0.21%
B. Others									
1. Ammonia production	1,130		1,130				6.48	1,130	1.57%
2. Metal (iron & steel)	208		208			0.01	0.01	208	0.29%

Table 12.3: (Contd.)

Source and Sink	CO₂ Emission	CO₂ Removal	Net CO₂	CH₄ Gg	N₂O Gg	NO₁ Gg	CO Gg	CO₁-Equipment (excluding CO₂ emissions from (TBB))	Percent of total CO₂ equivalent
3. Agriculture	2,384	2,384		1,363	0.11	3.84	97.30	28,667	39.82%
A. Enteric Fermentation				519				10,892	15.13%
B. Manure Management				73				1,534	2.13%
C. Rice Cultivation				767				16,107	22.37%
D. Agriculture Soils									
E. Prescribed Burning of Savannas (NA)									
F. Field Burning of Agricultural Residues	2,3841	2,3841		5	0.11	3.84	97	133	0.18%
G. Others (please specify)									
4. Land-use Change and Forestry	23,162	3,425	19,738					19,738	27.41%
A. Changes in Forest & Other Weedy Biomass Stocks	21,391	3,326	18,066					18,066	25.09%
B. Forest and Grassland Conversion (NA)	1,771		1,771					1,771	2.46%
C. Abandonment of Managed Lands (NA)		99	-99					-.99	- 0.14%
D. Others (please specify)									
5. Waste				44				918	1.27%
A. Solid Waste Disposal on Land				44				918	1.27%
B. Wastewater Treatment									
C. Others (please specify).									

Notes: CO₂ emissions from traditional biomass burning (TBB) are not included in subtotals and the national total. CO₂ equivalents based on global warming potentials (GWP) of 21 for CH₄ and 310 for N₂O, NO₂, and CO are not included since GWPs have not been developed for these gases.

Bunker fuel emissions have already been accounted for in the energy sector as per IPCG guidelines.

a. Difference between reference approach and detailed approach equal to (12,863-12,088) = 775 G₅'

Source: Asia Least-cost Greenhouse Gas Abatement Strategy, ALGAS, Bangladesh

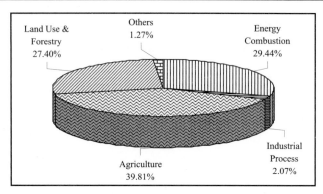

FIG. 12.3: Sectoral Emission of GHGs (1990)

15% due to enteric fermentation from livestock including cows and goats. Much of the methane (CH_4) emission (82%) in Bangladesh comes from rice cultivation and livestock management. The high percentage of methane in total emissions from agricultural activities (rice cultivation) signifies the subsistence nature of the economy.

There was a net emission from the forestry sector. Forestry and land-use change contribute a net emission of CO_2 as the rate of deforestation exceeds the rate of afforestation and new plantation. In 1990, the net contribution to total emissions by forestry and land use change was 27.41% or 19,738 kt of CO_2 equivalent.

Projections of Future Emissions
Future emissions of GHGs will be affected by population growth, growth of GDP, structural change in the economy, and the demand-supply balance in the energy sector. In addition, future emissions will also depend on the trend in energy efficiency.

The future projection of GHGs up to the year 2020 by the ALGAS study assumes that the population of the country will increase from 109 million in 1990 to 177 million in 2020. In addition, projection of sectoral shares of GDP up to 2020 is assumed to be as shown in Table 12.4.

Table 12.4:

Sector	1990	2000	2010	2020
Agriculture	38	30	26	11
Industry	10	12	16	25
Service	32	38	43	44
Others	20	20	20	20

Based on the above assumptions, the total emission of GHGs is projected to be 1,22,000 kiloton, of which 54,300 kt will come from energy and

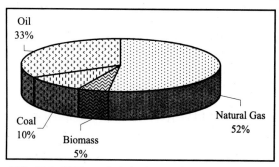

FIG. 12.4. GHGs emissions from different types of fuels

transformation industry, 63,300 from the demand subsector, 6,600 from traditional biomass for burning energy, and the remainder from fugitive fuel emission (Fig. 12.4).

There will be a sevenfold increase in emission from the energy sector in the year 2020 compared to the base year 1990. The reason behind this large increase is the expansion of the industrial and electricity generation subsectors. The share of natural gas in total emission from different types of fuel will increase from 44% in 1990 to 52% in 2020 and the share of coal will rise from 5 to 10% between 1990 to 2020. The share of biomass will decrease from 23% to 5% by the year 2020 (Fig. 12.5).

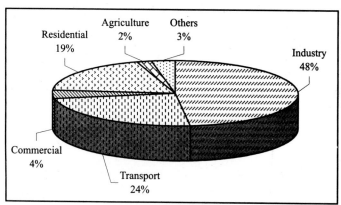

FIG. 12.5: Energy demand GHGs inventory by sector in CO_2 equivalent in 2020

With regard to sectoral emissions, the share of agriculture in total emission will decline from 6% to 2% between 1990 to 2020. The share of the residential sector will decline from 47% to 19%. A sharp rise in emissions will occur from transport and industrial sectors. The share of the industrial sector in total emissions will increase to 48% in 2020 from 26% in 1990 and that of the transport sector will rise to 24% from 16% between the years 1990 and 2020.

Vulnerability to Climate Change

The geographical location and sociodemographic feature of Bangladesh makes it one of the most vulnerable countries to climate change 12.6. Its long coast line, vast low-lying landmass, high population density and nature-dependent traditional agricultural practices would be impacted upon due to climate change. It is likely that vulnerability of the disadvantaged and poor community would be worse than the nonpoor and better-off strata

Prepared by: GIS Division, BCAS.

FIG. 12.6: Areas vulnerable to climate change in Bangladesh

of the society. The issue of climate change relating to vulnerability and adaptation for Bangladesh has been addressed through several studies.

Climate change scenarios were developed for Bangladesh using a General Circulation Model (GCM) for two projected years 2030 and 2075 with 1990 as the base year. The GCM outputs in combination with the observed time series data reveal that there would be more precipitation in the summer-monsoon (July-September) while winter (November-February) precipitation would decrease considerably (Alam, 1999). Average temperature would increase by 1.3°C and 2.1.6°C for the years 2030 and 2075 respectively. It was found that precipitation would decrease at a negligible rate in 2030 and there would be no appreciable rainfall in the winter in 2075. There would be excessive rainfall in the monsoon causing flooding and very little to no rainfall in the winter leading to drought. The GCM outputs lead to the conclusion that moderate changes regarding climate parameters would take place for the projected year 2030, while severe change would occur in the year 2075.

Impact on Crop sector

The impacts of climate change on production of rice and wheat, the staple food for the country's 130 million population, were assessed by studies using specific models developed for the purpose. One such study (Karim et al., 1999) led to the conclusion that production of rice would reduce for all the three seasonal categories e.g. Aus, Aman and Boro. Production of Aus rice would reduce by 19% for a rise in temperature of 2°C and reduce by 38% for a rise of 4°C. Production of Boro rice would go down by 4% and 7% for increase in temperature of 2°C and 7°C respectively. Production of HYV Aman rice would reduce by 13% for a rise of 2°C and by 25% for a rise of 4°C. The adverse impact of global warming and climate changes would be more severe on wheat than on rice. For a 2°C rise of temperature, wheat production would reduce by 37% and for 4°C the loss would be 68%. Although the extent of losses in production differs with the models applied, they are sufficiently substantial by all accounts to affect the livelihood of the rural poor who are dependent on agriculture and generally have no alternative source of income or living.

Soil Salinity and Erosion

Climate change and allied effects including sea-level rise would lead to soil salinity and consequent loss of agricultural crops (Karim, 1999). It was found that under moderate climate change scenario, 10% of the present nonsaline area would transform into saline area whereas under severe climate scenario, 45% of the present nonsaline area would transform into saline area in the coastal region. The loss of crop and food grain production would be 2 to 4 times higher under severe climate change scenario.

The impacts of climate change induced sea-level rise on beach erosion and land loss estimated using Brunn's formula found that 5,800 hectares would be lost and 11,200 hectare recessed in the long-term along the eastern coast line of Bangladesh (Huq. et al.). About 95% of the land exposed to the risk of erosion is currently under cultivation of various crops. The loss of agricultural land would further aggravate the poverty situation in the densely populated coastal area.

Impacts on Forest

The forest resources of Bangladesh, especially the Sundarban Mangrove forest, is likely to be adversely affected due to the impact of climate change. Climate change would be accompanied by an increase in rainfall in the summer leading to higher flood incidence. As a consequence, there could be relatively prolonged inundation in the Sundarban areas in the monsoon months (July-September). A different environmental condition, on the other hand, is likely to prevail in the winter months. There would be significant lowering of freshwater discharge in the rivers coupled with high rate of evaporation. As a consequence, salinity intrusion into the forest would increase. It is likely that increased salinity would have discernible adverse effects on forest regeneration and succession. Experts are specially concerned that the productivity of the mangrove forest might decline as more grassy plants may replace woody trees (Ahshan, 1999).

Effects of Sea-Level Rise

Sea-level rise would have far-reaching adverse effects on the economy, infrastructure, ecosystem, production practices, and livelihood system of the people of Bangladesh. It has been estimated that more than 25,000 km^2 (about 18% of the total area of Bangladesh) would be affected by inundation due to sea-level rise. About 80 to 100% of the land area of some southwestern coastal districts (Patuakhali, Barisal, Noakhali, Comilla) would be under threat of a one-meter sea-level rise. Other infrastructures, including 850 km of road, 28 km of railway, 85 small towns/municipalities, 4000 km of coastal embankment, and one seaport would be affected by the event of one-meter sea-level rise (Ali and Huq). Other adverse effects would include increased flooding, salinity intrusion, frequent tidal surges in the cyclone-prone coastal zone, etc.

Effects on Fisheries

Although the impacts of climate change on marine and estuarine fisheries are not explicitly known, freshwater fisheries are likely to be negatively affected due to climate change. Sea-level rise and intrusion of saline water into rivers and other water bodies in the coastal region would lead to a deterioration of the fish habitats for freshwater capture and culture fisheries

(Ali, 1997). As a result, production of freshwater fisheries is expected to decline in the coastal zone of Bangladesh.

International negotiations and what needs to be done

Bangladesh has been taking a somewhat passive posture in the international negotiations on climate change. As a member of the Group of Seventy-Seven (G77) it takes a common position with the rest of the G77 group but seldom takes a leadership role on any issue.

So, under these circumstances what is there that Bangladesh can actually do either to prevent the worst from happening or coping with it when it happens? The answer is indeed we can much do if we put our minds to it and start developing a strategy and implementing it right now. The major elements of such a strategy could be as follows: .

Recognizing the scale of the problem: The first step is for our leaders and policymakers to recognize the scale of the problem and how it is likely to affect us. Unfortunately, while the rest of the world has recognized that Bangladesh is going to be one of the worst affected countries, our own leaders do not seem to have grasped the scale of the problem that is likely to be created. The recent IPCC/TAR report and the UN Secretary General's visit and warning will (hopefully) have alerted them to this issue.

Realizing that it is a multisectoral developmental problem: So far the issue of climate change has been considered an environmental one, left to the Ministry and Department of Environment to handle. This includes attending the Conference of Parties (COP) of the United Nations Framework Convention on Climate Change (UNFCCC). This is clearly a mistake as the problem is not just an environmental one, but also a developmental one, which requires the attention of all relevant ministries and departments. Thus for example, the Ministry of Finance needs to be involved (especially with regard to possible investment opportunities through Carbon Trading), Ministry of Planning (in order to incorporate climate change impacts into national level planning, which is not done at present), Ministry of Water Resources (as water flows are likely to be one of the most adversely impacted by climate change, leading to both increased drought as well as flood), Ministry of Relief (as there will be increased likelihood of more frequent and more severe cyclones and floods), Ministry of Agriculture (as the entire coastal zone is likely to become increasingly more saline), and Ministry of Foreign Affairs (as the international negotiations need capable and senior diplomats to conduct them) as well as others.

Creating a high-level, multisectoral, multistakeholder body: Once the scale of the problem and its multisectoral nature is recognized by the highest policy-makers, they may set up a suitable body drawing on the different ministries (including those mentioned above) as well as others from NGOs and universities who have knowledge and experience of these issues. There already exists an Inter-Ministerial Committee on Climate Change headed by the Minister for Environment but it has not met for a long time and in reality is moribund. It may well be revived to serve this purpose.

Develop analytical and scientific capacity and knowledge: Despite being a poor developing country and having weak scientific and technical institutions, Bangladesh does have some leading scientists and experts on climate change in different institutions, e.g. BUET, BCAS, BUP, BIDS, BARC, IUCN, SPARRSO, EGIS, SWMC, etc. Indeed, of the hundreds of experts involved in the IPCC/TAR, Bangladesh has probably had the most scientists as Lead Authors (after India) among the developing countries. This is an important recognition of their capability and expertise at an international level (IPCC members are chosen purely for their internationally recognized expertise and there is no quota for countries). These human resources need to be nurtured and developed further to enable Bangladesh to develop a proper strategy to deal with the problem at all its levels (including adapting to changes in the country as well as improving our negotiation capabilities internationally).

Create better awareness about the issue: This needs to be done at different levels comprising policy-makers (including bureaucrats, technocrats, parliamentarians, etc.), scientists and technical people in different sectors (e.g. in water, agriculture, disaster management, etc.), and the general public (through NGOs, colleges, schools, media, etc.). The focus of this awareness raising must again be on how we can better cope with any adverse consequences of climate change as well as support our diplomatic efforts on the international negotiations front.

Develop a high-powered negotiating team: International negotiations on climate change related issues take place at the annual COPs. The last (sixth) COP was held in the Hague in the Netherlands last November 2002 but as it was unable to come to a satisfactory conclusion it was postponed and will now be re-convened next July 2003 in Bonn, Germany. The next (seventh) COP will be held next November 2003 in Morocco. Bangladesh usually sends a delegation with members drawn from the Ministry and Department of Environment (headed by the Minister for Environment) to attend the COPs. Over the years individual members of Bangladeshi delegations (including Ministers) did their best at these meetings but it must be acknowledged that Bangladesh has had very little impact so far for a number of reasons. First of all, seldom do the same people go to successive meetings (the rule is generally followed of allowing different members of the ministry to go on these trips, so if one person went last time, it is someone else's "turn" this time). This means that there is no development of knowledge and skills, as each new delegate has to learn the issues all over again each time.

Develop opportunities for carbon trading: The UNFCC and the Kyoto

Protocol (signed at the third COP in Kyoto in 1997) have opened the way for countries to trade their carbon emissions. As Bangladesh is one of the countries with the lowest carbon emissions but with a high population, we have an opportunity to use this fact to our advantage. The Clean Development Mechanism (CDM) is one such means of carbon trading which also allows inward foreign investment through the private sector. Bangladesh could easily attract such investments if it learned how to do so. Unfortunately so far only a few developing countries (e.g. India, China, Brazil, etc) are aggressively trying to use this mechanism. If we don't actively seek to take advantage of these mechanisms we shall definitely be left behind.

Focus on building adaptive capacity: The IPCC/TAR has identified the notion of "Adaptive Capacity" as a key parameter of how countries can cope with the adverse impacts of climate change. Thus, for example, the Netherlands with its low-lying deltaic topography, similar to Bangladesh's, is equally at risk to the impacts of sea-level rise but has the adaptive capacity to cope with the consequences by simply raising their sea walls and paying for it themselves (which they have already decided to do). In contrast Bangladesh has neither such simple solutions nor the financial and human capacity to undertake such coping mechanisms. Furthermore, the problem of climate change and its adverse impacts is not one of Bangladesh's making but one we have to suffer for. Hence the notion of international responsibility (if not legal liability) is one that we can usefully pursue to obtain financial assistance (if not compensation) at the international level. At the national level there is much we need to develop in terms of capacities and knowledge and public awareness to enable our people to better cope with the impacts in future.

Obtaining funding: The climate change arena has opened up a number of new funding mechanisms internationally which Bangladesh has (so far) been unable to access. These include carbon trading and the CDM mentioned above and other mechanisms such as the Global Environment Facility (GEF), which has been dispensing billions of dollars in funding environmental issues including climate change. So far Bangladesh has received a few small grants from GEF, but only for biodiversity and nothing for climate change. A recent COP and GEF Board decision was taken to allow GEF to fund projects on adaptation to climate change (in the past only mitigation projects were funded) and hence Bangladesh has a good opportunity to capture some of these funds. However, this will not be easy as GEF funds are not given on the basis of simple need or poverty, but on how well a project is designed and presented to them (which requires good technical skills which we have to develop). There are other funding opportunities opening up in the climate change arena (including, for example, capacity building), which Bangladesh (as one of the most vulnerable countries) can make a strong case for. These will not come to us as a matter of right, however, but rather only if we can make a good technical and logical case for them.

Conclusion

It is clear that climate change is very bad news for Bangladesh in the long term. Nevertheless in the short (and medium) term there are also some opportunities for the country to seize if we are clever and far-sighted enough to realize them and take the appropriate actions. A few possible actions which we could take to turn this looming (long-term) disaster to some (short and medium-term) advantage are described above.

References

1. ADB, GEF, UNDP, 1998. Bangladesh, Asia Least-Cost Greenhouse, Gas Abatement Strategy. Asian Development Bank (ADB), Manila, Philippines.
2. Ahmed, AU and Alam, Mazaharul. 1999. Development of climate change scenarios with general circulation models. In: Vulnerability and Adaptation to Climate Change for Bangladesh. S. Huq. et al. (eds.). Kluwer Acad. Publ., Dordrecht, Netherlands.
3. Ali, Youssouf M. 2001. Fish resources vulnerability and adaptation to climate change in Bangladesh. In: Vulnerability and Adaptation to Climate Change for Bangladesh. S. Huq. Et al. (eds.). Kluwer Acad. Publ. Dordrucht, Netherlands.
4. Bangladesh Bureau of Statistics. 2001. Statistical Year Book of Bangladesh 1999, Statistics Division, Ministry of Planning, Government of Bangladesh, Dhaka, Bangladesh.
5. Berk, M.M. et al. 2001. A strategic vision on near-term implications of long term climate policy options. Keeping Our Options Open, NRP, Netherlands.
6. Biagini, B (ed.). 2000. Confronting Climate Change: Economic Priorities and Climate Protection in Developing Nations. A Climate of Trust Report. National Environmental Trust (NET), USA.
7. Finance Division. 2001. Bangladesh Economic Review 2001. Ministry of Finance, Government of Bangladesh, Dhaka, Bangladesh.
8. Habibullah, M., Ahmed, A.U. and Karim, Z. 1999. Assessment of foodgrain production loss due to climate induced enhanced soil salinity. In: Vulnerability and Adaptation to Climate Change for Bangladesh. S Huq. and et al. (eds.). Kluwer Acad. Publ. Dorducht, Netherlands.
9. Islam, R., Huq, S., and Ali, A. 1999. Beach Erosion in the Eastern Coastline of Bangladesh. In: Vulnerability and Adoptation to Climate Change, for Bangladesh. S. Huq. and et al. (eds.). Kluwer Acad. Publ. Dorducht, Netherlands.
10. Rashid, Haroun Er. 1977. Geography of Bangladesh. Univ. Press Ltd., Dhaka, Bangladesh.
11. World Bank 2000. World Development Indicators. World Bank, Washington, DC.
12. Zahurul. K, Hossain, G., and Ahmed, A. 1999. Climate change vulnerability of crop agriculture. In: Vulnerability and Adaptation to Climate Change for Bangladesh. S. Huq. Et al. (eds.). Kluwer Acad. Publ. Dordrecht, Netherlands.

CHAPTER 13

Seeking the Middle Ground Between More and Less: A Canadian Perspective

Heather A. Smith
Associate Professor and Chair, International Studies Program, UNBC, 3333
University Way Prince George, BC, V2N 4Z9. e-mail: smith@unbc.ca

Introduction

"Where do we fit" and "do we matter" are perennial Canadian questions. Observed by some as lacking self-confidence, political navel gazing is a national Canadian obsession which may seem odd to those outside of Canada. Given our neighbor to the south, however, the obsession is not so odd. By virtue of geography, the United States is what Kim Richard Nossal calls "a relative invariant"[1] in Canadian foreign policy. As Nossal writes, "Canada has four neighbors, but demography, culture, language, economic structure, and ideology have combined to make the United States the most important focal point of Canadian foreign policy"[2].

The shadow of the US looms large over the Canadian political scene. Especially since the end of World War II, US orientations and behaviors have influenced the planning and execution of Canadian foreign policy. At times, and often depending in part on the relationship between the Prime Minister and President, our relationship has been strained. At other times, such as when Brian Mulroney and Ronald Reagan held office, our relations were warmer. But the tensions between solidarity with the United States or independence from the US; between being seen as different from them or being regarded as a US lackey; between regionalism and multilateralism – remain constant. And now, in the post 9-11 era, reference has been made to Canada as a fading power,[3] as a country hobbled by past poor decisions, and increasingly irrelevant to the US. For a country that prides itself on making a difference through the promotion of internationalism and good international citizenship as well as active participation in multilateral institutions, suddenly feeling you do not matter is difficult to accept.

Why describe this Canadian angst? The tensions identified above are integral to understanding the Canadian approach to climate change and at least some of the reasoning behind the Canadian ratification of the Kyoto Protocol. Whether the ratification decision was brave or foolhardy remains to be seen. The fact is that Canada is at least symbolically out of step with the United States. Yet, the US is not really out of the game, it is just sitting (albeit noisily) on the multilateral sidelines and for Canada, the US is never on the sidelines. Its presence is always felt. Regardless of the future of the Kyoto Protocol without the US, Canada will still cooperate with the United States—bilaterally, trilaterally, and perhaps less directly, multilaterally.

Rather than choose between the regional and the multilateral, solidarity and independence, difference and sameness, Canada is going to try to do both. Canada will remain active multilaterally. This choice was made with the ratification of the Kyoto Protocol. Yet, Canada will not become a leader in the promotion of new and innovative ideas designed to salvage the environmental integrity of the Protocol. Rather, it will take a position consistent with past behavior. This can be described as the "less" path—"the less I have to do to reduce emissions reductions, the better". Bilaterally and trilaterally as many of the activities taking place include Mexico, Canada will advocate the "more" path—"the US needs to do more to reduce emissions". The desired end is to encourage the crafting of international and continental rules and regulations that bridge the gap between the international and the US. It is not so much about bringing the US back in but ensuring that the playing field is level and that Canadian interests are served both multilaterally and continentally.

This chapter begins with a brief analysis of some competing perspectives on Canada and multilateralism. It will be seen that we must do more than explain why Canada engages in multilateralism, we must also investigate the practice and consequences of multilateralism. We then turn to Canada's international activities as related to climate change, with the objective of examining US influence on Canada's orientation. This section begins with an analysis of Canada's activities in the early stages of the climate issue, then moves on to the period between the first Conference of Parties in Berlin and the US withdrawal from the Kyoto Protocol. We then investigate Canada's behavior after the US withdrawal from the Kyoto Protocol in 2001, including Canadian participation in the Sixth Conference of Parties resumed, Canadian ratification and bilateral activities between Canada and the US. The historical overview allows us to trace Canadian policy development and Canada's position vis-à-vis the United States over a fourteen-year period. The overview provides us a means by which to assess how the tensions, noted above, played out in practice. We are also offered insights into the type of world order Canada is promoting through its international activities. Finally, the conclusion will reflect on the lessons of case study in light of the American withdrawal. While Kyoto may provide the "architecture" for the continuation of future negotiations, one should not expect the US to return to the fold in the immediate future. Rather, the United States will strike bilateral and trilateral deals with countries such as Canada, Mexico and Australia, intervene in future international discussions when it feels its interests are undermined, and rely on its allies who have ratified the Kyoto Protocol to further ensure that future negotiations do not threaten American interests. Canada will be one such ally.

Canada and Multilateralism

It is commonly accepted by mainstream scholars such as Tom Keating[4] and Kim Nossal[5] and critical theorists such as David Black and Claire Sjolander[6] that multilateralism has been a central component of Canada's foreign policy. Tom Keating identifies both international and domestic sources of multilateralism.[7] The international sources include a desire for autonomy from Great Britain (associated with the interwar period) and a desire for a rule-based international system which was, and is, seen as providing some security and stability. A rule-based international system, and formal international institutions, were also perceived as tools for constraining and limiting American unilateralism. Multilateral activities and institutions were also viewed as a means by which to promote Canadian interests and at the same time function as a counterweight to the dominance of the United States. The degree of connection to the US was such that Canadian policy-makers sought to devise means by which to regulate the bilateral relationship "but also to view international organizations as a forum in which a distinctly Canadian presence could be manifested".[8] With regard to domestic sources of multilateralism Keating argues that the commitment to multilateralism will "vary depending on domestic political coalitions".[9] Finally, Keating notes that multilateralism has consistently had the support of the Canadian public. For Keating, multilateralism has been and remains a hallmark of Canadian foreign policy. However, Keating does not explicitly problematize multilateralism while other scholars readily comment on the character and implications of Canadian multilateralism.

Kim Nossal agrees that multilateralism remains an important element of Canada's orientation toward the world, but he is sceptical about the values that drive Canadian foreign policy. In his analysis of internationalism, a dominant idea of Canadian foreign policy which he believes comprises multilateralism, good international citizenship, voluntarism and community, he argues that while multilateralism remains important, other elements of internationalism have mutated. Nossal argues Canada, since the election of Jean Chretien in 1993, has become increasingly unwilling to adequately resource foreign and defense policy. He describes this orientation as "pinchpenney diplomacy", of which he states "pinchpenny diplomacy is more than diplomacy that seeks to be as cheap as possible. Rather, it suggests a particular attitude toward international activity, an essential meanness of spirit that underwrites an overly frugal foreign policy conducted by a rich and secure community in a world that continues to be marked by poverty and insecurity".[10]

David Black and Claire Sjolander bring together the practice, meaning, and implications of multilateralism in their thoughtful neo-Gramscian analysis. They agree with Keating and Nossal that Canada's multilateral endeavors promote world order but they are sceptical of claims that it promotes a "better" world order; rather Canada's activities promote a hegemonic world order that is highly unequal. They also observe a gap between discourse and practice and ultimately conclude that multilateralism is becoming more exclusive. Essentially, they counsel us to query as to the kind of world order Canada is promoting and direct us to investigate the gap between the rhetoric of policy-makers and practice.

The Black and Sjolander analysis is particularly useful for our purposes here. It looks at multilateralism as more than a style of diplomacy. Both authors understand that attempts to build and promote particular types of world order, as is the case with international environmental negotiations, come with implications and consequences. This does not render Keating's analysis null since his explanations of Canadian diplomacy are insightful and Black and Sjolander have strengthened these insights that lead to this question: "For whom is multilateralism"? As such, their work complements Robert Cox's analysis of multilateralism and his query: "Who speaks for the biosphere?"[11]—an important question, given the topic of this chapter. Finally, Black and Sjolander connect multilateralism and regionalism. Multilateralism and regionalism are not dichotomised, but rather are treated as reinforcing styles and practices of diplomacy. As such, they offer us a platform from which to focus on Canadian international environmental diplomacy on climate change, with an eye directed toward the United States.

The Early Stages: Canada as Leader?

At the early political stages of the climate change issue Canada adopted a leadership role, consistent with the notion of technical/entrepreneurial leadership when it acted as a catalyst and facilitator[12]. There was also an element of a moral leader—promoting ideas and concepts that are "good" for the earth. This moral leadership was similar to the trend identified by Fen Hampson and Dean Oliver in Canada's recent international behavior. They describe Canada as a member of the moral minority "that distinguished (and self-styled) group of states whose moral multilateralism is predicated on their faith that the enunciation of a new set of global norms will lead inexorably to the creation of a just and more equitable international order".[13] The particular time period of leadership, practically speaking, lasted about a year, beginning in June 1988.

Canada's role as catalyst and facilitator is most evident in its promotion of the law of the atmosphere in June 1988 at the conference on "Our Changing Atmosphere: Implications for Global Security", otherwise known as the Toronto Conference, co-sponsored by the Canadian government, the United Nations Environment Program (UNEP), and the World Meterological Organization (WMO).

At this meeting Canada promoted the idea of a law of the atmosphere. The origins of the idea can be found in the Atmospheric Environment Service (AES) of the Department of Environment (DOE). The essence of the idea was to take a holistic approach to environmental issues that recognized and reaffirmed the linkages between major atmospheric issues.[14]

Typically the Toronto Conference is noted for the emission reduction targets that were a central element of the conference statement. Less well known is the inclusion of reference to a law of the atmosphere which emerged in the conference statement as a call for a comprehensive global convention and which was promoted by both Prime Minister Mulroney and Environment Minister McMillan. In an attempt to capture the momentum from the Toronto Conference, and perhaps in an attempt to take advantage of media attention, the Canadian Prime Minister proposed a follow-up meeting to consider the rapid development of a Law of the Atmosphere. The follow-up meeting, the meeting of Legal and Policy Experts, was held in Ottawa in February 1989. The communiqué from this meeting includes reference to the law of the atmosphere but essentially the concept died at that meeting. Statements from subsequent conferences on climate change focus on a framework convention and/or a protocol specific to climate change as opposed to the umbrella convention envisioned by proponents of the law of the atmosphere.

The end of the law of the atmosphere initiative can be accounted for, in part, by opposition from the head of UNEP, Mostafa Tolba, who feared a replay of the Law of the Sea. Moreover, the participation of the US in the issue was deemed necessary because it was the largest emitter of CO_2, and the law of the atmosphere lacked the support of the United States.[15] Even though the blocking power of the United States put an end to the concept of a law of the atmosphere, it did not undermine the status Canada had obtained as a champion in the fight against climate change.[16]

Canada continued to participate in the numerous meetings that led up to the beginning of the negotiations of the Framework Convention of Climate Change (FCCC) in early 1991. During this time Canada's claims of leadership were tempered as Canada began to develop sets of regulative rules, reflective of Canadian interests. We adopted a commitment "to stabilize emissions of CO_2 and other greenhouse gases, not covered under the Montreal Protocol, at 1990 levels by the year 2000"[17] in the December 1990 Green Plan. The comprehensive commitment, in the view of policy-makers, allowed the Canadian government the flexibility to deal with the gases that can be most readily reduced. This position placed us ahead of the George Bush-led United States who was seen as an international laggard, but it also divided us from the European Union who called for a CO_2 only convention. More ambitious emission reduction targets such as the Toronto target, are viewed as requiring further investigation prior to any Canadian commitment.

> The Government of Canada believes that further reduction in greenhouse gas emissions are required and that these should be based on a program of targets and schedules agreed upon internationally. In this context, the technical feasibility and cost and trade implications of further reductions in emissions will be examined, including the 20-percent reduction in CO_2 emissions called for by the 1988 Toronto Conference.[18]

It was also evident at that time that the preferred options for reductions would be market-based instruments. The 1990 Green Plan states very clearly that "we must meet our environmental goals in ways that promote economic prosperity"[19] and that "we must make use of forces of the marketplace and allow industry as much flexibility as possible in meeting specific environmental goals".[20]

Canada's self-described position during the negotiations of the FCCC was facilitator. Aggressive behavior was shunned and Canada worked to try to narrow the divisions between the developed states. The United States was typically pitted against the European Union. For Canada, the middle road was the comprehensive stabilization commitment, one that they saw the United States potentially moving toward in the latter part of the negotiations and a position they felt the Europeans would also accept.[21]

Early Stages: Canada as Leader Assessed

In the beginning of the period 1988 to 1992, Canada attempted to fashion itself as an international leader on climate change. This self-assigned perception of leadership is not unique to climate change. For example, in the late 1980s Canadian leaders ensured others knew Canada was the first to sign the 1985 Vienna Convention on Ozone Depleting Substances.[22] In relation to climate change, Canada sought to promote new ideas as

Prime Minister Mulroney engaged in rhetoric that clearly pointed to anthropogenic sources of climate change. He also sought to promote a law of the atmosphere. The prestige and media attention arising from the Toronto Conference was welcome because it acknowledged that Canada was different from the United States, more environmentally progressive, and a good international citizen. Practically, the claim to leadership was supported by a wealth of technical expertise housed in the Atmospheric Environment Service of Canada. Generally speaking, Canada's activities during this period represent support for multilateralism. There appeared to be an understanding that the issues of the global commons could not be solved by one country. Collective solutions were the preferred route. But during this period we also observe the power of the United States to squash ideas it does not like, as was the case with the law of the atmosphere initiative. Canada's multilateral adventurism and normative leadership was constrained and limited by the US.

As we move into the FCCC negotiations it can again be observed that Canada sought to function as facilitator. Canada wanted to ensure that the United States was part of the agreement. This is reflected in the fact that by the time of the FCCC negotiations, normative statements about the need to protect the well-being of the environment were not sufficient and it was time to start crafting some rules. Recall that Canada has a proclivity for a rule-based international system, but a clear preference for ensuring that those rules apply to the US as well as Canada. In seeking to facilitate between the US and the EU the aim was to ensure the US was "in". This is not to suggest Canada compromised its position by adopting the facilitator role. Canada did promote a comprehensive target in part because it was attractive to the United States but also because it was in keeping with Canada's interest. Canada did not want and has never wanted a CO_2 only climate change agreement. Canada's economy is very energy intensive. Flexibility in emission reductions became a watchword for Canada as early as the 1990 Green Plan. Market-based reduction strategies were and continue to be the preferred option. Canada, on the one hand, sought to present itself as different and separate from the US through the adoption of positions such as "leader" and "facilitator", while, on the other, once engaged with international negotiators, articulating a set of principles consistent with its own economic realities and by extension the realities of the United States.

Berlin to US Withdrawal From Kyoto: Canada and Coalition Politics

The FCCC came into force in March 1994. At COP-1 in Berlin and COP-2 in Geneva, Canada was part of a coalition of countries known as JUSCANZ (Japan, US, Canada, Australia, New Zealand). At COP-I, JUSCANZ rejected all attempts by the EU who wanted Annex I countries to cap their emissions at 1990 levels at 2000. Canada sought "a process of negotiations for post-2000 commitments without predetermining the outcome of these negotiations".[23] JUSCANZ was generally seen as complicating the negotiations with demands for emission reductions by more advanced developing countries and until COP-2 was seen as having an implicit coalition with the oil-rich Organization of Petroleum Exporting Countries (OPEC). The coalition between OPEC countries and JUSCANZ appeared to break down at COP-2 over the Intergovernmental Panel on Climate Change (IPCC) Second Assessment Report.[24]

The integrity of the JUSCANZ coalition was challenged in the lead-up to the Kyoto meeting with the American declaration in support of a legally binding protocol. Canada, caught with its guard down, stated in November 1997 it would do better than the American commitment to stabilization of greenhouse gases at 1990 levels by 2008-2012. The position negotiated with the Canadian provinces[25] and announced that same month agreed to seek to reduce GHGs back to 1990 levels by 2010[26] did not meet the requirements of beating the US and therefore on December 1st, the Minister of Environment, Christine Stewart announced Canada's position at Kyoto would be for industrialized countries to seek a 3% reduction of GHG emissions to 1990 levels by 2010 with further reductions of 5% by 2015[27].

The emission reduction targets noted above were the centrepiece of Canada's position at Kyoto. But, just as important was the emphasis placed on economic instruments. Canada was committed to and strongly promoted joint implementation, market-based instruments, and emission reduction credits that included sinks.

At Kyoto, as is well known, the EU, frequently in tandem with developing counttries, faced off against the United States and its partners in the JUSCANZ[28] coalition. The division was initially over emission reduction commitments. "The EU and others supportive of an ambitious target, such as the G-77/China and AOSIS, decided to hold out until the US signaled willingness to improve its offer of stabilization at 1990 levels".[29] To break this deadlock the Chairman proposed differentiated targets for industrialized countries. But to accept these differentiated targets counries such as Canada and the US expected "maximum flexibility" in implementation. As reported in the *Earth Negotiations Bulletin*: "The EU resisted conceding to the US and JUSCANZ members flexibility in implementation, notably on emissions trading and sinks criteria... but the US and JUSCANZ required commitments on these very issues".[30] An attempt was made to bind developing countries (non-Annex I parties) to some sort of voluntary commitment. On December 5th "New Zealand said that Annex I Parties' constituencies needed assurances that developing countries would adopt binding emissions limitation commitments"[31] in future. New Zealand was supported in this intervention by the US, Canada, Australia, Japan, Poland and Slovenia. This position was rejected by the EU and vehemently rejected by the G-77/China and a host of other developing countries. Ultimately, differentiated reductions for Annex I countries were adopted.[32] For Canada this meant accepting a reductions target of 6% GHGs from 1990 levels by 2008-2012. The United States accepted a target of 7% reductions, thus making it hard for Canadians to claim they beat the US.

Market-based mechanisms such as joint implementation and emissions trading, and the new Clean Development Mechanism (CDM) were all included in the Kyoto Protocol, thus providing the flexibility required by the American-led coalition. Provisions were also made for the inclusion of carbon sinks, associated with forestry, into the calculations of emission reductions. The inclusion of sinks was a highly desired element of the Protocol for Canadian negotiators because Canada has huge tracts of forest that can act as carbon sinks. No conditions imposed voluntary commitments for developing countries.

At the Fourth Conference of Parties, held in Buenos Aires in November 1998 and the Fifth Conference of Parties in Bonn, JUSCANZ – now the Umbrella Group—and the EU continued to collide over a number of issues. At Buenos Aires, the Umbrella Group squared off with the G-77/China over voluntary commitments, as the United States, leading the Umbrella Group, demanded "meaningful participation" from developing states if it were to ratify the protocol. The American led coalition expressed a "strong interest in moving quickly on the elaboration of guidelines and principles for the flexibility mechanisms".[33] There was also a very significant debate on the issue of whether or not the proportion of emission reductions a country can count from the flexibility mechanisms should be capped. The EU and developing states insisted on a cap while the Umbrella Group, Canada included, "steadfastly opposed it, stressing the need for maximum flexibility in meeting targets".[34] At Bonn, Canada continued to push on sinks, while other parties continued to disagree on ceilings for the mechanisms and compliance.

Canada went to the Sixth Conference of Parties with four specific objectives. According to Lloyd Axworthy, Head of the Delegation at the Hague: "First and foremost, we want to ensure that the global environmental objectives can be met; second, we want to meet our Kyoto targets at the lowest possible cost; third, we believe that it is important to ensure that there is a level playing field among major economies and competitors; and finally, we want to maximize opportunities for Canadian business." [35] These objectives are consistent with the neo-liberal economic philosophy that has underpinned Canadian climate-change policy since the 1990 Green Plan. These objectives drove Canadian behavior at the Hague meeting.

The EU was concerned that the flexibility mechanisms could be used to avoid significant domestic actions. They wanted limits on the use of the mechanisms. The Umbrella Group remained resistant to such caps. The Europeans, willing to move this matter along, seemed ready to accept the notion of qualitative limits on the mechanisms. Canada was willing to accept this compromise "as long as it felt that it had room to meet much of its target through credits from carbon sinks".[36] The US refused the European proposal. The American-led Umbrella Group also wanted to include sinks as activities acceptable under the mechanisms and pushed for simple procedures to define those mechanisms. The inclusion of sinks in the CDM, in particular, was a sticking point between the Umbrella Group and the EU. Similarly, the Umbrella Group wanted to include additional activities, such as land use, under the sink provision of the Protocol and the EU opposed this. "Countries such as the US and Canada believe it is in their best interest to include as many carbon sinks as possible to enable them to meet their Kyoto targets without drastic emissions cuts. They therefore sought a broad interpretation of the sinks provision as a precondition for ratifying the Protocol."[37] There was some accord on sinks but ultimately disagreement prevailed. Canada also wanted the right to gain credit for selling nuclear power to developing countries through the CDM, although indicating a willingness to remove this demand as negotiations progressed. The talks ultimately collapsed, however, with resumption scheduled for May 2001. The Umbrella Group was accused of sabotaging the talks with their intransigence over sinks and other crunch issues.

In an effort to restore some sense of hope, the EU and Umbrella Group met in Ottawa December 7th 2000 to continue discussions regarding international climate-change-negotiations. Environment Minister Anderson noted Canada's facilitation role in these meetings and stated regarding Canada's behavior at the Hague: "Canada was instrumental in presenting constructive solutions and was ready to accommodate the concerns expressed by other parties. Canada championed a comprehensive approach that would ensure the environmental integrity of the Kyoto Protocol."[38]

In March 2001, the United States pulled out of the Kyoto Protocol negotiations, but stated it would remain a member of the FCCC. In the mind of the new American President, Kyoto would have negative economic implications for the United States. Without the US there was genuine concern the requirements for ratification could not be met because the United States represents 36% of the 1990 emissions of GHGs.

Coalition Politics Considered: Multilateralism of Indifference

Between the First Conference of Parties in 1995 in Berlin until the American withdrawal from the Kyoto Protocol in March 2001, Canada allied itself with the United States and was an integral member of the US led coalition. There was an apparent aberration around Kyoto targets, but that aberration, to be addressed shortly, was more superficial than substantive. What is most striking about this period is that Canada stayed in the negotiations after the US left.

During this period, Canada continued to engage in multilateral activity and yet given the coalition-based nature of the climate-change negotiations one would be hard pressed to suggest that Canada functioned independent of the US. Thus one finds a contradiction wherein multilateralism traditionally provided for independence and autonomy from the United States and supported Canada's promotion of itself as separate and different from the US. Yet, Canada worked in tandem with the United States throughout much of this period, thereby forgoing many of the assumed benefits of multilateralism. Canada's place in the coalition is in part explained by the dynamics of the international environmental negotiations themselves wherein almost all countries adopted coalition-based positions and therefore it would be naïve to assume Canada could "go it alone". The coalition, however, was not simply the result of structural imperatives, it was grounded in a commonality of purpose.

In a document produced by the Federal/Provincial/Territorial Ministers of Energy and Environment in February 2000, the common objectives of the Umbrella Group, at least from a Canadian perspective, were outlined.[39] It is noteworthy that members of JUSCANZ came together at Kyoto to explore forming an emissions trading bubble and as it evolved into the Umbrella Group, the group became a negotiating bloc. "The Umbrella Group shares the basic strategic approach originally developed by JUSANZ, namely, the pursuit of Kyoto mechanisms to ensure the most cost-effective implementation possible."[40] The document also confirms EU fears about the Umbrella Group countries seeking to use the Kyoto mechanisms as their primary means of reducing emissions. "It is likely that most Umbrella Group members intend to maximize their use of the Kyoto mechanisms in meeting their commitments. The United States, for instance, has indicated that as much as 65-70% of their commitments could be met through the mechanisms".[41] The document also makes it very clear that the coalition is opposed to any limits on the mechanisms.

The coalition members may argue they were seeking to reduce global concentrations of CO_2 and thus reduction through the Kyoto mechanisms or sinks is perfectly acceptable. However, one commentator on Kyoto noted the inclusion of sinks in the Kyoto Protocol could be the undoing of the Protocol.[42] Emissions trading, the CDM and JI are all seen as means by which to export pollution while potentially maintaining a business-as-usual strategy at home. The difficulty (or perhaps unwillingness) to reduce at home, thus compelling countries such as Canada, the United States, and Australia to promote sinks and the Kyoto mechanisms, is highlighted by their per-capita emissions. In 1998 three key members of the Umbrella Group were also the three highest emitters of CO_2 per capita (the US, Australia and Canada respectively)[43]. By the measure of historical CO_2 and methane contributions by region, OECD North America represented approximately 29% of global emissions.[44] These figures, in conjunction with the discourse on competitiveness, suggest that assumptions about negative economic impacts underpinned the efforts of the Umbrella Group. The concern for negative economic impacts was reinforced by domestic constituencies both in Canada and the United States. Essentially, Canadian and American activities call into question the extent of the commitment to sustainable development and environmental integrity.

As noted earlier, Kyoto appeared to be an aberration during this period. There is a sense that Canada sought to "beat the US". Recall that the federal government jettisoned with the national consensus and brokered with the provinces in terms of emission reduction targets. The break with the provinces and the effort to "beat the US" was impelled, in part, by Canada's Prime Minister, Jean Chretien. Recalling the legacy of environmental leadership, shamed by missed FCCC commitments and very negative publicity about Canada's participation in the international negotiations on climate change, the Prime Minister pushed for a commitment better than the US, at least prior to the Kyoto meeting. For a brief moment, Canada claimed the status of difference, once again. The difference was more superficial than substantive, however,

Ultimately, the reduction targets adopted were more significant than the position with which Canada entered the negotiations. Canada, and the US, accepted the more stringent targets in return for maximum flexibility in the Kyoto mechanisms. For Canada, the additional flexibility was imperative as there had been an agreement with industrial stakeholders in advance that any reduction targets would be compensated for by international mechanisms, thus reducing the impact on industry at home. Multilateralism over this period did not promote difference, but rather sameness as regionalism and multilateralism reinforced each other.

Yet, Canada stayed in the negotiations after the US withdrawal. One can speculate continued participation was motivated by the desire to ensure the rules being designed would be in Canadian (or Canadian and American) interest. Perhaps it was a symbolic gesture with the aim of presenting an image of a good international environmental leader, or at least refurbishing the lost image. The decision to stay could also be viewed as "independent" action by Canada. Or perhaps, it was just a matter of wait-and-see – wait until the next set of negotiations and see what happens. The possibility or threat of withdrawal could be Canada's leverage. Indeed, the Minster of Environment, David Anderson, was not very optimistic about the future of Kyoto without the United States. Anderson reportedly stated, with reference to proceeding without the United States, in May 2001: "There's a certain amount of brave talk about 'Well, let's proceed, let's ratify, let's just proceed as though nothing has happened'. That's not realistic".[45]

The Minister changed his tune during the ratification debate in Canada. Why did Canada stay after the US departure and why did Canada ratify the Kyoto Protocol? And what does this mean for the future of the negotiations? These questions will be addressed after we examine the period between July 2001 and December 2003.

Bonn to Canadian Ratification: Leadership Reclaimed?

COP-Six resumed meetings in Bonn in July 2001, without the United States. Canada identified the four key issues at Bonn as follows: "First, efficient and accessible market mechanisms, including recognition for clean energy exports and technologies that reduce greenhouse gas emissions...Second, recognition of the critical role that managed forests and agricultural carbon sinks can play...Third, support for, and engagement of, developing countries...Finally, for Canada a successful Protocol must include a compliance regime that ensures that countries can and will meet their obligations".[46]

The Bonn COP resulted in a compromise agreement on key issues such as funding for developing states, flexibility mechanisms, and sinks, but the compromise came at the cost of the EU stance on "environmental integrity". For example, the Umbrella Group received the flexibility it desired in the mechanisms: "the [Bonn] agreements specify that domestic action should constitute a 'significant element of the effort' made by each party in implementing its obligations under the protocol, but without a quantitative limitation on the use of mechanisms".[47] The demands of the Umbrella Group resulted in an expanded range of sinks-related activities and members of the Umbrella Group worked to undermine a compliance regime that would have legally binding consequences. Compliance issues were not resolved and remain largely unresolved to this date.[48]

With the conclusion of the Bonn meeting there appeared to be almost a collective sigh of relief and some jubilation. David Wirth notes: "the Bonn Agreements have ... been widely cited as emblematic of the EU's capacity to exercise global leadership in the absence of US support for multilateral initiatives."[49] EU leadership is called into question, however, by the view that the agreement was reached because the EU gave into the rump of the Umbrella Group on key issues. It was about striking a deal – any deal, according to Thomas Legge.[50] Failure in the face of US withdrawal was unacceptable, and Canada, Japan, Australia and Russia used this to their advantage and played hardball. John Drexhage, a former

Canadian climate change negotiator, noted of Bonn: "The current deal contains a number of elements that, from the perspective of the global environment, are weaker than what could have been agreed to at the failed round of negotiations in the Hague last November. For example, Canada received substantially more credits for 'sinks' …than it would have received in the Hague, thereby easing the pressure to make direct emission reductions at home."[51] With the additional credits for sinks, flexibility in the mechanisms, and the unresolved compliance discussions, the Bonn Agreements, in the words of Canada's Prime Minister, opened "the way for Canada's ratification of the Kyoto Protocol next year."[52]

In the last two years there have been two subsequent Conference of Parties meetings: COP-Seven in Marrakech in November 2001 and COP-Eight, recently concluded in New Delhi. As noted above, since Bonn the push has been to get Kyoto ratified. Yet, there are some issues worth noting in the last two COPs, particularly related to Canada's positions on compliance and clean energy credits.

Marrakech was expected to be a meeting where there was some final tinkering on elements in the Protocol. Yet, according to one observer, Canada, Russia, Australia and Japan worked to undermine the compliance regime of the Protocol. At issue was whether or not the compliance regime would be legally binding. "Canada's chief negotiator Paul Fauteux said that although the Bonn agreement set out what the sanctions would be, ministers had not agreed whether or not they would be politically binding."[53] The official Canadian position was support for legally binding consequences for noncompliance with the caveat that the decision on compliance consequences be left to the first meeting of parties – after the entry into force of the Kyoto Protocol.[54]

In addition, Canada pushed the issue of credit for clean energy. The argument was that Canada should get credit for exports to the United States because the exports were cleaner than the fuels that would otherwise have been used. Officially, Canada offered to hold a workshop on the global benefits from cleaner energy and the Minister of Environment stressed: "Our Prime Minister has repeatedly stated the importance he personally places on ensuring appropriate recognition for the contribution made to addressing climate change through trade in clean energy"[55]

COP Eight was held in November 2002 in New Delhi and as such corresponded with the Canadian ratification debate. A statement by David Anderson at the conclusion of the New Delhi COP expressed disappointment over the lack of progress on credit for clean energy exports and indicated a determination to push the issue at the next conference of parties. In addition, he stressed Canada was vulnerable to the impacts of climate change. Finally, given the Prime Ministerial commitment to Canadian ratification by the end of 2002, the Minister of Environment implied Canada would again take a leadership position on this issue: "…we will be on track to meeting our Kyoto target. We are well-positioned to accelerate the international momentum towards achieving global climate change goals."[56]

In December 2002 Canada ratified the Kyoto Protocol. At the World Summit on Sustainable Development in the summer of 2002 the Prime Minister had stated Canada would ratify the Kyoto Protocol before Christmas 2002. It was an intense debate, pitting the provinces against one another and against the federal government. It was marked by hostile exchanges between Prime Minister Chretien and Ralph Klein, the premier of the province of Alberta, Canadas largest oil and gas producer. Industry and environmental nongovernmental organizations also participated in the debate. At the heart of the debate were concern's about negative economic impacts, both generally and as a result to being out of step with the United States, appropriate and proper consultation between the provinces and the federal government, and the depth of the analysis regarding implementation strategies. In spite of intense opposition, the Prime Minister was determined he would leave a "green legacy" in light of a decision to depart from Canadian federal politics, which he did in late 2003. Subsequently, the federal government produced an implementation strategy. The Minister of Environment argued the economic impacts would be minimal and the provinces were put off until a later date. In December 2002, closure was imposed in Parliament on the Kyoto debate, and the instrument of ratification was delivered to the United Nations in mid-December 2002. With ratification the Prime Minister reclaimed Canadian leadership and reinforced Canada's commitment to multilateralism stating "Canada has a proud tradition of working with other nations towards common goals. We are committed to leadership on international challenges."[57] In the view of the Prime Minister, "with this signature, we are doing the right thing for Canada, for the global environment, and for future generations."[58]

Bonn to Canadian Ratification Considered

What does the period between the Bonn meeting in 2001 and the New Delhi COP tell us about Canada and multilateralism and Canada's behavior in the face of the American withdrawal? Indeed, it is intriguing that Canada remained in the international negotiations after the US withdrawal given the apparent commonality of purpose and commonality of values that underpinned the Canadian and American positions. Clearly, the Canadians were thinking strategically and noting the dynamics of the negotiations worked to ensure that the best deal possible for Canadian interests would be struck. Those interests, as Drexhage noted above, paralleled American interests. Canada and its coalition members sought and got deals that were ironically consistent with American aims, but that may not have been possible had the Americans stayed in the negotiations. American intransigence fostered resentment. No doubt Canadian intransigence fostered resentment as well, but the conditions of the international arena had changed. In spite of using the negotiations in such a way as to potentially expand the loopholes, Canada was able to present itself, in the words of former Deputy Prime Minister Herb Gray as having been "instrumental... in negotiating a political agreement"[59] at Bonn. The mantle of "difference" was claimed once again.

The claims of leadership and difference, as in the past, are limited in substance. We must acknowledge Canada has ratified the Kyoto Protocol when its largest trading partner has not. Thus, at least the optics suggest the claim of difference has some merit. Yet, it is difficult to conceive of a complete turnabout in the Canadian philosophy regarding reductions of emissions. Recall Canada received more credit for sinks at Bonn and it has and will continue to push for clean energy credits. These loopholes are all factored in the Canadian implementation plan. The Kyoto commitment of 6% below 1990 levels is said to translate into a reduction of 240 MT from business-as-usual emissions.[60] To meet this reduction, the Canadian government counts numerous actions to date, but also anticipates counting on the clean energy credits. These credits would account for over one-quarter of Canada's emission reductions target, or 70 MT. Add to this a minimum of 10 MT from international market mechanisms and the anticipation of sinks credits. "Ongoing actions are expected to bring credits of 30 MT annually to Canada."[61] These numbers suggest there is a creative numbers game taking place where Canada will seek credit for business-as-usual activities, such as the export of natural gas to the United States and geography by way of sinks. Moreover, Canada's greenhouse gases rose between 1990 and 1999 by 15%.[62] This does not account for actions taken since 2000, but the trend is foreboding. Whether or not the Kyoto commitment can be met remains to be seen, but the implementation strategy leaves no doubt that the loopholes will be used to Canada's advantage. Preserving Canada's competitiveness will become paramount.

Given this scenario, we would be well advised to watch the Canadian position of compliance in future. It may be in Canada's interest to ensure there is a weak compliance mechanism in spite of support expressed for a legally binding compliance mechanism. Superficially, Canada may seek to position itself as a consensus builder, willing to forego a legally binding mechanism, in order to craft a deal on compliance that is ultimately weak. As a diplomatic maneuver this bolsters the Canadian image and places a Canadian ally in the line of fire over compliance but ultimately meets two Canadian objectives: refurbished image and weak compliance.

This may seem far-fetched but the Canadians have been wily and unpredictable negotiators. They put nuclear energy on the agenda twice at the Hague and Bonn and subsequently withdrew it both times, they threatened to leave COP-Six Resumed if their demands were not met, negotiated additional sinks credits at Bonn, and ratified the Kyoto Protocol without domestic consensus but to international applause – thus gaining more diplomatic credit in spite of a vague implementation plan. The multilateral negotiations have served many purposes.

The US withdrawal from Kyoto does not translate into no US action on climate change whatsoever. Nor does the US withdrawal from Kyoto translate into an absence of Canadian-American activities related to climate change. As noted earlier, from the Canadian perspective, the US is never really out of the game. It is always present. Given this reality, it should not be surprising there are activities in the US that affect Canada and cooperative ventures on many levels between Canada and the US.

The cooperation between Canada and the US at many levels is undertaken by many kinds of actors. For example, the US Department of Energy has become active in carbon sequestration research. In July 2001, it initiated a project with an international team of nine energy companies, including two Canadian energy companies: Suncor Energy and Pan Canadian. The US government has committed $5 million (US) to this project.[63] The three North American states, Canada, the United States and Mexico, through the institution of the Council of the Commission for Environmental Cooperation "pledged to explore further opportunities for market based approaches to carbon sequestration, energy efficiency, and renewable energy in North America."[64] As well, the five easternmost Canadian provinces, including Quebec, and six New England states have adopted collective goals of reducing total regional greenhouse gas emissions. Canadian business, the federal government, and provincial governments are all engaged with the United States in climate change initiatives.

Specifically linked to the international dimension are the efforts by Canada to encourage the Bush Administration to be more proactive in its climate change initiatives. The federal Minister of Environment, David Anderson, has undertaken on several occasions efforts to push the US to go further in its efforts. For example, upon release of the American Climate Plan, David Anderson stated that he believed the US made a mistake in rejecting Kyoto and that the American plan was not "all that is needed from the United States' side to deal with the issue of climate change."[65] In March 2002, Minister Anderson went to Washington with the objective of encouraging the United States to do more. The aim, according to Anderson, was to bring the US actions into greater conformity with the Kyoto approaches. He did concede, however, that he expected that Canada would be doing more than the US in the short term.[66] He also sought support of Canadian clean energy credits (which he did not get). He did concede that Canada will not get the US to change its mind. But the two countries did announce the intent to increase bilateral cooperation. These efforts can be regarded as Canada trying to undertake a facilitation role but they can also be seen as trying to encourage the US to "do more" – thus leveling the continental playing field.

The irony, however, is perhaps it is Canada that needs to "do more". Canada, and other countries, would be well advised to expand their focus on the US beyond the Bush Administration. A recent study by the Pembina Institute for Sustainable Development and the World Wildlife Fund compares the subnational activities of the US states and the Canadian provinces. The findings are compelling. It is claimed "governments in the U.S. have taken far more significant action to reduce GHG emissions than have governments in Canada."[67] It is further noted:

> State governments in the U.S. are far ahead of provincial governments in Canada in implementing GHG-reducing measures. State actions, while still far from sufficient to reverse the rising trend in total U.S. emissions, are having real impacts now and are gathering pace. This, combined with clear public support for action on climate change, will ultimately make it impossible for the U.S. federal government not to take action.[68]

The significance of this study for Canada is that it undermines the argument that there must be a level playing field between Canada and the US. There is no level playing field. The US or at least American states such as California are leading the way on greenhouse gases reduction and Kyoto or no Kyoto, this is the market in which Canada will have to operate. Furthermore, the reality of subnational government activities in the US challenges claims of Canadian leadership and the assumption that Canada will act to encourage the US along the "right" path.

For Canada the multilateral and the bilateral cannot be easily separated. Efforts to encourage the US to "do more" are completely consistent with multilateral efforts to limit the amount of Canadian domestic reductions. If Kyoto can be crafted in such a way as to meet the US expectations, perhaps through various avenues the US will engage in GHG reduction activities outside the Kyoto fold, but consistent with the rules of Kyoto. Therefore, the level playing field is created (in spite of the questionable nature of this claim) without the US being formally part of the Kyoto Protocol. This is not to claim significant reductions will be achieved, but that somewhere between doing less internationally and encouraging the US to do more domestically, Canada finds a middle ground between the desire to be different and the need to be the same.

Cynicism aside, it is important to note that Canada did stay in the negotiations when in fact they could have withdrawn from them. Part of the reason for staying may be a latent anti-Americanism that surfaces in Canada from time to time. For example, the view put forward by the American ambassador to Canada that Canada should not ratify Kyoto and should join a North American climate accord, would not sit well with Canadians who like to regard themselves as independent actors. In addition, one must return to the bilateral context, especially after September 11, 2001. Through a variety of initiatives ranging from "Smart Border" agreements, to the actions in Afghanistan, Canada appears to be increasingly drawn into the American orbit. Continentalism is running rampant. Also recall the earlier discussion of "Canada as a fading power". In the midst of the increased continentalism exists a sense that Canada is losing its influence – except perhaps multilaterally. Thus, staying in the negotiations may represent an attempt to reclaim the spotlight, at least in some small measure. In other words, it is not just about climate change, but Canada-US relations broadly.

Finally, we must consider the environmental impacts taking place in Canada. In his statement regarding the Bonn Agreements, Herb Gray noted: "Canada is already feeling the effects of global warming, particularly in the North."[69] Economic motives dominate the Canadian agenda on climate change, but the urge to resist continentalism in some small way, coupled with the recognition of environmental impacts compelled Canada to stay in the negotiations, and ultimately, to ratify the Kyoto Protocol.

Concluding Reflections

So, what lessons are to be drawn from the Canadian case? What does it tell us about the future of Kyoto given the US withdrawal? First, as suggested in the preceding section, we must pay attention to the numerous levels at which climate change emission reduction activities can take place. While the US federal government is viewed as obstructionist in its behavior internationally, efforts to reduce emissions are taking place in spite of the Bush Administration. Future initiative to cooperate with the US might best be focused not on the federal government, but the US states and business that can function to create pressure from within the US. Cooperation between Canadian and American and American and European business and subregional groups could be catalytic for change.

Secondly, the future must focus on the meaningful implementation of the Kyoto Protocol by countries that have ratified it, including Canada. As a result of international negotiations, several loopholes have been built into it and expanded in the interests of countries such as Canada and Russia. Canadian ratification would be meaningless were there no incentives domestically and internationally to implement the protocol with the end of achieving real reductions. There is some doubt as to whether Canada is committed to preserving environmental integrity or its international image. Ratification by Canada is not enough. Implementation must be carefully scrutinized.

Third, this analysis gives us pause to consider the meaning of leadership. Canadian leadership, at whatever stage of the negotiations, has been shown to be more superficial than substantive. European leadership was called into question at Bonn. This begs the question: leadership for whom? If Canadian leadership is about pushing for more sinks credits and forcing the hand of the Europeans to concede to its desires, then this is not leadership in the name of environmental integrity. It is "leadership" in its own interests and, one may argue, in the interest of the United States. In the past Canada may have exhibited technical leadership, but in future, one should expect actions consistent with Canadian diplomacy at Bonn and the Hague. Canada will remain central to a well-established blocking coalition and will use its status as a member of the Meeting of Parties (assuming Russia ratifies) to ensure that loopholes are maintained and flexibility in reductions is preserved. The commonality of purpose and commonality of ideology that bound the US and Canada has not dissipated simply because symbolically the two countries are out of step. Canada will endeavor to fashion itself as different and independent from the US, but observers would be well advised to watch Canadian diplomacy carefully and skeptically.

Finally, this chapter encourages us to revisit ideas about multilateralism, at the very least, the Canadian academic dialogue on this matter. Consistent with the insights of Black and Sjolander, multilateralism has served a purpose for Canada wherein it has been party to fashioning an international agreement that privileges economics over the environment. The world order Canada, and by extension other countries, such as the US and Australia, reinforce is one where competitiveness, economic well-being and the private sector dominate. It is a world order that we know to be terribly unequal characterized by multilateralism of the strong. Moreover, Black and Sjolander are accurate in their observation that regionalism and multilateralism reinforce each other. Canada always has an eye toward the US and while there are a few exceptions to this rule, on the matter of climate change Canada's behavior is conditioned by and influenced by the activities of the US. As the international negotiations on climate change continue, the Canadian reality of the relative invariant of the United States must be factored into our analysis of Canadian climate change policy and diplomacy.

Footnotes

1. Kim Richard Nossal, 1997. *The Politics of Canadian Foreign Policy*, (Prentice-Hall, Scarbourough, Ontario, Canada, p. 23.
2. Ibid. p. 28.
3. See Norman Hillmer and Maureen Appel Molot, (eds). 2002. Canada Among Nations 2002: A Fading Power. Oxford University Press, Toronto, Ontario, Canada.
4. Tom Keating, 2002. Canada and World Order: The Multilateralist Tradition in Canadian Foreign Policy. Oxford Univ. Press, Toronto, Ontario, Canada (2nd ed.)
5. See Nossal, The Politics of Canadian Foreign Policy.
6. David Black and Claire Sjolander. 1996. Multilateralism re-constituted and the discourse of Canadian foreign policy. Studies in Political Economy, 49, (Spring): 7-36.
7. See Keating, Canada and World Order, pp. 9-16.
8. Ibid., p. 12.
9. Ibid., p. 14.
10. Kim Richard Nossal, 1998-9. Pinchpenny diplomacy: The decline of good international citizenship in Canadian foreign policy. Int. Journal (Winter): 104.
11. Robert Cox, 1997. Introduction. In: The New Realism: Perspectives on Multilateralism and World Order. Robert Cox (ed), N.Y. United Nations Press, New York, NY.
12. See Andrew F. Cooper. 1997. Niche diplomacy: A conceptual overview. In: Niche Diplomacy: Middle Powers after the Cold War. Andrew W. Cooper (ed.) (St Martin's Press, New York), N.Y.
13. Fen Osler Hampson and Dean F. Oliver. 1998. "Pundit diplomacy: A critical assessment of the Axworthy Doctrine. Int. Journal LIII, (3): 381.
14. Howard Ferguson. 1988. "Draft–World Conference on the Changing Atmosphere: Highlights of the Planning Process.", Atmospheric Environment Service, Toronto, Ontario, Canada, p.2.
15. Heather A. Smith. 1998. "Canadian federalism and international environmental policy making: The case of climate change, Working Paper 1998 (5), Queen's Institute of Intergovernmental Relations, Kingston, Ontario, Canada.
16. See for example, Fred Pearce. 1989. Turning Up the Heat: Our Perilous Future in the Global Greenhouse, Paladin, London: UK, p. 7.
17. Canada's Green Plan for a Healthy Environment. 1990. Minister of Supply and Services), Ottawa, Ontario, Canada, p. 100.
18. Ibid., p. 101.
19. Ibid., p. 15.
20. Ibid.
21. Canada, Climate Change Advisory Committee Reports, January 1991-January 1992.
22. Heather A. Smith. 1993. A new international environmental order? An assessment of the impact of the global warming epistemic community, Unpublished Ph.D. diss., Queen's University, Kingoton, Ontario, Canada, P. 131.
23. Canada, Canadian Delegation Report. 1995. Climate Change First Conference of Parties, p. 5.
24. See Sebastian Oberthur and Hermann Ott. 1995. "The first conference of parties. In: Environmental Policy and Law 25 (4/5): 144-156; and Sebastian Oberthur.

1996. The second conference of parties: In: Environmental Policy and Law 26(5): 195-201.

25. In the context of climate change the Canadian provinces have a very significant role to play. The Canadian constitution is such that the federal government has the right to represent Canada abroad, and negotiate and sign international agreements. However, in areas of provincial jurisdiction, such as natural resources, the federal government is required to cooperate with the provinces on implementation, where implementation affects provincial jurisdiction. This demands a type of two-level game for the federal government, which works multilaterally and then must come home and negotiate with the provinces in a manner whereby it is another member at the table. The role of the provinces is extremely significant to this case.

26. Canada, Environment Canada.1997. News Release: Canada's Energy and Environment Ministers Agree to Work Together to Reduce Greenhouse Gas Emissions, 12 November, p. 1.

27. See Douglas MacDonald and Heather Smith, 1999-2000. Promises made, promises broken: Questioning Canada's commitments to climate change: Int. Journal LV (1): 105-124.

28. The coalition expanded to include Switzerland.

29. International Institute for Sustainable Development. 1997. Earth Negotiations Bull. 12, (76): 41. On-line: http://www.iisd.ca/linkages/download/asc/enb1276e.txt

30. Ibid.

31. Ibid., 35.

32. Ibid.

33. International Institute for Sustainable Development, 27. 1998. Earth Negotiations Bull. 12 (97): 16, 27.

34. Victoria Kellett and Chad Carpenter. 1998. Inching forward at Buenos Aires. Insti. Sustainable Development. website, p. 2.

35. Government of Canada. 2000. Speaking points for the First News Conference by Lloyd Axworthy, Head of Delegation, November 20, 2000. on-line. http://climatechange.gc.ca/english/whats new/001120 s e.htm.

36. International Institute for Sustainable Development. 2000. Climate Canada: A Canadian lens on global climate change, 5 December, p.2.

37. Ibid.

38. Environment Canada. 2000. Statement by Environment Minister David Anderson: Climate change officials make progress, 7 December; p. 1.

39. See Federal/Provincial/Territorial Ministers of Energy and Environment. 2000. National implementation strategy on climate change, Block 2 – International Context, February. 2000.

40. Ibid., p. 24-25.

41. Ibid., p. 28.

42. Hermann E. Ott. 1998. "The Kyoto Protocol: Unfinished Business", in Environment, 40, 6, (July/August), 20.

43. Federal/Provincial/Territorial Ministers of Energy and Environment. 2000, p. 20.

44. Ibid., p. 21.

45. Planetark US stance key to global warming talks – Canada. May 14, 2001. (Planetark.com, 14 May 2001, accessed 10/09/2001).

46. Environment Canada. 2001. News Release: Deputy Prime Minister Herb Gray leads Canadian delegation to climate change negotiations in Bonn, Germany, 13 July 2001, pp. 1-2. On-line: http://www.ec.gc.ca/press/2001/010713 n e.htm.

47. David A. Wirth 2002. The Sixth Session (Part Two) and Seventh Session of the Conference of Parties to the Framework Convention on Climate Change. Ameri. J. Inte. Law 96: 653.
48. Ibid., pp. 654-655.
49. Wirth. 659.
50. Thomas Legge, 2001. "The unexpected triumph of optimism over experience. Centre for European Policy Studies Commentary, 27 July 2001, p. 20. (http://www.ceps.bc/Commentary/July01/unexpected.htm accessed 3/12/02).
51. International Institute for Sustainable Development, Commentary: Canada and Cop-6, p.1 . On-line:. http://www.iisd.org/bonnoped.htm
52. Canada. Office of the Prime Minister. 2001. Statement by the Prime Minister 23, July 2001. On-line: http://pm.gc.ca.
53. Robin Pomeroy 2001. Canada Says Kyoto doesn't need legal sanctions; 2 November 2001. ClimateArk.org, 1.
54. Government of Canada. 2001. Intervention by Canada on compliance" 30 October, 2001. On-line: http://www.climatechange.gc.ca/cop7/compliance statement 10 30 e.htm.
55. Government of Canada, 2001. "Canada Country Statement by the Honourable David Anderson, Minister of the Environment of Canada", 7, November, http://www.climatechange.gc.ca/cop7/Canada Country Statement 1107 e.htm.
56. Government of Canada, 2002. Statement to the Media by the Honourable David Anderson, Minister of Environment, Government of Canada, 1 November 2002. on-line: http:www.ec.gc.ca/press/2002/021101 s e.htm.
57. Government of Canada, Office of the Prime Minister. 2002. Statement by the Prime Minister, 16 December 2002. On-line: (http://www.pm.gc.ca).
58. Ibid.
59. Government of Canada, 2001. Statement by Deputy Prime Minister Herb Gray at Part II of the Sixth Conference of Parties to the United Nations Framework Convention on Climate Change, 23 July, 2001
60. Government of Canada. 2002. Climate Canada Plan for Canada. November 2002, p. 11. On-line : www.climatechange.gc.ca.
61. Ibid., p. 12.
62. Environment Canada. 2001. Canada's Greenhouse Gas Emissions: 1990-1999. On-line: http://www.ec.gc.ca/press/2001/010711 b e.htm
63. Government of the United States, The White House. 2001. Action on climate change review initiatives, p.3. On-line. http://www.gcrio.org/OnLnDoc/climate initiative010713.html.
64. Ibid.
65. Environment Canada. 2002. Media teleconference by Minister Anderson following the US Proposals on clean air and climate change 14 February, http://www.ec.gc.ca/minister/speeches/2002/020215_t_e.htm.
66. David Ljunggren, 2002. Canada to press US to do more on global warming. World Environment News Planetark.org p. 1.
67. Matthew Brambley, Kirsty Hamilton and Leslie-Ann Robertson, 2002. A Comparison of Current Government Action on Climate Change in the U.S. and Canada. Pembina Institute and World Wildlife Fund Canada, Ottawa, Ontario, Canada, p.1.
68. Ibid.
69. Ibid.

CHAPTER 14

Costa Rica and Its Climate Change Policies: Five Years after Kyoto

Ana V. Rojas
CEDARENA, Apto 134-2050, Sc. Pedro Costa Rica
e-mail : anarojas@cedarena.org, ana-v-rojas@hotmail.com
www.cedarena.org

Introduction

An explanation is given of how emission reduction projects have influenced policy-making processes in Costa Rica and how the Activities Implemented Jointly (AIJ) Pilot Phase was a kick start for the Costa Rican Government to experiment and help develop market-based mechanisms capable of achieving environmental benefits at a lower cost for Annex I countries. As a consequence of the experience gained, Costa Rica has had a strong participation during international negotiations. This effort was recognized during COP-7, when the Head of the Costa Rican Delegation was appointed as member of the Executive Board on behalf of Non-Annex I Parties. Therefore, it is only reasonable to expect that Costa Rica will continue to have a strong voice during the climate change negotiation process, especially when project-driven issues are to be discussed.

At the national level, the relevance of Clean Development Mechanism (CDM) projects is evident, and it may have even overshadowed the efforts directed toward vulnerability and adaptation issues. This probability is reflected here in a comparison of the emphasis given to these issues vis-a-vis that given to CDM projects and activities.

Costa Rican Framework for Climate Change

The Ministry of Environment and Energy (MINAE)[1] is the national institution in charge of complying with the United Nations Framework Convention on Climate Change (UNFCCC) and the Kyoto Protocol (KP) mandates. To achieve a better approach to the subject, climate change issues are shared by two offices: the National Metereological Institute (IMN)[2] and the Costa Rican Joint Implementation Office (OCIC).[3]

The IMN is the technical office and represents Costa Rica at IPCC meetings. It is also the institution in charge of: national communications, GHG national inventories, analysis of the country's vulnerability before climate change, as well as development of adaptation and public awareness projects. Meanwhile, OCIC is the office responsible for AIJ pilot phase projects and future CDM projects. OCIC also has the responsibility to follow up international negotiations (COPs and SBSTA/SBI meetings) on behalf of the Costa Rican government.

There is also a Climate Change Consultative Commission at the national level. This working group was created by Executive Decree in 1998,[4] with the objective of coordinating at the national level the efforts to confront climate change. However, at the moment this Commission is not active as its functions, membership, and constitution are under review in order to reflect more accurately the new dynamics—national and international scenarios—surrounding climate change issues.

Costa Rica has translated its climate change commitments into two working areas: establishment of national policies and legislation, and development of emission reduction projects. In the political area, it has introduced different incentives in order to implement a development model that incorporates climate considerations in its equation. For example, the Forestry Law[5] establishes the possibility for the Government to charge for the environmental services from which the population benefits. Mitigation of GHGs is among the environmental services recognized by this legislation.

How does it enter into force? How do you charge for environmental services? One way is through the establishment of a selective tax. This happens in Costa Rica with fossil fuels. Therefore, when a person goes to a gas station to fill the car's tank, a percentage of what is paid for the fuel goes to the National Forestry Finance Fund (FONAFIFO).[6] FONAFIFO distributes the funds among small and medium owners of forest and forestry plantations, thereby compensating the owners for the environmental services their activities produce: either for the emissions avoided avoided from precluding deforestation or from sinking GHG emissions in their plantations.

On the other hand, efforts have been made in the energy sector, both at a rational use level and by emphasizing use of renewable generation sources. In this sense, Law No. 7200[7] and its reform[8] stimulate the participation of private generators and usage of renewables. These legal instruments support the use of biogas (obtained from solid and organic wastes), hydraulic force, and geothermic and eolic sources for energy generation. The legislation contained in the Law for Rational Use of Energy, and the creation of a National Commission for Energy Conservation (this office is part of MINAE) supports these same objectives.

Regarding the second working axis, it is important to mention again that Costa Rica has invested strongly in project development issues. Not only has the country kept an important spot at international negotiations, but OCIC has also gained important experience through the approval and development of emission reduction projects, be they AIJ pilot phase projects or future CDM projects.

Berlin Mandate and AIJ Pilot Phase

In 1995, the Berlin Mandate was adopted, allowing signing parties to develop pilot AIJ projects. The objective of this pilot phase was to create practical examples to be analyzed, and through these experiences improve the procedures and requirements needed for the emission projects. In other words, the pilot phase is a practical laboratory.

Straight away Costa Rica got fully involved in the pilot phase. OCIC was created by the end of the year, and started promoting individual AIJ projects. For this reason, Costa Rica was able to put together four different projects early in the pilot phase: ECOLAND, KLINKI, CARFIX and BIODIVERSIFIX. The first is an AIJ project between Costa Rica and the United States and has been wholly executed. The objective of ECOLAND is to consolidate the limits of the Piedras Blancas National Park (2,340 ha), located in the Osa Peninsula. The environmental benefits of the project have been calculated in 366,2000 metric tons of carbon, with a total cost of US $ 1,000,000.[9]

The Klinki Forestry Project is located in Turrialba and was approved in 1995. It is a klinki pine (*Araucaria hunsteinii*) reforestation project in an area of 6,000 ha. For this project, the benefits are calculated in 1,966,495 metric tons of carbon, and the cost of the project US $ 3,800,000[10]

BIODIVERSIFIX is a restoration and consolidation project for both the dry and humid forests located in the Guanacaste Conservation Area (ACG).[11] For practical reasons the project was divided into two zones: restoration of 45,000 ha of abandoned pastures into regeneration zones for tropical dry forest (DRYFIX), and a restoration area of 13,500 ha of abandoned and regeneration lands for tropical humid forest (WETFIX).[12]

CARFIX is a project of sustainable forestry management in the Central Volcanic Range Conservation Area (ACCVC).[13] Among its activities it includes regeneration of natural forest (10,670 ha), forest sustainable management (20,502 ha) and forestry plantations (5,533 ha). The main objective of this proposal is to preserve the forestry cover of the ACCVC, while supporting activities which increase carbon sequestration. This project is under the administration of FUNDECOR.[14]

It was with this project-based experience that Costa Rica entered the international negotiations held in Kyoto. By this time, the Costa Rican Government was very fond of the market and project-driven approach. Therefore, it is no wonder that the national delegation strongly supported the introduction of flexible mechanisms in the KP. These mechanisms were viewed as an alternative for combining funds for domestic environmental protection and yet contributing to emission reductions worldwide.

Article 12 and its CDM were born as a symbiosis between two different proposals.[15] The first was the Brazilian proposal for the creation of a Clean Development Fund. The driven idea was to establish a punishment/compensation system, by which noncomplying Annex I countries had to contribute financially to mitigation and adaptation actions. The Fund was to managed by the GEF (Global Environmental Facility).[16]

The second proposal was to introduce financial mechanisms that would ease in an economic sense the commitments of Annex I countries and enable them to still achieve their emission reduction targets. This proposal concerned the flexible mechanisms Emissions Trading and Joint Implementation.

Hence, before and during COP-3, both the Brazilian and Costa Rican delegations worked on texts that could combine the existence of a fund for adaptation and mitigation projects with trade mechanisms that would allow financial and technology transfer.[17] The result was a double purpose mechanism. On the one hand, it is an instrument for Non-Annex I countries to achieve sustainable development, for it increases financial and technology transfer. On the other, it helps Annex I countries to achieve part of their emission reduction targets at a lower cost. Moreover, it establishes that a percentage of the funds produced by CDM projects (credits) has to be destined for adaptation projects in Non-Annex I countries.

Beyond Kyoto

Costa Rican Portfolio: avoided deforestation, LULUCF and renewable energy projects

As stated above, OCIC is focused on developing AIJ and CDM projects. Therefore, it is also the office entrusted with development of the legal framework requisite for implemantation of AIJ and CDM projects, and the institutional framework required to facilitate the issuance and selling of the credits generated by the projects.

From its creation, OCIC's activities at international negotiations for project promotion had focused on forestry projects. Due to the international scenario and pressure against the inclusion of avoiding deforestation projects, OCIC had to shift its tactics.

In order to achieve its goals, OCIC then established alliances with different organizations at national level. Among them, the most important is FUNDECOR, which provides incentives for forestry projects that capture or reduce CO_2 emissions. OCIC also has cooperative agreements with the Costa Rican Coalition for Development Initiatives (CINDE)[18] and the Costa Rican Electricity Producers Association (ACOPE).[19]

It may be recalled that the early development of forestry projects allowed OCIC to "learn by doing" the needed requisites and procedures for a project to effectively achieve avoidance or reduction of GHG emissions. After the individual projects were created, OCIC began a second phase constituted by the so-called umbrella projects. This means that the projects developed included a series of small projects or components. This is how the Reforestation and Forest Protection Project (known as PFP) and the Territorial and Financial Consolidation Project for National Parks and Biological Reserves of Costa Rica (known as PAP) came to life.

PAP is an effort to consolidate protected areas under the absolute conservation category in the country (national parks and biological reserves). Its implementation is to be translated into conservation of 530,498 ha, equivalent to a total of 18,000,000 metric tons of carbon. It is important to highlight that the BIODIVERSIFIX project was included under the PAP.

Through the PFP, the Costa Rican Government acquires the compromise to implement the "payment for environmental services", and recognizes the quantified emission reduction benefits generated by the private conservation and plantation activities. In other words, the PFP is a framework for FONAFIFO's activities regarding payment of environmental services. Moreover, this project incorporates the CARFIX project mentioned before.

Once these umbrella projects were constituted, a financial instrument was developed. It is a certification of the amount of GHGs mitigated by a project and is known as a Certified Tradeable Offset (CTOs). PAP obtained a certification from the Societe Generale de Surveillance (SGS), due to which the first CTOs of emission for this project were achieved. On the other hand, PFP has also generated CTOs, which have been traded with the Norwegian Government and one of its national enterprises.

Regarding the energy sector, Costa Rica has already presented several renewable energy projects to the UNFCCC's Secretariat[20] (Table 14.1). This energy portfolio is mainly constituted by hydroelectric and aeolic projects.[21]

OCIC also has the Energy Export Project to Central America. This particular project is supported by ACOPE and offers the possibility of expanding its coverage and becoming a national project. Its design includes an installed capacity of 268 MW, with an annual generation of 1,400 GWH, producing a reduction of 1,400,000 metric tons of CO_2.[22]

Table 14.1: Renewable Energy Projects presented to the UNFCCC Secretariat by Costa Rica

Project Name	Type of Project	Installed Capacity (MW)	Emission Reductions (tm C)
Doña Julia	Hydroelectric	16	562,020
El Encanto	Hydroelectric	8	300,000
Tierras Morenas	Aeolic	20	562,020
Aeroenergía	Aeolic	6,4	146,000
Aeolic plants	Aeolic	20	506,720
Tejona	Aeolic	20	800,000
ICAFE	Biomass	No data	17,323

The last big project in which OCIC has been involved is ECOMERCADOS.[23] This is a package that includes a loan for US $32,600,000 from the World Bank, and a GEF donation for US $8,000,000. This project was designed by OCIC, in collaboration with MINAE and the Ministry of Treasury.

The ECOMERCADOS objective is to develop markets for the environmental services produced by the forest ecosystems, and to contribute to national development through forestry and energy generation projects. Therefore, it contains the following components:[24]

* Strengthen development of the environmental services markets
* Increase the number of certified emission reductions (CERs)
* Strengthen administration and field supervision of the payment for environmental services program (PSA)[25]

As for the distribution of the funds, this will be done in the following way: the GEF funds will be used to pay for biodiversity conservation services; US $14,000,000 will be destined for the payment of previous PSA contracts; US $9,000,000 will be directed to forest conservation in priority areas of the Mesoamerican Biological Corridor; and US $10,000,000 used to promote a CDM portfolio based on renewable energy generation.[26]

Regarding the project portfolio of emission reductions financed by ECOMERCADOS, it is constituted as shown in Table 14.2.[27]

International Negotiations and the Costa Rican Position: LULUCF and CDM

Costa Rica was an active participant in the negotiations of KP rules. Its scope of action can be separated into two different but related issues: (Land use, landuse change and to Forestry) projects and rules on CDM. This is evident when reviewing the national submissions presented before the UNFCCC Secretariat, especially during the previous COP-6 meetings.

From the Costa Rican perspective, an exclusionary interpretation for Article 12 was (and is) inconsistent with the mandatory accounting

Table 14.2: Projects Financed by Ecomercados

Project Name	Type of Project	Installed Capacity (MW)	Emission Reductions (tm C)
Chocosuela II	Hydroelectric	16	158,400
Chocosuela III	Hydroelectric	5	49,500
Brasil II	Hydroelectric	30	291,060
Río Azul	Bio-thermal	3	88,000
Cote	Hydroelectric	6	26,532
General	Hydroelectric	39	392,040
Jiménez	Hydroelectric	50	435,600
San Gabriel	Hydroelectric	0.375	3,372
Caño Grande III	Hydroelectric	3.3	24,413
PE Chorotega	Aeolic	6.6	NA
PE Vara Blanca	Aeolic	9.6	NA

framework for Annex I Parties established under Article 3.3 of the KP.[28] According to this rule, Annex I countries must take into account certain LULUCF activities in order to achieve their commitments under Article 3. Since the objective of article 12 is to help Annex I Parties to achieve their commitments, it is therefore only logical that "... the scope of projects eligible under Article 12 should correspond to the activities eligible under Articles 3.3 and 3.4."[29]

The exclusion of LULUCF projects from CDM was (and is) considered as inconsistent with the purpose of Article 12 to assist Non-Annex I countries to achieve sustainable development. The sustainable management of natural resources is critical for the achievement of sustainable development as well as for addressing vulnerability to climate change.[30] Furthermore, experience has shown that LULUCF projects do help local communities to develop, as they present new ways of increasing local income.[31]

During the negotiations it was argued that since Article 12 used the term "emission reductions", only projects which actually reduced emissions were eligible under CDM. However, the term "emission reductions" is never defined in the Protocol's text or in the Convention. In fact, this term is used in reference to units of account rather than particular types of activities. Costa Rica's position was that, whenever the drafters wanted to distinguish between different types of activities, they explicitly named those which were eligible or those which were excluded from certain kinds of projects. And in the case of CDM there is no explicit distinction among types of projects.

Going further, even if the term "emission reductions" is taken literally for Article 12, it is important to note that not all LULUCF projects are sinks. It has been recognized by the IPCC that forests can be sources, sinks or reservoirs. Therefore, LULUCF projects can slow, reduce or avoid deforestation. These projects reduce anthropogenic emissions by sources.[32] Hence, Costa Rica advocated from two different fronts, first the inclusion of LULUCF in general, and if this failed, at least for the inclusion of avoided emissions through avoided deforestation projects (the so-called "conservation projects").

In a different order of things, Costa Rica used the experience gained during the pilot phase in order to propose its position regarding the CDM structure. The various contributions made by the country can be tracked through the early submissions to the UNFCCC Secretariat,[33] as well as in the different consolidated texts produced by the Chair of the Mechanisms Negotiating Group.

Costa Rica's contributions in the development of these rules were recognized during COP-6 since it was included in the "Friends of the Chair Working Group". This was a reduced negotiating group which helped the Chair of the Mechanisms Working Group to consolidate the different positions still held during the Hague convention.

International scenario and its outcome

Rules on CDM and its impact for Costa Rican projects

During COP-6 bis, although consensus couldn't be achieved in all the negotiation items, a political agreement was produced.[34] This result was almost a miracle. After the failed negotiations in the Hague, and the withdrawal of the USA from the KP, most people thought that the Protocol and the Convention were dead.

The so-called Bonn agreement became the framework for further negotiations on the KP rules during COP-7. However, the text left very small room for further negotiations in some cases. This agreement held the KP issues in four different areas: financial issues, LULUCF projects, flexible mechanisms, and compliance.

Regarding CDM, several agreements were achieved. Perhaps the most important one from a Costa Rican perspective was the definition of LULUCF activities eligible under Article 12. Contrary to the logic of the article, "conservation" projects (avoided deforestation) were excluded from CDM, while reforestation and afforestation projects (sinks) were included as eligible projects under Article 12.

During COP-7 several other issues were defined. For example, the fact that eligibility of "conservation" projects for the second commitment period would be defined in further negotiations, contrary to what some countries held; (if excluded from the first commitment period, they should be excluded from the following periods as well.)

The definition of rules for the design, operation, and monitoring of LULUCF projects will be studied by the IPCC and a report presented for COP-9, when rules on this matter should be adopted. Therefore, there will be some uncertainty, at least for the next two years, regarding the regulations required for a LULUCF project to be included as a CDM activity.

For Costa Rica, as well as for the other countries willing to participate in such activities, this means that either they start developing projects now, trying to address leakage and permanence issues in the best way possible, or they wait until the rules are defined. Both options are risky. The first requires a great amount of investigation, and a strong monitoring of the project, with the danger that when the rules are decided, their measures would not comply with the requisites established. However, if the procedures adopted are similar to those ruled by COP-9, these projects would be able to market their CERs soon after their approval. On the other hand, waiting until COP-9 to design and create a project based on the rules decided by the Parties means that the developers might be years behind their competition.

For Costa Rica, it seems that the OCIC will not initiate or propose new LULUCF projects itself. However, it will not stop any private developer from presenting new project ideas in this field.

The agreement achieved in Marrakech establishes other institutional regulations, such as the environmental additionality and reductions in environmental and social impacts due to the development of a project. Besides, Parties agreed on the retroactivity of CDM eligible projects. This means that a project's emission reductions can be calculated from January 2000 and forward. Any emission reductions obtained prior to this date will not be considered as a CER. In order to be granted this privilege, the CDM project has to be validated and registered before December 2005. This decision opens an opportunity for Costa Rica to present its AIJ projects as CDM projects.

During COP-7 the CDM Executive Board was established. Its main objectives include development of procedural rules and registry systems for accounting and information, establishing methodologies for baselines and monitoring procedures, and development of fast-track procedures for small projects. The work of this Board will be crucial during the next few years in order to kick start CDM, and properly define the procedures and guidelines projects should follow.

Costa Rica understood the importance of participating first hand in this phase. As a result of its participation during the whole KP process and its negotiations, its Ambassador in Special Mission before the UNFCCC, Mr. Franz Tatenbach, was appointed by consensus of the G77 and China as one of its two representatives before the Executive Board. For the country it is a great recognition of the effort put in during the different negotiations, and an invaluable opportunity to further contribute to the development of guidelines and methodologies for CDM projects.

Rising issues: vulnerability and adaptation

So far, at international and national levels, most of the efforts toward climate change have focused on project development. It could be said that there is a State policy to develop mitigation projects. However, this is not true for vulnerability and adaptation.

Although OCIC is a governmental office, it has the profile of a private company. Perhaps this is a consequence of its own nature—the development and support of projects. But the National Meteorological Institute (IMN), the body responsible for vulnerability studies, is a technical one. Big efforts have been undertaken in this area, and the Costa Rican National Communication was one of the first to be ready in Central America. But this is not enough.

Central America is very vulnerable to climate change. So far, the region has suffered extreme climate events: hurricanes followed by droughts. Although these events have affected Costa Rica in a lesser way, they are an indicator of the possible future. These experiences should be studied in order to develop an adaptation strategy.

There is a growing need for the IMN to take a stronger position and to start evolving from a technical body into an information center, and a discussion and policy-making entity. Part of the need to strengthen or highlight this functions is the reason why the decree No. 26964-MINAE-MIDEPLAN is being studied and restructured. Attention should be drawn to the future reality of the country.

Steps are being taken regarding this matter. At this point, the IMN is strongly reflecting on the need to work in a collaborative way with research institutions, universities, ministries, and NGOs.[35] Hopefully in a short time Costa Rica will be able to have a participatory policy-making process on vulnerability and adaptation.

Conclusion

As explained throughout this chapter, the Costa Rican government placed a bet on a "learning-by-doing" process. And the outcome has been a strong presence during the UNFCCC and KP negotiations. It seems that after 5 years the bet is paying off, and its effort has been recognized. Participation in the CDM Executive Board ensures that the work to develop project rules will be sustained at an international level for at least the next two years.

At the national level, development of CDM energy projects will increase early in the process. However, it is important that these projects not overshadow the vulnerability and adaptation aspects of climate change. This will only be obtained with a strong participation of NGOs and research entities. A strong lobby work should be done in order to produce and implement national policies and strategies regarding the country's vulnerability.

Footnotes

1   Ministerio de Ambiente y Energía (in Spanish).
2   Instituto Metereológico Nacional (in Spanish).
3   Oficina Costarricense de Implementación Conjunta (in Spanish).
4   Executive Decree No. 26964-MINAE-MIDEPLAN, 1998
5   Law No. 7575, 1996
6   Fondo Nacional de Financiamiento Forestal (in Spanish).
7   Law No. 7200, 1990
8   Law No. 7508, 1995
9   COSTA RICAN GOVERNMENT, 1999. National Report on Activities Implemented Jointly, p. 7
10 Ibid
11 Area de Conservación Guanacaste (in Spanish).
12 CHACON, CORDOBA and MACK. 1998. Pilot Phase Joint Implementation Projects in Costa Rica, p. 42
13 Area de Conservación de la Cordillera Volcánica Central (in Spanish).
14 CHACON, CORDOBA and MACK, p. 43. FUNDECOR is the Fundación para la Conservación de la Cordillera Volcánica Central. In 2001 they won the International Award for Development, King Baudouin, for their conservation work.
15 CURTIS and AMIN ASLAM. 1998. "The Clean Development Mechanism".
16 Executive Summary of the Brazil Proposal for Definition of the Protocol to the Convention, 1997.
17 Franz TATENBACH, Interview, 2001.
18 Coalición Costarricense de Iniciativas de Desarrollo (in Spanish).
19 Asociación Costarricense de Productores de Electricidad (in Spanish).
20 COSTA RICAN GOVERNMENT, p. 12.
21 This matrix was generated with information gathered from the following documents: COSTA RICAN GOVERNMENT, National Report on Activities Implemented Jointly, and MINAE, Annual Report, 2001.
22 MINAE, Annual Report, 2001.
23 Law No. 8058, 2000.
24 ECOMERCADOS PROJECT. 2000. Summary Description of the Project, p.1.
25 Pago por servicios ambientales (in Spanish).
26 BALTODANO. 2000. Benefits and weaknesses of the Ecomercados project.
27 MINAE Annual Report, 2001. pp. 73-74.
28 COSTA RICA, FCCC/SB/2000/MISC.1/Add. 2, p. 2. This document was also signed by the following Latin American countries: Argentina, Bolivia, Chile, Colombia, The Dominican Republic, Ecuador, Guatemala, Honduras, Mexico, Nicaragua, Panama, Paraguay and Uruguay.
29 Ibid., p. 3.
30 Idem.
31 Franz TATENBACH. 1998. The Costa Rican Experience with Market Instruments to Mitigate Climate Change and Conserve Biodiversity, p.16.
32 COSTA RICA, FCCC/SB/2000/MISC.1/Add. 2, p. 2
33 For detailed information on the Costa Rican proposal, consult the following document: REPUBLIC OF COSTA RICA, Partial Proposal: Clean Development Mechanism (article 12). Kyoto Protocol. United Nations Framework Convention on Climate Change, 1999.
34 UNFCCC, Doc. FCCC/CP/2001/L7, 2001.
35 Rita, CHACON, Interview, 2001.

References

Laws and executive decrees
Law No. 7200, 1990
Law No. 7508, 1995
Law No. 7575, 1996
Law No. 8058, 2000
Executive Decree No. 26964-MINAE-MIDEPLAN, 1998
Articles and books
BALTODANO, 2000. Benefits and Weaknesses of the Ecomercados Project.
COSTARICAN GOVERNMENT. 1999. National Report on Activities Implemented Jointly.
CHACON, CORDOBA, and MACK. 1998. Pilot Phase Joint Implementation Projects in Costa Rica.
CURTIS AND AMIN ASLAM. 1998. The Clean Development Mechanism.
ECOMERCADOS PROJECT. 2000. Summary Description of the Project.
MINAE. 2001. Annual Report.
TATENBACH. 1998. The Costa Rican Experience with Market Instruments to Mitigate Climate Change and Conserve Biodiversity, p.16
UNFCCC Documents
Executive Summary of the Brazil Proposal for Definition of the Protocol to the Convention, 1997.
COSTA RICA, FCCC/SB/2000/MISC.1/Add. 2
REPUBLIC OF COSTA RICA. Partial Proposal: Clean Development Mechanism (article 12). Kyoto Protocol. United Nations Framework Convention on Climate Change, 1999.
UNFCCC, Doc. FCCC/CP/2001/L7, 2001
Interviews
Rita Chacón, 2001
Franz Tatenbach, 2001

Chapter 15

India and the Climate Convention: The Challenge of Sustainable Development

Joyeeta Gupta
Institute for Environmental Studies, Vrije Universiteit, Amsterdam, The Netherlands

Introduction

During the last few decades a large number of environmental treaties have been negotiated at the international level. In the days of globalization, it is inevitable that nations all over the world will be involved in the process of treaty negotiation. But while these countries are drawn into the negotiating process, many are less than prepared to deal with the complex issues involved. India, too, has been caught up in the global commitment to address global environmental issues and has been participating in a number of international treaties. India has signed and ratified, among others, the Montreal Protocol on Ozone Depleting Substances,[1] and its London and Copenhagen amendment, the Basel Convention,[2] the United Nations Convention on the Law of the Sea,[3] the Convention on Biological Diversity,[4] the CITES Convention,[5] and the United Nations Framework Convention on Climate Change (UNFCCC)[6]. From a generally defensive role[7] in international environmental treaties, India is moving very slowly towards a proactive policy. In October 2002, India hosted the Eighth Conference of the Parties to the Climate Change Convention. The act of hosting the Conference is viewed as "an important capacity building exercise in the country and will also provide an opportunity to showcase efforts made by India in the environmental arena."[8] Against this background, this paper analyzes the role of India in relation to the Climate Change Convention.

Increasingly a common thread through many of these negotiations is the notion of sustainable development. It plays an important role in Agenda 21,[9] the Rio Declaration,[10] the UNFCCC, and several others. It is against the background of the evolving discussion of sustainable development that this chapter analyses the role of India in the climate change negotiations. First the legal commitment to the concept of sustainable development in the climate change regime is highlighted. Next the concept of sustainable development from a broader multidisciplinary perspective is examined. Then the historical evolution of India's negotiating strategy in relation to the climate change regime is presented, followed by a brief examination of the domestic policy of India on climate change. Based on this information, the strengths and weaknesses of India in the international negotiating arena, and the threats and opportunities facing the country are analyzed, after which some conclusions are drawn.

Sustainable development and climate change

Legal commitment to sustainable development in the climate regime

Climate change negotiations began in 1990. The key landmark dates in the negotiations are the adoption of the UNFCCC in 1992, adoption of a decision to seek an agreement that included quantitative commitments for the developed countries at the first Conference of Parties in 1995,[11] adoption of the Kyoto Protocol to the Convention in 1997[12], and adoption of the Marrakech Accords in 2001[13] (for details see Gupta and Lobsinger, this volume).

A cornerstone of the climate change agreements is the concept of sustainable development. This section first reviews what the agreements have to say about sustainable development. The Climate Change Convention states clearly that Parties have a right to and should promote sustainable development.[14] However, in the rest of the text, the term "sustainable development" does not appear; instead the expressions "sustainable social and economic growth"[15], and "economic development should proceed in a sustainable manner" prevail.[16] The obligation of countries to sustainably manage sinks and reservoirs of greenhouse gases is then highlighted.[17] Such sustainable economic growth is not just included to satisfy the developing countries who have been traditionally skeptical of the concept, but also the developed countries.[18] The way sustainable development is included in the Climate Change Convention highlights the controversial commitment to the concept. At that time, Parties understood that sustainable development would imply a change in development patterns which could bring costs with it. The choice was thus sustainable development or development, and in order to integrate the concepts, it was chosen to see this in terms of a linear development from development to sustainable development.

By the time the Kyoto Protocol was adopted in 1997, it appears the ambiguity had disappeared. The Protocol emphasizes that developed countries must aim at sustainable development through a range of measures focusing on energy efficiency, sinks and reservoirs, agriculture, transport, renewable energy, etc.[19] It emphasizes the need to achieve sustainable development in achieving the quantitative goals of the convention.[20] The Protocol specifies the need to ensure sustainable development in relation to financial assistance and technology transfer to developing countries,[21] and that Clean Development Mechanism projects should focus on sustainable development.[22]

In the Marrakech Accords of 2001, the term sustained economic growth has almost entirely disappeared from the text. The Ministerial Declaration emphasizes the relationship between climate change and the other conventions and that through these means one should try and achieve sustainable development.[23] The Declaration states: "The problems of poverty, land degradation, access to water and food and human health remain at the center of global attention; therefore, the synergies between the UNFCCC, the Convention on Biological Diversity, and the United Nations Convention to Combat Desertification in those Countries Experiencing Serious Drought and/or Desertification, Particularly in Africa, should continue to be explored through various channels, in order to achieve sustainable development."[24] The articles on capacity building, vulnerability and adaptation programs all emphasize the need for sustainable development and that these should be in line with national sustainable development strategies.[25]

Concept of sustainable development

What then is sustainable development and why does it appear so attractive and yet so controversial? From the legal perspective it is useful to distinguish between the principle or right of sustainable development and the law of sustainable development. The principle of sustainable development calls on the global community to focus on protecting the needs of future generations while meeting the needs of current generations.[26] It has thus a strong equity and a precautionary dimension[27]. The law of sustainable development, a term that has recently acquired currency, refers to seven sets of principles according to the International Law Association. These include the responsibility of states to ensure sustainable use of natural resources, the principles of equity and eradication of poverty, common but differentiated responsibilities, precautionary approach, public participation and access to information and justice, good governance and integration with human rights and social, economic and environmental objectives.[28]

At the same time, sustainable development is based on the idea that countries can continue to grow while they reduce their emissions per unit product. This is based on the optimism that technologies will become available that will gradually increase the efficiency of production processes. However, these technologies will be developed and/or become affordable as and when societies become richer. This is what economists refer to as the

Environment Kuznets curve or the inverted U curve.[29] This curve shows that as countries become more developed their pollution increases, but beyond a certain point in the developmental process pollution decreases. This leads Southern nations to argue that they should be allowed to grow before sustainable development becomes affordable for them. The process can be accelerated if Northern nations make such modern technologies financially accessible to the developing countries.[30] This perception underlies both the angst towards sustainable development in developing countries and their persistent demand for access to technologies and aid in a number of international treaties.[31]

India's evolving role in climate change negotiations

Negotiating in the international arena

India is a large country (the seventh largest in the world) with the second largest population in the world. It is also among the poorest countries in the world but with a huge and rapidly developing middle class. It is not surprising that India's total emissions of greenhouse gases is quite high. In some estimates India's total emissions of CO_2 equivalent emissions was 164 Mt C in 1990 and might increase to 693 Mt C in 2020. India's emissions per capita are low about 0.2 t C - 0.3 t C[32]. The Indian Institute of Science in Bangalore estimates that although net emissions from forests were negative in 1990, this has possibly increased to 29 Mt in 2000 and may rise to 77 Mt by 2020 because of forest degradation and reductions in forest carbon intensity.[33] However, its per capita emissions remain very low; the literature indicates that the per capita emissions of carbon are 0.2-0.3 metric tons.[34]

As far back as 1989, the Government of India's position on climate change was that the problem was global and hence the solutions too must be global. Furthermore: "It may be counterproductive to lay down targets for countries which are still striving to raise the living conditions of their masses. It may be equally counterproductive to reach agreements to combat climate change, without devising mechanisms to ensure global participation".[35] In India's preparation for the 2002 World Conference on Sustainable Development, the Government reiterated that the climate change policy of India is based on three ideas: first, developed countries are primarily responsible for the emissions; second, developing countries need to focus on fulfilling the primary developmental needs of their populations; and finally, that the developed world should transfer resources to the developing world in order to assist them in meeting these goals.[36] This argument is based on the principle that India's per capita emissions are low.[37] Thus, to the outside world, the core ideas of India have not changed. But is India quite as defensive as the policy suggests?

Although the issue has now been ten years on the formal domestic agenda as an imported agenda item,[38] there is not yet a formal official evaluation of the greenhouse gas emissions in India. While in the initial years the data on emissions and impacts, and appropriate policy measures, was far from adequate to support negotiations, the data since then has improved somewhat but is not yet definitive. Meanwhile, researchers have been investigating a wide variety of policy options for India. The preliminary data suggested that India's per capita emissions were minimal compared with those of the developed countries and this therefore reduced the urgency for action within India. But because of limited domestic information, negotiators were often taken by surprise by scientific information circulated by the West, such as the reportedly high methane emission levels of India.[39] This has made it difficult for the Government of India to take a strong but constructive negotiating stand in the negotiations.[40]

By 1992, the Government was negotiating in favor of emission reduction in the North and the need for the North to pay the agreed fixed incremental costs of taking action to reduce emissions in the South.[41] The then Minister of Environment, Kamal Nath, emphasized in a number of speeches that the technologies transferred to India should be "clean" and not "cleaning", affordable, accessible, and without strings attached—a point that the then Prime Minister reiterated in his speeches also.[42] There was also support for the idea of the previous Prime Minister, Rajiv Gandhi, that there should be a planet protection fund established by a 0.1% of GNP from all countries except the least developed.[43]

The Climate Change Convention adopted in 1992 was acceptable to the Government of India, but did not reflect many of its concerns. Recognition of the per capita emissions was limited to a reference in the Preamble,[44] the Global Environment Facility was entrusted with the responsibility to be the interim financial mechanism though India and other developing countries were not initially in support of this,[45] and the concept of Joint Implementation was introduced, despite developing country skepticism of the concept.[46] At the same time, the target for the developed countries was nonbinding in nature[47] and there were no quantitative and, hence, measurable obligations to assist developing countries. The lack of specific quantitative commitments for India made it easy for the Government to sign and ratify the Convention in quick succession.[48] India ratified the Convention on November 1, 1993.

Immediately after the negotiations of the Convention, it was clear that the developed countries should have legally binding quantitative commitments. While this had been a key point in India's negotiating position, it appears that India began to have second thoughts; it was feared that if the developed countries had binding targets, it would not be long before such targets would also be applied to developing countries. But possibly because of pressure of non-state actors, the Government finally came up with a green paper which emphasized the need for the developed countries to reduce their greenhouse gas emissions by 20%.[49] Professor Mwandosya, who was chairing the G77 in relation to the climate change negotiations in 1995, recorded with admiration the way the Government of India negotiators set about convincing the rest of the developing countries to push the developed countries into taking action. In doing so, India isolated the oil-exporting countries who had been opposing strong measures out of fear that this would reduce their export income.[50] The so-called 'green-G77' was born, and this was to be a reminder to the OPEC countries that the G77 would negotiate, if necessary, without them. (OPEC of course eventually retaliated by offering to chair the G77 with Nigeria as Chair in 2000 and Iran in 2001!). The pressure of the green G77 and the European Union led to the eventual adoption of the Berlin Mandate in 1995.[51]

At the same time, it is unclear as to what precisely was the Government's role in the negotiations over the concept of Joint Implementation. While documents and interviews indicated that the Ministry of External Affairs and the Ministry of Environment and Forests opposed the concept, there are diplomatic rumors that the Indian government eventually supported the process that led to the adoption of the concept of Activities Implemented Jointly (AIJ) in 1995, which was a voluntary, pilot phase of Joint Implementation without credits. In 1997, at a conference on joint implementation, the then Minister of Power, S. Venugopalachari, emphasized that for the pilot phase to be successful a number of projects in different sectors and countries needed to be initiated,[52] and the conference concluded that AIJ projects would only succeed when they would also take into account local priorities and not when local priorities were sacrificed in favor of global priorities.[53]

While preparations for the Kyoto Protocol negotiations were in progress, the US Senate adopted a resolution that the US would not be willing to sign and ratify the Kyoto Protocol until key developing countries promised to participate in a meaningful manner.[54] The pressure thus began to build up on India to demonstrate meaningful participation. Meanwhile Brazil tried to push the idea of the Clean Development Fund, which unlike Rajiv Gandhi's idea would be financed by fines from countries in noncompliance with the Protocol.[55] With increasing developing country opposition for the idea of Joint Implementation, the Clean Development Mechanism took a more centerstage role in the negotiations, and by the end of the negotiations this concept was adopted in a revised form in which it is arguably a new name for Joint Implementation. It was supposed to be different in that it would focus on sustainable development issues, i.e. it would take the message of the Indian Joint Implementation Conference on board. But, in fact, since Joint Implementation projects under Article 6 are also subject to the general obligations in Article 2 and 3 of the Kyoto Protocol, such projects must also meet the criteria of sustainable development. Instead the Clean Development Mechanism is bureaucratically more cumbersome and subject to an adaptation tax.[56] India also accepted the concept of emission trading at the Kyoto Protocol, but discussion with the negotiators revealed that at best they had no choice but to accept the concept and, at worst, they did not understand the implications of introducing this concept. Later, in a private comment in a newspaper, the Indian negotiator spoke of the (belated) attempts of the South to express its concern about emission trading in 2000.[57] Ultimately, the Government of India has delayed signature of the Protocol as a means to express its annoyance with the link to meaningful participation, and because of the metamorphosis of the Clean Development Mechanism and the introduction of emissions trading.[58]

The following year at Buenos Aires, Argentina tried to include the issue of voluntary commitments for developing countries, but because of opposition the item was dropped. However, Argentina[59] and Kazakhstan[60] decided to go ahead with voluntary commitments; both sensing that there would be financial rewards for doing so. Since then, Argentina has been facing economic crises and has not followed up on its earlier statements, but Kazakhstan has gone ahead. India continues to take the defensive role.

Meanwhile India has participated in several international research programms: with UNEP in relation to the development of a methodology for costing of emission abatement studies, with the Global Environment Facility, and with the Asian Development Bank. In 1999, India and the United States signed a Joint Statement on Cooperation in Energy and Related Environmental Aspects. The following year, the Prime Minister of India, Atal Bihari Vajpayee and US President Bill Clinton adopted a "Joint Statement on Cooperation in Energy and Environment between India and the United States" with the intention to cooperate in relation to the Kyoto Protocol and the Clean Development Mechanism; to this end they established a Joint Consultative Group on Clean Energy and the Environment. Protocol of Intent was signed on September 15, 2000 between the US Government and India to promote cooperation between the US Agency for International Development and the Department of Energy with the Ministry of Power in India in relation to modern power technologies.

India is under an obligation to respect the principles in the Climate Change Convention, to implement the obligations in Article 4.1 of the Convention and Article 12 of the Convention, subject to the availability of financial resources and technological help and in accordance with the principle of common but differentiated responsibilities. Under the Kyoto Protocol India is under an obligation to implement Article 10 of the Protocol and is invited to participate in the Clean Development Mechanism. A key obligation is to submit a National Communication under Article 12 of the Convention. At the same time, it is specified in the Convention that this obligation is subject to the availability of funds. It is reported that the Global Environment Facility has a limit to the amount of funds ($350,000) it can make available for preparing such Communications and that such funds are inadequate to cover the costs of preparing a National Communication for India.[61] Official sources also feel that diverting huge resources to conduct a study when the country has a number of priorities is difficult to justify domestically.

At Marrakech, funds have been established and processes launched to stimulate capacity building and technology transfer. Since Marrakech, the EU and Japan have ratified the Protocol. Russia may ratify in the near future. India, too has ratified the Protocol. The question is: How best can a proactive policy on climate change be designed? But before addressing this question, it may be useful to understand the domestic policies in India.

Domestic context

Under the Indian constitution, the state shall promote international peace and security and foster respect for international law and treaty obligations.[62] However, international treaties do not automatically become part of national law; there has to be a national legislative process to incorporate relevant treaty provisions in municipal (national) law and national law even if it is contradictory to international law has to be respected.[63] Given that India signed and ratified the Climate Change Convention, it would not be too far-fetched to imagine that climate change as an issue would have featured in the Eighth Five Year Plan (1992-1997).[64] But this did not happen. In 1991, a Coastal Zone Regulation Notification was passed calling on the provinces to prepare appropriate policies. In 1992 the Conservation Strategy of the Government of India did, however, mention the need to develop coping mechanisms to deal with climate change. The Ninth Five Year Plan (1997-2002) makes no recommendations with respect to climate change but does integrate the CO_2 discussions within its energy policy.[65] Few of the Annual Reports of the Ministry of Environment and Forests have provided any real coverage on the climate change issue.

Having said that, one can argue that there are other policies that may have a direct or indirect influence on the national greenhouse gas emissions. The national forestry policy calls for one million hectares to be planted annually. But, it is the energy policy that is most relevant. Since 1990, the process of liberalization of the electricity sector has been launched in India. The Indian Electricity Act was amended in 1991 to allow the private sector to participate in the generation of electricity. Renewable energy was given a major boost with the establishment in 1992 of a Ministry for Nonconventional Energy Sources. This Ministry is run with passion if not adequate funds since the goals of the Eighth Five Year Plan have been exceeded, despite some initial teething problems.[66] In 1998, the Electricity Regulatory Commission Ordinance was promulgated to establish a tariff structure that was independent of political pressure and economically rational. A Power Sector Reform Bill was passed to allow the State Electricity Boards to be unbundled into generation, transmission and distribution utilities, and to promote corporatization and privatization. Parliament is also considering the Energy Conservation Bill, the Renewable Energy Bill, and the Electricity Bill. These legal developments are remarkable by any standard and the Renewable Energy Bill's goals are comparable with that of the European Commission's Directive on the Promotion of Electricity from Renewable Sources of Energy in the Internal Electricity Market.[67] The Indian Judiciary has also taken a prominent role in the discussions and has judged that commercial vehicles in New Delhi must switch to Compressed Natural Gas. Although there are delays in the implementation process, the Court orders have to be followed and have set a precedent for other states in India.

In addition, the liberalization process itself is expected to lead to national and international competition and this is leading to increased efficiency in production in energy consuming sectors and in energy generation sectors. Liberalization of the banking sector may stimulate increased investments in energy efficiency, and all these tendencies may lead to a decoupling between growth and greenhouse gas emissions.

At the same time, there is a gap between theory and practice. While the Indian electricity sector has adopted the theory of liberalization and has translated it into policy measures and laws, there are problems in implementation. The literature reveals that the subsidies for the windmills were not well designed,[68] contracts with international companies not necessarily protecting India's legal and economic interests,[69] privatizing State Electricity Boards may lead to undervaluing these assets, and over-valuing the ability and willingness of the consumer to pay higher tariffs. The question is, are these problems merely beginner's problems or are they reflections of a structural problem in importing a Western ideology in a society not yet ready for it?

Without attempting to answer that question, what is clear is that in the energy sector at least, there are ample opportunities to promote policies to reduce emissions. The social actors too seem to have become increasingly engaged in the international discussion and see the Clean Development Mechanism as a possible solution to the problem of resource crunch. The Journal of the Confederation of Indian Industry, Global Climate Change: Emerging Green Opportunities, highlights the opportunities for industry in the climate change regime. Visits to a number of stakeholders in this sector in 1999, 2000, and 2001 indicated strong interest coupled with a healthy skepticism in the new mechanism. There are a range of divergent views in industry about how best to use the CDM funds, and while some are very pessimistic that if Indian industry does not reach out for every opportunity to modernize quickly, it will never be able to compete; others argue that India should bargain for the best possible deal. Meanwhile there appears to be a growing consensus in the literature that supercritical boilers, coal gasification technology, water pumps, renewable energy projects etc. could all benefit from international cooperation.[70]

Meanwhile intellectual input into the debate is coming from a number of nonstate actors in India. Several NGOs and scientists are increasingly involved in the debate, and while the NGOs and scientists are making a substantial contribution to the international debate, it is unclear to what extent they are communicating their results to the public and raising public demand for accountability and action from the government.

A SWOT Analysis

The above documentation brings me now to the question: What then are the strengths and weaknesses of India in the negotiating process and what threats and opportunities face the country? These are discussed below, then summarized in Fig. 15.1.

Strengths

The strengths of India in the negotiating process are highlighted in terms of three levels: international, national, and the actor level.

Fig. 15.1: Strengths, weaknesses, threats and opportunities facing India in the climate change regime.

International level

At the international level, India has arguably three key strengths:

- *Diplomatic experience*: The diplomatic service in India has considerable experience in dealing with international issues and can be seen as a strength for the Indian government. The intellectual quality and the negotiating ability of the diplomatic service are considered high by interviewees, and should there be a constructive change in policy, the diplomatic service has possibly the means to implement their skills.
- *Moral and Market power*: Since India has a combination of low per capita emissions and high gross emissions, it is possible for India to take a high moral and economic stand. Moral, on the grounds that its emissions are low, and economic, on the grounds that it has a huge market for foreign technologies and that it is in the economic and environmental interest of the West to engage with India. The power is also strong because there are other such countries in the developing world that are in a similar situation (e.g. China). This provides India with negotiating power which has a high moral content and economic allure at the moment. This power will, however, begin to decrease as India's per capita emissions grow, and especially in the context of decreasing emissions elsewhere. It will also diminish as other countries with similar positions out-compete India and join the developed world.[71]
- *Influence within G77*: India, by virtue of its size, ideas, and diplomatic skills, has the ability to influence and shape the policies of the Group of 77 and can thus influence the way the developing countries deal with the problem of climate change. In the past it has had ambitions to lead the G77 and has successfully pushed issues within the framework of the G77.

National level

Implementation potential : At the national level, the key strength of India is its implementation potential. According to Article 21 of the Constitution of India, no person can be deprived of his life or liberty except according to the procedures established by law. Case law indicates that life, public health and ecology are prioritizd over unemployment and loss of revenue,[72] and that this fundamental right also includes the right to pollution-free air.[73] Article 48 A of the Constitution also specifies that the state shall endeavor to protect and improve the environment and to safeguard the forests and wildlife.[74] As mentioned earlier, there are also a number of legal initiatives in the energy policy sector. Clearly, the legal framework is in place and an active judiciary promotes environmental protection.

At the same time, there is huge potential to reduce emissions at not very high costs. While India's electricity sector is expected to grow by 5 times in 2020 relative to 1990, there are indications that existing policies and trends may lead to only a fourfold increase in related greenhouse gas emissions. There are a number of policy options that could be used to reduce the emissions from the power sector which include fuel substitution and increased efficiency in generation and reduced loss in transmission and distribution. The latter includes the reduction of electricity theft which is as high as 47% in New Delhi. Furthermore, there are a number of options to reduce the rate of growth of demand for electricity in India, and thereby reduce the growth of emissions.[75] The IIS Bangalore estimate that 700,000 Mt of carbon can be abated by dedicating 69 million hectares of land to agroforestry, enhanced natural regeneration, long and short rotation forestry, and social forestry.[76]

Actors

Number of engaged actors : India also has a number of active NGOs, qualified scientists, an informed and communicative press and these actors can see the importance of dealing with the climate change problem as well as identifying the critical policy options for the government. The Centre for Science and Environment publishes Down to Earth to inform the public about various environmental issues, among other initiatives. The Tata Energy Research Institute also undertakes research and policy-oriented work to support the decision-making process in India. Researchers in different universities and organizations are also working on the issue of climate change.

Weaknesses

But if India has so many strengths, why is it unable to effectively influence international policy? This brings us to a discussion of its weaknesses.

International

- *Poor neighborliness*: While a key international law principle is that of good neighborliness, one could argue that India practices the opposite. Unlike countries in other regions of the world, Indian negotiators do not form coalitions with negotiators from neighboring countries and rarely with the rest of Asia in preparation for the G-77 and international negotiations on climate change. Diplomats explain that the interests of India are very different from those of the neighboring countries, nor can India entrust its neighbors to represent its interests at the simultaneous multiple meetings taking place and that "we are Indocentric".
- *Sustainability dilemmas*: I have argued earlier that India faces a number of sustainability dilemmas in relation to a number of international regimes. These can be summed up as the development dilemma (modernizing without westernizing), the poverty dilemmas (begging without mortgaging resources, surviving

without squandering resources), the privatization dilemma (how to empower the private sector to solve public problems), the ecospace dilemma (how to demand equity internationally without guaranteeing it domestically), economic dilemma (how to go for short-term profit without incurring long-term loss); and the negotiation dilemmas (how to negotiate pragmatically without being corrupted, how to cooperate with G77 without being weakened by G77).[77]

- *Defensive strategy*: Because of the lack of a detailed negotiating mandate (see below), India tends to take a defensive strategy at international negotiations.[78] This focuses on the responsibilities of the North and in taking the moral high ground while neglecting the economic dimensions. The defensive strategy is aimed at protecting India's right to grow. India's problems in preparing its national communications is also attributed to the fear that "such inventories may also be used to further the political demand from the North for advancing the commitments of India to the Climate Convention".[79]

- *Handicapped negotiating power:* As argued extensively earlier, India also has handicapped negotiating power in the international arena. This is because of the hollow mandate, the handicapped coalition building power, the lack of staying power in the negotiations, the small negotiating teams that have to cope with multiple simultaneous negotiating sessions and corridor lobbying where the decisions are, in fact, precooked.[80]

National

At the national level, one can identify four key weaknesses:

- *Lack of negotiating mandate*: At the national level, the key problem is the lack of a clear negotiating mandate that goes beyond platitudes and rhetoric. This is because of ideological doubts, lack of detailed scientific information and societal debate that can support decisions, and lack of any weighing mechanism to assess the importance of the interests at stake. As a result, there is neither a detailed, fine-tuned mandate, nor is there an effort to integrate climate change into other policy areas. At the Planning Commission, the think tank of India, there are only two officials working on environmental issues. The Ministry of Environment and Forests does not have much clout or influence and inevitably the negotiators appear to have an implicit mandate to discuss North-South issues, the platform of the developing countries in all negotiations, but not domestic policy.

- *Lack of legitimacy*: While the mandate tends to build on pseudoindicators of legitimacy,[81] the lack of a well-developed domestic process to involve all stakeholders in the development of a strategy for the country as a whole. The lack of time and resources, and not lack of interest, is the key reason and excuse of officials for not consulting the stakeholders and reading and integrating their input. Officials are seen as representing the "elite interests of a hierarchical society" with an immediate political and not substantive response : "The developed countries are trying to stop us from developing".

- *Lack of enforcement and incentives*: India also has the problem of poor enforcement of environmental policies and ill-structured incentives to encourage environmental reform. As a federal state, only some policies fall under the jurisdiction of the central government. While foreign policy and implementation of international treaties is a union government responsibility,[82] environment, forestry and electricity are on the concurrent list.[83] At the same time, India has more than 6 dozen ministries, of which environment is just one. Coordinating with other ministries is thus also a major challenge. This and the lack of political will mean that implementation will always be the Achilles heel of India and that since India can't ensure implementation, it should be careful to take on legally binding commitments, explained a former ambassador in an interview with the author. "We know that we have to clean up our coal and we have to try and get action in the country. But we cannot guarantee that it will happen and we cannot make it happen."

- *Political instability*: Although India is a relatively stable democracy, the last ten years have been characterized by political instability. The country has been ruled by the National Front Government (1989-90), a breakaway National Front Government (90-91), the Congress (91-96), the Bharatiya Janata Party (for thirteen days), the United Front (1996-1998), Bharatiya Janata Party led coalition (1998-1999), and then the National Democratic Alliance. All political attention is thus focused on efforts to stay in power as opposed to governing.

Actors

General apathy : At the domestic level, given the domestic political turmoil and economic stress, the general public is apathetic to environmental issues, especially when they are esoteric such as climate change. There is also considerable bureaucratic inertia. While there are a large number of scientists in India, many of these scientists lag behind the scientific developments in the North for the purpose of international negotiation. Domestic scientists have been able to demonstrate that foreign science is irrelevant or biased,[84] and to come up with policy recommendations, they have yet to put forward constructive proposals that could help the Government take a proactive stance in the international negotiations. "Multi-disciplinary team work is rare, and public involvement through education and awareness is even rarer".[85]

Threats

The key problem lies in the perceived threats to India of participation in the international negotiations:

International

At the international level there are arguably three key threats that shape India's negotiating strategy:

- *Development cap and evaporating sovereignty*: A key potential threat to India is the fear of the eventual development cap—that the developed countries will prevent the further development of India by restricting its ability to use and exploit its own natural resources in accordance with the principle of permanent sovereignty over natural resources. Principles to restrict sovereignty[86] so as not to cause harm to others have been supported by India in the past, but the climate change issue is a test case. The issue becomes specially sensitive in relation to energy. With nuclear energy seen as environmentally dangerous, with the World Commission on Dams[87] making recommendations to limit dam construction, and now with the Climate Change Convention possibly leading to a restriction on the use of coal, the Government of India sees its control over its own natural resources decreasing. This leads it to a defensive position.
- *Dependency and debt*: At the same time, there is fear that India may be forced to import resources (e.g. gas) and modern unaffordable technologies to reduce the rate of growth of its greenhouse gas emissions, which could lead to increased technological dependency and financial insolvency.
- *Fear of future noncompliance measures*: Constructive participation in international agreements also implies that if a country fails to comply with its commitments, non-compliance mechanisms may be imposed on it.

National

- *Technological lock-in*: A major domestic threat is that if India ignores the global developments and invests in currently affordable long-life technologies (which is mostly the case in the electricity generation sector), then it may be locked into a technological trajectory that will restrict its ability to participate constructively in the climate change regime in future, even if it were able to accelerate its own development process considerably.

- *Potential impacts of climate change*: The potential impacts of climate change can be quite severe for India. The glaciers in the Himalayas have already started to melt and this is likely to affect the hydrological cycle in India. The monsoon patterns may be affected. Sea-level rise threatens the coastal regions of India and rising temperatures in a country that has high summer temperatures may exacerbate the existing climate to such an extent as to affect the basic necessities—availability of adequate drinking water, food, shelter and health.[88]

Actors

- *Legitimacy*: Whatever policy the government adopts, if it lacks legitimacy and support from the people, it might lead to further political disruption. Those who may suffer from the potential impacts of climate change may organize themselves to force change. Those who are in business realize that in the context of liberalization they have to compete with international companies and may demand a domestic institutional framework that promotes modernization. Others may resist the negative impacts of liberalization.

Opportunities

The opportunities can be defined as ways to maximize the strengths, minimize the weaknesses and address the threats. These include :

International

- *Leadership of the G77*: India has diplomatic skills and potential ability to influence the G77. It has for the present moral and market power. It has implementation potential (the potential to decouple growth from emissions) and an elite group of social actors that are well informed about the problem. It has to consider whether Indocentricity, poor neighborliness policy, defensiveness, and morality are going to serve its long-term goals. It has to face the threat that climate change could further impoverish the resource base in the country, and that without enlightened policy, India could be locked into a technological trajectory that would lead to its isolation in global environmental politics. Dogmatic defense of the right to grow without an understanding of the consequences is a foolhardy and shallow excuse for intellectual poverty. Clearly India wants to defend its right to grow; but this right needs to be articulated in considerable detail in the context of the scientific and social information we have and in a way that guarantees sustainable development for India and the rest of the developing world. This is where India can play a leading role in convincing other countries of how sustainable development needs to be defined in order to guarantee the survival of developing countries and in order to prevent their automatic marginalization or inevitable exclusion from the international environmental system. Without the rest of the developing world India's negotiating power is weakened considerably, and developing countries will be played against each other. There is also urgency in adopting such a position because of the declining power of India as emissions rise and technological lockin sets in. Diplomats argue that leadership has no benefits for India, especially after negative experiences in the trade negotiations, but the climate change negotiations are different and perhaps as challenging. Lessons need to be learned from the trade experience but not blindly transferred to the climate change regime. Stakeholders from outside the diplomatic sphere, who have less inside knowledge of the politics of negotiation, but who are more aware of international dynamics, feel that India alone would never be able to make an impression on the Western world; the only way to strengthen the negotiating power is to coordinate the power of the G77 and consolidate it by nurturing political synergism.

- *Bridge between North and South*: India could also try and present itself as a bridge between the North and the South, and focus on communication of key ideas and concepts to both sides. The North-South discussions have reached a deadlock not least because of India's own dogmatism and conservatism in this area. Of course, the United States' position does not help. It might be useful to recall an anecdote from the Kyoto negotiations. It appears that among many of Al Gore's activities in Kyoto prior to the adoption of the Protocol was a meeting with the leaders of the G-77. Mwandosya recorded the awe with which the developing country negotiators listened to an explanation of all the problems in the US which hampered the ability of the Clinton administration to take stronger steps. What is glaringly absent is whether the negotiators from the South also took the opportunity to explain the problems within their societies to the Vice-President of the US. There is an assumption in almost all the literature that capacity building needs to be focused on the South; people cannot believe the ignorance in the West about how developing societies actually (mal) function.
- *Improve its public relations*: As this chapter has indicated, India has taken a number of measures to deal with several related issues. It has, however, chosen thus far not to present these measures as part of its climate change policy. However effective such policies may be, the fact remains that the measures taken in India to modernize its electricity sector, for instance, are a trend-break and substantial by global standards. I would argue again, as I have done in the

past, that India must emphasize its *de facto* commitment to taking measures to deal with the climate change issue, even if it is unwilling to take on *de jure* commitments. This would remove the excuse of individual developed countries to argue that developing countries are unwilling to take measures.

National

- *Sustainable development*: It is clear that India faces sustainability dilemmas both at a national and at an international level. Avoiding discussion of this issue does not increase India's ability to deal with issues. Sustainable development is fast becoming an "inescapable logical necessity."[89] It is of vital importance for the government and civil society to come to terms with this concept and to translate it into practical policies and measures that can be implemented in India and are relevant for the Indian context. This calls for industrial transformation. It implies that affordable modern technologies need to be identified. If technologies are necessary and not affordable, India needs to be able to make a deal with the rest of the developing countries to see whether it can come up with an OPEC-like trade block: Environmental Technology Purchase Organization—and make mass orders for specific technologies if they are made available at low costs.
- *Make relevant issue linkages*: While India does not have much of a climate change policy, a lot of developments are taking place within the domestic context. A number of measures have been taken in the forestry sector and in the energy sector. Liberalization itself is expected to lead to the closure of small-scale industry, fast growth of the service and communication sector, and efficiency increases in large-scale industry if they wish to survive in the new competitive sphere. All this will lead to a decoupling of emissions from economic growth.

Actors

Stimulate public debate on sustainable development: This is necessary for a number of reasons. First, to create awareness of the climate change and environmental problem and to help people adapt to changing climate. Second, to fill the current intellectual deficit regarding how sustainable development can be defined. Third, to involve the public in the design of policies and approaches likely to have a high-compliance pull. Fourth, to minimize the threat of destabilization caused by actors whose views are ignored in the process. Fifth, to unleash social experiments that can help with social initiatives to deal with the climate change problem. Sixth, a mutually supporting process should be initiated wherein the public demands accountability and action from the government and the government expects implementation of its rules.

Conclusion

The decision to host the Eighth Conference of the Parties in New Delhi marked a major change in policy in India. But for that decision to prove effective and efficient in results, there must be a well-planned agenda. Otherwise, the conference will be registered as an expensive exercise in futility. The sphere around the negotiations has changed with the US withdrawal from the negotiations, followed by Australia. More than 100, other countries have ratified the Protocol. Nevertheless, with the exit of the largest emitter (USA), the stability of the regime is at stake and this is likely to have negative influences on all countries, given that climate change is a serious problem. Further, Europe is engaging more seriously in dialogue with its immediate neighbors. All this weakens India's negotiating position and ultimately the benefits it could seek from participating in the regime. It is time therefore to adopt a different strategy. This brings me to the following conclusions.

First, while sustainable development for the West might mean at the least incremental fine-tuning of the domestic systems since many are beyond the hump of the inverted U curve, this is not relevant to India. India is too much on the left side of the hump to be thinking in terms of incremental change. Nor can it afford to wait and see how others define this concept. It has to elaborate on this concept in order to operationalize it for the benefit of the domestic society. This has to be done through a nation-wide dialogue on the issue in order to capture the intellectual capital of the country without compromising on the future for the poor and thereby securing legitimacy for the content of the concept.

Second, India must follow a policy of capitalization, and not capitulation, in the area of climate change, because this is the only viable policy for India in the longterm. Threatened as it is by the impacts of climate change and technological lockin which implies long-term political isolation, decline in moral power, and increasing domestic environmental impacts, India cannot afford to take a rhetorical, moralistic defensive stance. This is especially in the context of the withdrawal of the US from the negotiations and the fact that the former East and Central European countries are competing for assistance. This means that India has to go far beyond its advertised three-prong policy on climate change and demonstrate that it has already *de facto* taken action, if not *de jure*, and that it is willing to continue to do so.

Third, India must find a way of effectively linking climate change to other international treaties so that it can minimize contradiction and repetition and consolidate the beneficial impacts. Concomitantly, it must find a way to link international with local priorities in order to keep both domestic and international stakeholders engaged in the discussion. Interviewees have argued that India is constantly falling back in each round of the negotiations, and that opting out is not an option for India. If that is the case the only way for India to deal with the climate change issue is to adopt a constructive and proactive approach.

Acknowledgements

This paper is based on ongoing research work undertaken in the context of the Vrije Universiteit Amsterdam project on the Law of Sustainable Development and Climate Change.

Footnotes

1 Protocol on Substances that Deplete the Ozone Layer (Montreal), September 16, 1987, in force January 1, 1989, 26 I.L.M. (1987), p. 154
2 Convention on the Control of Transboundary Movement of Hazardous Wastes and Their Disposal (Basel) March 22, 1989, in force May 24, 1992; 28 I.L.M. (1989), p. 657.
3 United Nations Convention on the Law of the Sea, (Montego Bay) December 10, 1982, in force November 16, 1994, 21 I.L.M. (1982) p. 1261.
4 Convention on Biological Diversity, (Rio de Janeiro) June 5, 1992, in force December 29, 1993, 31 I.L.M. (1992).
5 Convention on International Trade in Endangered Species of Wild Fauna and Flora (Washington), March 3, 1973, in force July 1, 1975, 993 UNTS p. 243.
6 United Nations Framework Convention on Climate Change, (New York) May 9, 1992, in force March 24, 1994; 31 I.L.M. (1992).
7 Gupta, J. 2001. India and climate change policy: Between diplomatic defensiveness and industrial transformation. *Energy and Environment*, 12, (2&3) 217-236.
8 Govt. of India, Ministry of Environment and Forests. 2001. *Annual Report*, New Delhi.
9 UN Doc. A/CONF.151/26 (1992).
10 UN Doc. A/CONF.151/26/Rev.1 (Vol.I) (1992).
11 UN Doc. FCCC/CP/1995/7/ Add.1, Dec. 1/CP.1, para 4
12 UN Doc. FCCC/CP/1997/7/Add.1 (1997)
13 UN Doc. FCCC/CP/2001/13 (2001)
14 Article 3.4 of the UNFCCC, see note 6 above.
15 Preamble of the FCCC, see note 6 above, para 22: "Recognizing that all countries, especially developing countries, need access to resources required to achieve sustainable social and economic development and that, in order for developing countries to progress towards that goal, their energy consumption will need to grow taking into account the possibilities for achieving greater energy efficiency and for controlling greenhouse gas emissions in general, including through the application of new technologies on terms which make such an application economically and socially beneficial."
16 Article 2 of the FCCC, see note 6 above.
17 Article 4.1 (d) of the FCCC, see note 6 above: "Promote sustainable management, and promote and cooperate in the conservation and enhancement, as appropriate, of sinks and reservoirs of all greenhouse gases not controlled by the Montreal Protocol, including biomass, forests and oceans as well as other terrestrial, coastal and marine ecosystems."
18 Article 4.2.a of the FCCC, see note 6 above: "Each of these Parties shall adopt

national policies and take corresponding measures on the mitigation of climate change, by limiting its anthropogenic emissions of greenhouse gases and protecting and enhancing its greenhouse gas sinks and reservoirs. These policies and measures will demonstrate that developed countries are taking the lead in modifying longer-term trends in anthropogenic emissions consistent with the objective of the Convention, recognizing that the return by the end of the present decade to earlier levels of anthropogenic emissions of carbon dioxide and other greenhouse gases not controlled by the Montreal Protocol would contribute to such modification, and taking into account the differences in these Parties' starting points and approaches, economic structures and resource bases, the need to maintain strong and sustainable economic growth, available technologies and other individual circumstances, as well as the need for equitable and appropriate contributions by each of these Parties to the global effort regarding that objective. These Parties may implement such policies and measures jointly with other Parties and may assist other Parties in contributing to the achievement of the objective of the Convention and, in particular, that of this subparagraph."

19 See Article 2 of the KPFCCC, see note 12 above.
20 See Article 3 of the KPFCCC, see note 12 above.
21 See Article 10 of the KPFCCC, see note 12 above.
22 Article 12 (2) of the KPFCCC, see note 12 above.
23 Decision 1/CP7.
24 Para 3, Decision 1/CP7.
25 Annex A of Decision 2/CP.7, para 4; Annex A of Decision 3/CP.7, para 4, Annex A of Decision 2/CP.7, Para 17.c.; Annex A of Decision 2/CP.7, para 18.
26 See World Commission on Environment and Development (1987). *Our Common Future*, The World Commission on Environment and Development, Oxford University Press, Oxford, UK.
27 Freestone, D. and E. Hey (eds.) 1996. *Precautionary Principle: Book of Essays*, Environmental Policy and Law Series, Kluwer Law International, Hague, The Netherlands.
28 Resolution 3/2002 of the International Law Association: The New Delhi Declaration of Principles of International Law Relating to Sustainable Development.
29 The inverted U curve or the Environment Kuznets Curve refers to the thesis that as countries get richer their pollution continues to increase to a certain point after which the pollution per capita begins to decrease.
30 See, for example, Gupta, J. 1997. *The Climate Change Convention and Developing Countries—From Conflict to Consensus?*, Environment and Policy Series, Kluwer Acad. Publ., Dordrecht, Netherlands, 256 pp.
31 Such as the treaties over desertification and biodiversity.
32 Gupta, J., J. Vlasblom and C. Kroeze with contributions from K. Blok., M. Hisschemoller, C. Boudri and K. Dorland 2002. An Asian Dilemma: Modernising the Electricity Sector in China and India in the Context of Rapid Economic Growth and the Concern for Climate Change, NOP, RIVM, Bilthoven, NOP report no. 410200097; Sharma R. 2000. Status of national communications to the UNFCCC, in K. Ramakrishna, B. Bamberger and L. Jacobsen (eds.) *Asia: Looking Ahead, Initial Stages of National Communications Reporting*. Woods Hole Research Center, Woods Hole, MA, pp. 71-81, at p. 77.
33 Quoted in Sharma R. 2000. See note 32 above at p. 77.

34 See Gupta et al., 2002, note 32 above and Sharma, 2000, note 32 above.

35 Minister Mahesh Prasad made this clear at the Noordwijk Conference of 1989; Prasad, M. 1990. Statement of Mahesh Prasad, India. In: Vellinga, P., P. Kendall, and J. Gupta (eds.) Noordwijk Conference Report, Vol. II. Ministry of Housing, Physical Planning and Environment, Hague, Netherlands, pp. 73-74.

36 GOI, 2002. India and International Agreements, India Towards World Summit on Sustainable Development, http://www.wssindia.ord/agreement.html. Site visited in June 2002.

37 See Dasgupta, C. 1994. The climate change negotiations. In: Mintzer, I.M. and Leonard, J.A. (eds.) *Negotiating Climate Change: The Inside Story of the Rio Convention*, Cambridge Univ. Press, Cambridge, UK, pp. 129-148.

38 See Gupta (1997), note 30 above.

39 See Agarwal A. and S. Narain 1991. Global Warming in an Unequal World: A Case of Environmental Colonialism. Centre for Science and Environment, New Delhi.

40 When constructive positions were prepared, they were not always accepted. A draft negotiating text prepared in 1991 was not accepted by the developed countries. See Agarwal, A., S. Narain and A. Sharma 1999. Green Politics: Global Environmental Negotiations. Centre for Science and Environment, New Delhi.

41 See Article 4.3 of the FCCC, note 6 above.

42 See Nath, K. 1993. *Selected Statements on Environment and Sustainable Development*, Government of India, at p. 71, and Rao, P.V. Narasimha 1993. *Selected Speeches*, Government of India Publications Division, New Delhi, Vol.I at p. 3.

43 See Prasad (1990), note 35 above.

44 See para 3 of the Preamble of the FCCC, note 6 above.

45 For details regarding the initial resistance to the GEF, see Gupta, J. 1995. The Global Environment Facility in its North-South context, *Environmental Politics*, 4 (1): 19-43.

46 See Gupta (1997), note 30 above, and Maya, S. and J. Gupta (eds.) 1996. *Joint Implementation: Weighing the Odds in an Information Vacuum*, Southern Centre on Energy and Environment, Harare, Zimbabwe, 164 pp.

47 See Article 4.2a7b of the FCCC, note 6 above

48 See Gupta (1997), note 30 above.

49 See Agarwal et al (1999), note 40 above.

50 See Mwandosya, M.J. 1999. *Survival Emissions: A Perspective from the South on Global Climate Change Negotiation*; DUP 1996. LIMITED and The Centre for Energy, Environment, Science and Technology (CEEST-2000), Dar es Salaem, Tanzania.

51 See Mwandosya (1999), note 50 above.

52 Venugopalachari, S. 1997. *Inaugural address*, on the occasion of the Conference on Activities Implemented Jointly, New Delhi, January 8.

53 Conference Statement 1997. Statement of the Conference on Activities Implemented Jointly: Developing Country Perspectives, New Delhi, 8-10 January.

54 See Byrd Hagel Resolution, Congressional Record: October 3, 1997 (senate) pp. S10308-S10311; available at http://www.microtech.com.au/daly/hagel.htm; also Clinton, W.J. 1997. Remarks by the President on Global Climate Change, National Geographic Society, October 22, 1997.

55 See Werksman, J. 1998. The Clean Development Mechanism: Unwrapping the

'Kyoto Surprise', *Review of European Community and International Environmental Law*, 7 (2) : 147-158.

56 See Article 12 of the KPFCCC, note 12 above.

57 "The South was strident about the emission trading and transnational 'carbon offset' projects. It feared that the rules for 'buying' and 'selling' greenhouse gas (GHG) allowances could very well gift rights to entitlements, freezing North-South disparities" Sharma, V. 2000a. Global Warming: the North-South Factor. *The Economic Times*, 22/12/00.

58 Interviewees explain: "But there is arm twisting by the West—the CDM was expected to be a non-compliance fine and now it has turned out to be quite different"; "We will ratify the Kyoto Protocol unless there are unwarranted linkages to meaningful participation and any consideration of contingency on domestic actions. What was the need for such linkages?"; "The reality of the 1992 agreement has been deformed by the 1997 agreement."

59 Argentina appeared willing in 1998 to adopt a commitment. This commitment is expressed as a dynamic target based on the relation between emissions and GDP. The target is expressed as $E = I * vP$ where E is emissions measured in Carbon equivalents, P is GDP in Argentine pesos, and I is the index value which is equivalent to 151.5.

60 Kazakhstan signed the FCCC in June 1992 and ratified it in 1995. Its emissions have fallen considerably since 1990 and it expects to suffer from the potential impacts of climate change. In 1998, Kazakhstan ratified the Protocol and also expressed its willingness to take on commitments. In 1999, the Permanent Mission of Kazakhstan requested the FCCC to upgrade its status to Annex I country.

61 See Sharma, note 32 above at 74.

62 Article 51 of the Constitution of India.

63 Bakshi, P.M. 1998. *The Constitution of India: With Selective Comments by P.M. Bakshi*. Universal Law Publ. Co., Delhi.

64 Government of India 1992. *Eighth Five Year Plan*, 1992-97, Planning Commission, New Delhi, Vol. I and II.

65 Government of India 1997. *Ninth Five Year Plan*, 1992-97, Planning Commission, New Delhi. Vol. I and II.

66 See, for example, Rajsekhar, B., F.V. Hulle and J.C. Jansen. 1999. Indian Wind energy programe: performance and future directions. *Energy Policy*, 27, (11) 669-678.

67 COM (2000) 279; Directive 2001/77/EC

68 See Rajsekhar et al., note 66 above.

69 See Mehta, A. 1999. *Power Play: A Study of the Enron Project*. Orient Longman, New Delhi.

70 See Gupta, J. et al., 2002, note 32 above; TERI 2000. Climate Change Research at TERI, Brochure, Tata Energy Research Institute, New Delhi; CII 2000. Potential CDM projects. *Global Climate Change: Emerging Green Opportunities* (2) 6.

71 See letter to Prime Minister Atal Behari Vajpayee dated October 19, 1998 from Anil Agarwal and Sunita Narain of the Centre for Science and Environment.

72 M.C. Mehta vs Union of India (1987) Sup SCC 131, AIR 1987 SC 1086.

73 Subhash vs State of Bihar, AIR 1991 SC 420, para 7.

74 Article 48 A was inserted by Constitution (Forty-second Amendment Act) 1977, sec. 10.

75 See, *inter alia*, Gupta et al., note 32 above.

76 See Sharma notes 32 and 33 above, at 77.

77 See Gupta, J. 2000. Global Environmental Issues: Impact on India. In: S.N. Chary and V. Vyasulu (eds.) *Environment Management: An Indian Perspective,* Tata McGraw Hill, New Delhi, pp. 252-281; and Gupta, J. 2002. Environment and Development: Towards a fair distribution of burdens and benefits: In: J.J.F. Heins and G.D. Thijs (eds.). Ontwikkelingsproblematiek: *The Winner Takes It All? Verdelings Vraagstukken in de Wereld,* Themabundel Ontwikkelingsproblematiek no. 12, pp. 35-50. Vrije Universiteit Amsterdam Press, Amsterdam, Netherlands.

78 Gupta, J. 2001. India and climate change policy: Between diplomatic defensiveness and industrial transformation. *Energy and Environment* 12, (2&3) 217-236.

79 See Sharma, note 33 above at 74

80 See, for details, Gupta (1997), note 30 above.

81 When negotiators have no clear mandate, they often develop a mandate based on proxy indicators of legitimacy such as compatibility with precedents, principles and positions in other issue areas; see Gupta, J. 2000. North-South aspects of the climate change issue: Towards a negotiating theory and strategy for developing countries. *Inter. J. of Sustainable Development* 3 (2): 115-135.

82 See Article 14 of List 1 of the Seventh Schedule of the Constitution of India.

83 See Articles 17A&B and 38 of List 1II of the Seventh Schedule of the Constitution of India.

84 The dispute between CSE and WRI, see Agarwal and Narain (1991) note 39 above, Parikh, J. 1992. IPCC strategies unfair to the South, *Nature*, vol. 360, pp. 507-508; Kandlikar, Milind and Ambuj Sagar 1999. Climate change research and analysis in India: An integrated assessment of a North-South divide, *Global Environmental Change*, Vol. 9. pp. 119-138; Gupta, J. 2001. Effectiveness of air pollution treaties: the role of knowledge, power and participation in Matthijs Hisschemöller, Jerome Ravetz, Rob Hoppe and William Dunn (eds.) *Knowledge, Power and Participation*, *Policy Studies Annual*, Transaction Publishers, New Brunswick, New Jersey, USA. pp. 145-174.

85 See Sharma note 33 above at 73.

86 See Principle 2 of the Rio Declaration, note 10 above.

87 See www.dams.org: In 1997, the World Bank and IUCN initiated talks leading to the establishment of the World Commission on Dams. This Commission consists of 12 Commissioners whose job was to evaluate completed dams and lessons learned from such dams. The two key objectives were: (1) review the development effectiveness of dams and assess alternatives for water resources and energy development, and (2) develop internationally acceptable criteria and guidelines to advise future decision-making in the planning, design, construction, monitoring, operation, and decommissioning of dams.

88 See Watson, R.T., M.C. Zinoera, R.H. Moss, and D.J. Dokken 1998. *The Regional Impacts of Climate Change: An Assessment of Vulnerability.* Cambridge Univ. Press, Cambridge, UK.

89 Case concerning the Gabcikovo-Nagymaros project (Hungry/Slovakia), Separate Opinion of Vice-President Weeramantry, at p. 88

CHAPTER 16

Indonesia: The Suffering Nation

Harry Surjadi
Pesona Depok C 9, JL. Margonda Raya Depok - 16432, Indonesia

Anyone who has ever been to a rain forest knows that it is among the noisiest places on earth. You can hear the cacophony of a thousand insects, birds, monkeys and deer, added to the roar of waterfalls.

But stillness hangs over the Kutai National Park, the 1,98,000-hectare natural forest in Kalimantan—the largest of the 17,000 islands of the Indonesian archipelago and home to most of the tropical rain forests in the region.

In April 1998, the forests of Kutai were devastated by fire. And although it had begun to rain in June, it was still so quiet there, you could almost hear yourself breathe.

Almost every year Indonesia experiences forest fires. The greatest forest fire in history was in 1982-1983, which affected over 3.5 million hectares of forest and land of East Kalimantan, an area 56 times the size of Singapore.

A team, that investigated the vegetation affected by the 1982/1983 fires, concluded that they had destroyed 800,000 hectares of primary forest, 1,400,000 hectares of logged-over forest, and 750,000 hectares in areas identified as secondary forest, swidden gardens or settlements, and 550,000 hectares of peat swamp forests.

The ecological and economical damage was tremendous. The fires altered the composition of vegetation, the diversity of species, and the structure of the forest. Long-term impacts on hydrological cycles, soil composition, and wildlife continue to this day (State Ministry for Environment, 1998).

The fires produced heavy smoke and haze that peaked in April 1983, halting air transportation not only in East Kalimantan, but also in Surabaya, Jakarta, and Singapore.

Economic losses due to the fires 1982/1983 were estimated at more than US $9 billion (Schindele et al., 1989). For comparison, this amount is three times the total annual revenues from the forestry sector in 1989, which amounted to US $3 billion. Another researcher estimated that damages to pepper crops planted by local people alone amounted to US $2 billion.

This was precisely what happened in 1982-1983 when El Niño resulted in protracted drought in East Kalimantan. The dry season began in June 1982 and lasted through May 1983. Forest fires began in November/December 1982 and although initially not very intense, by early 1983 the fires were spreading more rapidly and burning more intensely.

Again, in 1997 and 1998 Kalimantan and other parts of Indonesia experienced fires, which in many ways were very similar to the 1982/1983 fire destruction of Indonesian rain forests. The 1997 land and forest fires, which occurred between February and October, were considered the worst in the last fifteen years. Cumulatively, fires erupted in 23 provinces across the country (Indonesia consisted of 26 provinces at that time, including East Timor) leaving only Jakarta, Yogyakarta, Bali, and East Timor fairly unaffected.

According to a study done by the State Ministry for Environment and UNDP in 1998, the total area burned in 1997 was approximately 838,869 hectares (State Ministry for Environment, 1998). But, the Worldwide Fund for Nature Indonesia reported the total area burned as over five million hectares. Data for the WWF report was obtained from a survey conducted by the Center for Remote Imaging Sensing and Processing in Singapore through satellite imaging in Sumatra and Kalimantan. The European Union Fire Response Group in cooperation with the Ministry of Forestry estimated the area burned as 3.6 million hectares.

WWF estimated that the total economic losses due to forest fires in 1997 amounted to Rp 9.5 trillion (US $95 million). The State Ministry for Environment and UNDP study estimated total national economic loss at Rp 5.96 trillion or about US $ 59.6 million (State Ministry for Environment, 1998). This represents lost revenues for the country which could have been used for other productive purposes. Economic losses included in the study were direct losses borne by the plantation, forestry, health, transportation, and tourism sectors as well as the direct costs incurred in trying to control the fires. The amount of losses did not include all ecological effects, but only the tangible ecological losses.

F. Siegert from Ludwig Maximilian University, Department of Biology, München, Germany, (Siegert et al., 2001) estimated that a total of 5.2 million hectares, including 2.6 million hectares of forest, were burned with varying degrees of damage in the 1997-1998 forest fires in Kalimantan.

According to a study published in the December 10, 2001 issue of Science by an international research team led by assistant professor of tropical ecology, University of Michigan, Lisa M. Curran, this ecological and economic resource is being destroyed by human activity, which has intensified the effects of regional climate change.

From 1985 to 1999, Curran and her colleagues studied Dipterocarpaceae — the main family of rain forest canopy trees in Indonesian Borneo. Field research covered a 57-square-mile area, but focused on six square miles in the Gunung Palung National Park. The research team also surveyed timber concessions throughout the surrounding province of West Kalimantan and studied 30 years of export records to determine the impact of logging in the region and on the park.

Curran's study is the first to document an ecosystem with a rare reproductive strategy called masting. More than 50 different species of Bornean dipterocarp trees synchronize their reproduction—limiting fruit and seed production to brief, intense periods. Curran discovered that these bursts of reproduction are initiated by the arrival of the ENSO, a periodic shift in tropical Pacific circulation patterns that brings drought to Indonesia.

"We recorded four masting episodes from 1986 to 1999 with an average interval of 3.7 years," said Curran in a press release of University of Michigan December 7, 1999. "With the possible exception of one very minor event in 1994, they all occurred during ENSO years. Climatic conditions of an El Niño year trigger simultaneous fruiting in dipterocarps, and are essential for regional seed production."

According to Curran, masting gives canopy trees an important survival advantage. In a typical six-week masting period, her research team collected 180 pounds of seed—ranging in size from a chestnut to a pistachio nut—from the forest floor in every acre of the six-square-mile survey area.

"It's like Thanksgiving in the forest," Curran said. Wild boar, orangutans, parakeets, jungle fowl, partridges, and other animals congregate to stuff themselves. Local villagers collect baskets of seeds called illipe nuts to sell as a cash crop. Because so much seed is produced simultaneously over such a large area, however, there is still enough leftover to germinate and produce a carpet of new seedlings on the forest floor.

In an article in the December 10, 2001 issue of *Science*, Curran and her coresearchers describe how the forest in Gunung Palung is changing following a decade of intensive dipterocarp logging in huge timber concessions surrounding the park. From 1991 to 1998, production of mature, viable dipterocarp seed fell from 175 pounds per acre (196 kilograms per hectare) to 16.5 pounds per acre (18.5 kilograms per hectare). Despite a major fruiting event during the 1998 El Niño year, no new dipterocarp seedlings were found in the survey area.

"Even though the park is supposedly off-limits to logging, the forest is losing the ability to regenerate itself," said Curran, "because seed predators can't find food outside the park, they move inside to eat the dipterocarp seeds before they germinate." Massive forest fires on nearby logging plantations, which destroyed an area the size of Denmark or Costa Rica in 1997-1998, brought pollution and intensified El Niño's drought, killing the few remaining dipterocarp seedlings.

El Niño or ENSO (El Niño Southern Oscillation) has been attributed to the fires and has had a lasting effect in prolonged dry seasons from June 1997 to April 1998. Due to the strong El Niño effect, part of East Kalimantan received rain. Although rain did fall on several occasions during December 1997 and January 1998, there was no rain in East Kalimantan until May 1998.

If we correlate Southern Oscillation Index (May-October 1997) data with forest fire area, it is very clear that the stronger the El Niño, the larger the area burned (State Ministry for Environment, 1998). El Niño is a weather pattern characterized by a movement of warm water currents in the Pacific Ocean to areas farther north than usual, which brings radical changes to weather around the planet.

Indonesia has only two seasons, a dry season (June-September) and a rainy one (December-March). In April to May and October to November the climate is in transition from one type of climate to another. As a tropical country, humidity ranges, between 63-91% throughout Indonesia. And, given the various terrestrial situations, temperature varies significantly depending on elevation above sea level, distance from coast, and type of ground cover. During the same year, diurnal temperature varies between 23°-36°C while night temperature varies between 18°-30°C.

Prolonged drought has occurred in Indonesia at least once every ten years. Data on rainfall in Sumatra, Java, Kalimantan, Sulawesi and Bali since the early 1900s show that prolonged drought occurred in 1903, 1914, 1925, 1929, 1935, 1948, 1961, 1963, 1965, 1967, 1972, 1977, 1982, 1987, 1991, 1994, and 1997.

Drought in Indonesia generally occurs during the dry season, which in most parts of the country occurs between May and October. When the dry season in Indonesia occurs at the same time as El Niño, the result is a prolonged drought, which extends from June to November and can continue until May of the following year. Of the 17 drought years mentioned above, 11 took place at the same time as El Niño, these were 1903, 1914, 1925, 1965, 1972, 1977, 1982, 1987, 1991, 1994, and 1997.

In Indonesia a prolonged drought as a consequence of El Niño has affected different parts of the country differently, depending on the strength of El Niño and the monsoon winds sweeping past Indonesia. Nevertheless, areas south of the equator experience drought more frequently especially during El Niño years.

Yet, El Niño is not the cause of the forest fires. Human activities cause the fires, but El Niño has enhanced the provision of dry biomass fuels to burn and prolonged the drought season. It is also true that Earth's temperature is warmer than ever. According to NASA researchers, in 1997 and 1998 global temperatures reached their highest level since record-keeping began last century; 9 of the 11 hottest years in the 20th century occurred within the last 10 years. The global mean surface temperature has increased by 0.4ºC in the past 25 years, and climate scientists are becoming increasingly confident that the anticipated process of global warming has begun. And there is evidence that El Niño events have increased in magnitude since the mid-1970s (Trenberth and Hoar, 1996) and climate change may alter the frequency and magnitude of the El Niño cycle (Timmermann et al., 1999).

Forest fires in 1997-1998 primarily affected recently logged forests; primary forests or those logged long ago were less affected. Siegert et al. (2001) wrote that these results support the hypothesis of positive feedback between logging and fire occurrence. The fires severely damaged the remaining forests and significantly increased the risk of recurrent fire disasters by leaving huge amounts of dead flammable wood. Fires are further exacerbated by dry climate, such as the long drought caused by El Niño, the changing composition of vegetation due to monocropping, and the changing composition of soil brought on by mining practices that expose coal deposits in Kalimantan.

Indonesian tropical rain forests are not only stressed by fire, but also threatened by illegal logging and land-use change. Since the economic crisis faced by Indonesia, more trees have been cut not only by concessionaires but also by people living around the forests. Two main problems threaten sustainability of Indonesia tropical forest. First, the quality of the remaining forest area and the prospects of obtaining a sustainable supply of timber from this resource in future. Second, and linked to the first, the demand for logs greatly exceeds the official supply. The big demand for logs or raw woods fuels widespread illegal harvesting in order to meet the shortfall in raw materials for the wood-processing industry. Table 16.1 shows that industry capacity exceeds wood supply.

Annual deforestation rates were 1% during 1990-1995. This rate increased to almost 2.5% per year during 1996-1999. Recent estimates of forest loss caused by logging and forest conversion are 1.3 million hectares per year. Because of economic crisis and lack of law enforcement under the corrupted and unstable government, annual deforestation rate can be estimated as more than 2.0 million hectares per year. And the rate is increasing year by year. If no action is taken to stop logging activities, illegal or legal, Indonesian forest will be gone in 30-50 years.

Table 16.1: Long-run supply and demand for roundwood in Kalimantan

Province	Pulpwood		Plywood and sawnwood	
	Supply (m³)	Industry capacity (m³ rwe)	Supply (m³)	Industry capacity (m³ rwe)
West Kalimantan	1,211,574	158,400	880,671	5,501,187
Central Kalimantan	3,004,310	0	3,916,645	3,329,710
South Kalimantan	354,316	0	676,828	4,310,161
East Kalimantan	1,419,115	2,800,000	2,806,834	7,128,333

Source: Ministry of Forestry, 1999

According to the ALGAS (Asian Least-cost Greenhouse Gas Abatement Strategy) study on Indonesian greenhouse gases inventory in 1994, forestry and land-use change accounted for 74% gas emission. Table 16.2 and 16.3 present a summary of Greenhouse Gases Inventory in 1990 and 1994. These tables clearly show us that forestry and land-use change contributed most of the CO_2 emissions. It is clear that activities from forestry and land-use change emit the most greenhouse gases, contributing 64% and 74% in 1990 and 1994 respectively. And within four years, the forests' capacity to absorb emissions deteriorated significantly by 74%.

Table 16.2: Summary of Indonesian greenhouse gases inventory, 1990

Sources and Sinks	CO_2 Uptake	Emission (Gg)				
		CO_2	CH_4	CO	N_2O	NOx
TOTAL NATIONAL CO_2 EQUIVALENT: 552,648 Gg						
I. Energy and Transport	—	140,102	1,886	29	82	313
II. Industrial Processes	—	17,901	—	—	—	—
III. Agriculture	—	—	3,388	31	14	1,071
IV. Forestry and Land-Use Change	1,537,686	280,607	277	69	2	2,428
V. Waste (Landfill)	—	—	288	—	—	—
Total Emissions and Removals by Gas	1,537,686	438,610	5,839	413	97	3,499

After 1994, there has been no follow-up inventory study of Indonesian greenhouse gas emissions. No doubt the deforestation rate has increased every year since Indonesia faced economic crisis, even though no exact number of how much Indonesian tropical forest has disappeared is available. The carbon emission from forestry and land-use change increase in the same percentage.

Table 16.3. Summary of Indonesian greenhouse gases inventory, 1994

Sources and Sinks	CO_2 Uptake	Emission (Gg)				
		CO_2	CH_4	CO	N_2O	NO_x
TOTAL NATIONAL CO_2 EQUIVALENT: 875,871 Gg						
I. Energy and Transport	—	170,016	2,396	8,422	5.7	818
II. Industrial Processes	—	19,120	—	0.5	—	0.01
III. Agriculture	—	—	3,244	331	52.9	18.7
IV. Forestry and Land-Use Change	403,846	559,471	367	3,214	2.5	91.3
V. Waste (Landfill)	—	—	402	—	—	—
Total Emissions and Removals by Gas	403,846	748,607	6,409	11,966	61.1	928

After a long drought, contrarily, when the La Niña phenomenon took place, Indonesia suffered from extremely heavy rainfalls. Floods cost millions of rupiah, claim houses, and even thousands of life every year.

"*Pak* (Dad), I'm starving. Give me something to eat," five-year-old Darno whimpered. Raising his voice, Darno repeated his request several times, but his father, Kartanom, remained silent. Kastiyem, Darno's mother, embraced him and pulled him onto her lap. "We just ate a piece of boiled sweet potato each. We have nothing left to eat, nothing to cook. Our rice stock and all our belongings have been swept away by the flood," Kartanom, 45, said. This conversation was reported in the only national English newspaper, The Jakarta Post, October 27, 2001.

"The children have only eaten boiled sweet potato since morning. We are ashamed to ask other people for money or food," said Kartanom, pointing to his three children. Darno is the second child, the smallest is an 18-month-old boy. Kastiyem, 40, just wept, embracing Darno with her right hand while breastfeeding Darno's younger brother.

Kartanom said he was very upset. "What should or can I do? No money to buy something to eat. Asking neighbors to help is impossible as most of them have a similar story." Kartanom, Kastiyem and their three children have been spending their days on a wooden bed. The bed is propped up in such a way that it is 20 centimeters higher than the water level, which reaches more than 50 centimeters.

Their family is one of 100 families in the village of Sikampuh who refused to seek refuge, arguing that by staying at home they could do more for their families. At least 1,500 houses in the village of Sikampuh, in the Kroya district of Cilacap regency, have been engulfed by floodwaters since Tuesday. The floods have hit more than 20 districts in the regencies of Cilacap, Banyumas and Kebumen, which are located in the southern plain of Central Java (Jakarta Post, 2001).

At least 35,000 houses in Kebumen, Banyumas and Cilacap regencies were submerged in the latest floods, which forced at least 30,000 people to evacuate their homes. Flooding also hit other parts of Indonesia. According to the Meteorology and Geophysics Agency's forecast, this year's rainy season which started in early October 2001, will continue up to April 2002.

Floods in rural areas were caused more by land-use change (again especially converting forest to agricultural land), legal logging, and illegal logging. The World Bank has estimated the deforestation rate to be about 1.2 million hectares a year. Floods in urban areas or in big cities, such as in Jakarta (Capital City) and Semarang (Central Java), were mostly caused by bad drainage systems and overexploitation of groundwater.

In its Third Assessment Report, the Intergovernmental Panel on Climate Change (IPCC) acknowledged the significance of human contribution to climate change. "This is already stronger than its Second Assessment Report," said Prof. Daniel Murdiyarso, Deputy Minister for the Environment of Indonesia, in a statement to the COP-7 UNFCCC, in Marrakech, October 29-November 9, 2001.

The Report of Working Groups II on impact, adaptation and vulnerability as a part of the IPCC Third Assessment, stated that vulnerability of human societies and natural systems to climate extremes is demonstrated by damage, hardship, and death caused by events such as droughts, floods, heat waves, avalanches, and windstorms. While there are uncertainties attached to estimates of such changes, some extreme events are projected to increase in frequency and/or severity during the 21st century due to changes in the mean and/or variability of climate, so it can be expected that severity of their impacts will also increase in concert with global warming.

The Working Groups II also projected that climate changes during the 21st century have the potential to lead to future large-scale and possibly irreversible changes in Earth systems, resulting in impacts at continental and global scales. The effects of climate change are expected to be greatest in developing countries in terms of loss of life and nature. A country such as Indonesia is more vulnerable to climate change because it also under pressure from such forces as population growth, resources depletion, and poverty. And climate change would exacerbate threats to biodiversity due to land-use and land-cover change and population pressures. The Indonesian population was more than 200 million in mid-1997. Population growth is estimated at around 2% per annum.

Indonesian biological diversity richness, which vies only with Brazil for the title of richest country on Earth in terms of biological diversity, would be affected and be gone forever. Indonesia is home to 515 species of mammals (second on the world mammal list behind Brazil), 39% of them endemic; 511 species of reptiles (fourth in diversity), 150 endemic; 1,531 species of birds (fifth), 397 endemic; 270 amphibian species (sixth), 100 endemic; 75 species of psittacine birds (first), with 38 endemic; 35 species of primates (fourth), and including a tree kangaroo (*Dendrolagus mbaiso*), the new mammal species found in the 20th century in Irian Jaya. All these could well be extinguished.

This suffering country is also in the top five for plant diversity with an estimated 38,000 higher plant species. It heads the world list in palm diversity with 477 species, 225 endemic, and has over half of the 350 known species of dipterocarp trees, with 155 endemic in Kalimantan. Indonesia also ranks behind only Brazil and possibly Columbia in freshwater fish diversity, with about 1,400 species. If Indonesian tropical forests are lost, the habitat for many already endangered species, some endemic species will be lost forever. And the loss is irreversible.

Many species are already at high risk and are expected to be placed at greater risk by the synergy between climate change, rendering part of the current habitat unsuitable for many species, and land-use change, fragmenting habitats and raising obstacles to species migration. Without appropriate management, according to IPCC's Third Assessment Report, these pressures would definitely cause some species currently classified as "critically endangered" to become extinct and the majority of those labeled "endangered or vulnerable" to become rarer, and thereby closer to extinction, in the 21st century.

Large-scale impacts of climate change on oceans are expected to include increases in sea surface temperature and mean global sea level, decreases in sea-ice cover, and changes in salinity, wave conditions, and ocean circulation. IPCC has confidently projected that sea-level rise would put ecological security at risk, including mangroves and coral reefs. Many coastal areas would experience increased levels of flooding, accelerated erosion, loss of wetlands and mangroves, and sea water intrusion into freshwater sources as a result of climate change. The State Ministry of Environment (1998) using Semarang as a case study concluded that a total of 2,300 hectares in Semarang are vulnerable to flooding and 3,195 hectares are vulnerable to salt water intrusion.

More than 80% of Indonesia's 51,000 km^2 of coral reefs have already been threatened mainly due to destructive fishing practices such as using bomb and cyanide which bleached the coral reefs, according to UNEP's new World Atlas of Coral Reefs. The World Atlas of Coral Reefs was prepared by UNEP World Conservation Monitoring Center. The Atlas shows that Indonesia, followed by Australia and the Philippines are the three largest reef nations. Conservation International, a conservation non-governmental organization based in Washington, DC, stated that while Brazil surpasses Indonesia in terrestrial and freshwater species diversity,

Indonesia unquestionably leads in marine diversity and has very impressive terrestrial and freshwater diversity as well. Sea-level rise and increases in sea-surface temperature are the most probable major climate change-related stresses on coastal ecosystems. Coral reefs may be able to keep up with the rate of sea-level rise but could suffer bleaching from higher temperatures. Sea-level rise may also threaten a wide range of mammals, birds, amphibians, reptiles, and crustaceans living in coastal areas.

Not only would the rich Indonesian coral reef be gone, 110 million Indonesians living in the coastal areas, most of them poor fishermen, would be affected. Semarang and Jakarta would be gone with a sea-level rise of 1m. Indonesia's 81,000 km coastlines (42,530 km mangrove forests) would retreat by 1,500 meters. And about 116 small islands and groups of small islands would be gone. In its Third Assessment Report, IPCC projected a sea level rise of 9-88 cm if global temperature increases between 1.4-5.8°C (IPCC, 2001).

Changes in average climate condition and climate variability would have a significant effect on agriculture in many parts of the Asian region. Moreover, agricultural areas in Indonesia have already been affected by many environmental hazards such as frequent floods and droughts. Almost every year since 1987 to 1997, paddy fields were effected by extreme climate conditions. When La-Nina came they flooded and when El-Nino came they dried up.

Paddy field effected by climate change, flooded and dried up in 1988-1997

Year	El-Nino/La-Nina	Flooded (ha)	Dried up (ha)
1987	El-Nino	***	430.170
1988	La-Nina	130.375	87.373
1989	Normal	96.540	36.143
1990	Normal	66.901	54.125
1991	El-Nino	38.006	867.997
1992	Normal	50.360	42.409
1993	Normal	78.480	66.992
1994	El-Nino	132.975	544.422
1995	La-Nina	218.144	28.580
1996	Normal	107.385	59.560
1997	El-Nino	58.974	504.021

Source: Indonesian State of the Environment 2002

Are all these sufferings attributable to climate change? Only one study done by Dr Alex Wilson of the University of Arizona, sponsored by PT Freeport Indonesia (a mining company subsidiary of Freeport McMoRan

Copper and Gold Inc), has concluded the that ice-cap covered tip of Puncak Jaya in Lorentz National Park, a World Heritage Site in Irian Jaya, is decreasing due to increasing global temperature. The glaciers are in an alpine environment and at an elevation higher than 5,030 meters above sea level. Glaciers in Papua New Guinea (eastern half of the island of New Guinea) and two in Irian Jaya (Papua) have disappeared in recent decades due to this same warming trend.

Exploitation of natural resources associated with rapid urbanization and high population growth, industrialization, and economic development has led to increasing pollution, land degradation and other environmental problems. Climate change represents a further stress to Indonesia.

Despite some gains in curbing population growth, Indonesia remains the most populous nation in the subregion. Its population in 1999 was 207.4 million, growing at a rate of 1.6% between 1995 and 1999. The island of Java, where the capital Jakarta is, is one of the most densely populated places in the world. More than 15 million people live in Jakarta greater area. To a large extent, the demands of this big population have led to serious social and environmental problems in the country. In recent years, the incidence of poverty has risen, after years of decline. In 1996, 11.3% of Indonesians were still below the poverty line. Due to economic crisis, the number increased to 20.3% in 1999, i.e., about 40 million people lived in poverty (Suryahadi et al., 1999)

Yet, Indonesian greenhouse gases emission is increasing and Indonesia is a highly populous nation, fourth in the world. If climate change really happens because of lack of commitment from developed countries to reduce their greenhouse gas emissions, Indonesia will suffer first. As an archipelago of more than 17,000 islands with a coastline of around 80,000 km, Indonesia will be adversely affected by any small change in sea-level rise. More than half of Indonesia's employment is in the agricultural sector, one of the most vulnerable and climate-dependent sectors. Last year alone, Indonesia spent about US $10 million on climate-related disaster relief. "Indonesia will face threatened food security and water scarcity if we let climate change get worse," said Prof. Murdiyarso.

Indonesia has been involved in climate change negotiations since they began. In 1992 Indonesia signed the UNFCCC and ratified it in 1994 with a decree of Parliament. In 1998 Indonesia signed the Kyoto Protocol and has backed it since then. The Indonesian position on the Kyoto Protocol is clear. When the President of America, George W. Bush, opposed the Kyoto Protocol, the Indonesia Minister of Environment sent a clear message to the US Ambassador in Jakarta that the Indonesian government demanded the US to reconsider its position. The Minister of Environment was backed up by 17 environmental NGOs. The 17 NGOs sent a disappointment letter to President George W. Bush asking that he reconsider his opposition and take US responsibility to reduce emissions of greenhouse gases seriously.

Indonesia does not have legally binding commitments, but seriously considers the recommendation of the UNFCCC on implementing national climate policies in the context of common but differentiated responsibilities. In fact, Indonesia has gone beyond the requirements of the Convention and the Kyoto Protocol.

A study done by the Directorate General of Electricity and Energy Development and Energy Assessment of the University of Indonesia in 1999, projected that under the current crisis scenario, energy demand would drop and reach its lowest point in 2000. In fact, between 1998 and 2000 energy demand dropped by 0.7% per year. But, CO_2 emissions from the energy demand sectors are projected to triple between 2000 and 2020 as the share of coal in energy supply is expected to increase by a factor of ten.

Indonesia's greenhouse gas emissions are projected to increase rapidly after the current economic crisis is overcome. To help reduce fuel consumption through greater efficiency, in 2000 the State Ministry for Environment, Republic of Indonesia, with support from the Government of Germany and the World Bank, engaged the National Strategy Study on the Clean Development Mechanism (CDM), the only mechanism that enables developing countries such as Indonesia to participate in the Kyoto Protocol.

The Clean Development Mechanism (CDM) is one of the 'flexibility' mechanisms included under the Kyoto Protocol. It allows countries with greenhouse gas (GHG) emission limitation and reduction commitments (Annex B countries) to engage in project-based activities in developing countries, with the twofold aim of assisting them in achieving sustainable development and helping Annex B countries to meet their emission reduction targets. CDM projects produce GHG emission reduction units, called certified emission reductions (CERs), which must be verified and authenticated by independent certifiers.

The National Strategy Study (NSS) developed a strategy for attracting CDM investment and implementing CDM projects in Indonesia. It emphasized the technical potential for and cost of GHG emission reduction projects, using both a top-down, MARKAL-based (an energy system modeling tool), and a bottom-up approach. Analysis of sinks will be done in a separate report. The study also estimated the size of the CDM market and the factors that will affect Indonesia's share, defined international and national institutional settings for the CDM, and developed a project pipeline.

Under a standard pre-Bonn market scenario, the total volume of the CDM in Indonesia until 2012 was projected to be 125 Mt CO_2. At a price of US \$1.83 per ton CO_2, the CDM revenue would amount to US \$228 million. Under pessimistic and optimistic scenarios Indonesia's share in global CDM was estimated at 1.5% and 3.5%, compared to 2.1% under the standard assumptions; using Indonesia's (precrisis) share in FDI (Foreign Direct Investment) as a proxy would lead to a share of 4%.

The Indonesian CDM project pipeline identified in the NSS consisted of 10 projects with an overall reduction potential of about 3.5 Mt CO_2 equivalent. Development of this project pipeline was based on the following priorities: compatibility with national energy policy and regulations, concentration on projects perceived as "low" risk assessment of mitigation costs, and assessment of stakeholder views based on the above-mentioned questionnaire results. Another important consideration is the difficult financial situation of the state electricity company, PLN. PLN had stated that its short-term priorities were to restore its financial viability and to finance postponed projects. Any CDM projects linked to the supply of power (cogeneration, fuel switch, renewable energy, etc.) would have to consider this priority.

Without US participation in the Kyoto Protocol and including sink in CDM, the CDM global market will reduce 75%. Olivia Tanujaya, a policy researcher from Pelangi, belonging to a leading Indonesian NGO in climate change and energy issues, estimated Indonesian potential revenue from CDM would decrease from US \$224 million per year to only US \$56 million per year at a price of US \$1.83 per ton CO_2. If the Kyoto Protocol were implemented without US participation, hot air could swamp the global carbon market, crowding out the CDM. "Quantitative limits on hot air or sales of emission credits from EIT countries generally could ensure meaningful developing countries' involvement via the CDM," concluded Tanujaya.

High priority CDM projects proposed in the NSS include boiler improvements, gas turbines, cogeneration, utilization of flared gas, gas combined cycle, small hydropower and low temperature cogeneration. Based on a survey of stakeholders done through NSS, strategies identified as priorities include fast-tracking small-scale projects, developing a national CDM manual and guidelines, establishing a national CDM board and clearinghouse, conducting a comprehensive baseline study, fostering invesment relations, and maximizing the potential for no-regrets and low-cost projects.

To strengthen meaningful participation, Indonesia with Brazil, Bangladesh, South Africa, and the Netherlands joined South-South-North CDM pilot projects. Pelangi is the SSN team for project development in Indonesia. The SSN Project derives its name from cooperation between southern or developing countries, and the relationship between these southern countries and countries in the north, or developed countries. The SSN mission is to promote such cooperation. The SSN project is committed to the reduction of greenhouse gases, combating global warming and climate change, sustainable development, and environmental integrity. The principal aims of SSN are to learn by doing, host governments with CDM structures, build capacity, and promote Internet resources. The project website is: http://www.southsouthnorth.org. Eight pilot projects have resulted in the design of a useful set of criteria and indicators for CDM projects.

From the two concluded phases of the SSN endeavors, nine projects were identified as potential CDM projects. The types of projects considered included transportation mechanisms, developments for renewable sources of energy, and sequestration projects. Of the nine potential projects, two were identified for immediate consideration and chosen as the two CDM pilot projects to be developed under SSN. These are the emission reduction program for urban buses and a geothermal development project. Each of these projects meets the essential requirements necessary for CDM and both are ready to begin. In addition, both projects scored high in a variety of categories considered important in assessing the relative priority among possible CDM projects. These categories pertained to additionality, sustainability, and feasibility in various spheres. By quantifying the relative importance of the different criteria used as indicators for assessing the priority and viability of each scheme, the team developed a matrix for ranking the schemes assessed.

The emission reduction program for urban buses pilot project is basically concerned with establishing new and efficient engines for public transportation in Yogyakarta, Central Java. The project aims to restructure urban public transport management, improve the efficiency of the urban bus system, and provide cleaner engines for two-fifths of the buses currently in operation. The project scored high in all criteria considered relevant for a successful CDM project. There are no barriers to its implementation and it will be managed by Yogyakarta Urban Transport Alliance.

The other pilot project is to construct a Sarulla geothermal power plant in Sumatra. This project was designed to fulfill the growing demand for electricity on the island of Sumatra while producing less emission. It has the highest rank among other project candidates, based on indicators of a successful CDM project. UNOCAL Geothermal Indonesia, Pertamina, and PLN will carry out the project jointly. The local government of Sarulla subdistrict also officially supports the project. The project is not financially viable compared to the baseline project, a coal power plant. However, the project is ready to proceed to the implementation phase.

According to the ALGAS greenhouse gases inventory study 1994, activities in forestry and land-use change contributed 74% of Indonesian greenhouse gas emissions. Since COP-6 UNFCCC decided to include sinks from afforestation and reforestation in CDM, Indonesia is very keen to implement sink in CDM projects to help stop deforestation. But, said Muriyarso in explaining the Indonesian position: "The use of sinks in both domestic actions and in the Kyoto mechanism needs to follow a scientifically sound method."

On the issue of including sinks in the CDM, Indonesia attaches high importance to settlements of the definition of forest, afforestation, and reforestation. The use of sinks in the first commitment period should be allowed only when key issues such as baseline, leakage, permanence, and public participation are resolved. In this regard, "Indonesia supports the idea of giving a mandate to SBSTA to draft the terms of reference of the modalities. This should be discussed at the next SBSTA sessions and be adopted at COP-8," said Murdiyarso. "And the inclusion of sinks in the CDM projects should not violate other international treaties such as the Convention on Biodiversity, and other forest-related agreements and should consider their interrelationships."

Does this mean that Indonesia should accept forest sinks in CDMs? The logging industries are very keen to have such projects but some NGOs are afraid that sink projects would decrease Indonesian forests. Forest plantation companies would look for forests to be cut and transferred into sink projects. The companies would get money from the trees they log, reforestation funding from the government, and money from sink projects. NGOs think the sinks on CDM do not resolve forestry problems but, on the contrary, would raise more problems, especially problems related to indigenous people.

In the World Summit on Sustainable Development, took place in Johannesburg, South Africa, Indonesia played a central role. Prof. Dr. Emil Salim, a former and the first Indonesian Minister of Environment, chaired the Preparatory Committee for WSSD. He was actively "influencing" delegation through his Chairman Paper, which was finally adopted by all delegations in the Summit.

Indonesian delegation worked closely with Prof. Salim to draft the Chairman Paper. The issues in Chairman Paper were poverty eradication, changing unsustainable patterns of consumption and production, protecting and managing the natural resource base of economic and social development, sustainable development in a globalizing world, health and sustainable development, sustainable development of Small Island Developing States, sustainable development initiatives for Africa, means of implementation.

Indonesia supposedly ratified Kyoto Protocol by 2003 before COP9 in Milan, Italy, due to national situation that the Government has to deal with terrorism and has to prepare for a general election in 2004, it was postponed. The draft regulation for ratification has been discussed amongst sectoral departments and was ready to be submitted to the Parliament. To quit from its suffering, Indonesia put on hope to developed countries especially Russia to ratify Kyoto Protocol in order to be implemented and stopped climate change.

References

ALGAS (Asia Least-cost Greenhouse gas Abatement Strategy). 1997a. *Technical note on sources and sinks.* ALGAS Project.

Directorate General of Electricity and Energy Development and Energy Assessment University of Indonesia. 1999. *A Strategy for Long-term Energy Provision with Optimization Principle: Final Report.*

Forest management may mitigate global warming. 2001. (http://www.eurekalert.org/pub_releases/2001-11/uow-fmm111901.php)

Global warming 'may cause El Niño' Thursday, April 9, 1998. Published at 21:24 GMT 22:24 UK (http://news6.thdo.bbc.co.uk/hi/english/sci/tech/newsid_76000/76693.stm)

IPCC Special Report on the Regional Impacts of Climate Change: An Assessment of Vulnerability. 1997. (http://www.grida.no/climate/ipcc/regional/index.htm)

Jakarta Post, November 29, 2000. *Tropical storm 3B to drift away.*

Jakarta Post, June 21, 2001. *Java, eastern Sumatra to have longer dry season.*

Jakarta Post, October 27, 2001. *C. Java flood victims lack food, medicines.*

Rising Seas Threaten Cities, Erode Beaches and Drown Wetlands in Key Developing Countries. *Climate Alert,* Vol. 8, No. 2 March-April 1995. (http://www.climate.org/Climate_Alert/articles/8.2/)

Schindele, W., Thoma, W. and Panzer, K. 1989. *The Forest Fire 1982/1983 in East Kalimantan.* FR-Project, ITTO, GTZ, BPPK, and DFS.

Siegert, F. Ruecker, G. Hinrichs A. and Hoffmann. A. A. 2001. *Increased damage from area in logged forests during droughts caused by El Nino,* Letter to Nature, *Nature* 414: 437 - 440.

State Ministry for Environment. 1998. *Forest and Land Fires in Indonesia. Impacts, Factors, and Evaluation* (Vol. 1). State Ministry for Environment, Republic of Indonesia and United Nations Development Program.

State Ministry for Environment. 2001. *National Strategy Study on the Clean Development Mechanism in Indonesia.* Jakarta.

Suryahadi, A., Sumarto, S. Suharso, Y. and Pritchett. L. 1999. *The evolution of poverty during the crisis in Indonesia, 1996 to 1999.* SMERU Working Paper. Jakarta. SMERU.

Timmermann, A. Oberhuber, J. Bacher, A. Esch, M. Latif, M. and Roeckner, E. 1999. *Increased El Nino frequency in a climate model forced by future greenhouse warming.* Nature 398, 694-7.

Trenberth K.E., and Hoar T.J. 1996. *The 1990-1995 El Nino Southern Oscillation event: longer on record.* Geophys Res. Lett. 23:57-60.

CHAPTER 17

Trump Card in Kyoto Pact: Russian Federation's Interests and Positions on the Global Climate Change Regime*

Natalia Mirovitskaya

Terry Sanford Institute of Public Policy Box 90239, Durham, NC 27708-0239 USA   e-mail: nataliam@duke.edu

Introduction

The position of the Russian Federation in the international climate regime has become vital for its future. Russia is among the five countries that produce more than half the world's fossil carbon as well as among five countries that account for about half of all fossil carbon consumption. Its territory accounts for 21% of the world's forests[1] and hosts nearly 8 million sq km of "wildlife" areas – places virtually untouched by any economic activity.[2] Many Russian scholars and politicians allege that the tremendous capacity of Russia's environment to absorb carbon emissions makes the country an "environmental donor" to the world. As one of the major producers of fossil fuel emissions and the largest international "sink", the country plays an important role in the global climate system. By sheer virtue of the size of its territory, character of its ecosystem, and size of economy, Russia can strongly influence the structure and effectiveness of any international agreement related to climate change.

Currently, it is the destiny of the Kyoto Protocol to the United Nations Framework Convention on Climate Change (UNFCCC) that rests squarely on Russia's shoulders. While the Framework Convention, now universally accepted, spells out the principles and objectives for common climate

*This article was received while the book was in press – V.I.Grover.

policy, the Kyoto Protocol contains legally binding obligations for developed countries to reduce or stabilize their greenhouse gas emissions into the atmosphere. In particular, it calls for strict quantitative limits (or allowances) on emissions from 38 developed countries, which are scheduled to take effect during the so-called First Budget Period, from 2008 to 2012, with 1990 as a baseline. For the Kyoto Protocol to come into force, it must be accepted, approved, acceded to, or ratified by at least 55 signatories to the Convention, including developed countries that together in 1990 accounted for at least 55 percent of total carbon dioxide emissions. In effect, this provision means that to come into force the protocol must be ratified or approved by either the United States or Russia, which together accounted for over 50 percent of emissions from Annex I countries in that year. As of January 2004, the Kyoto Protocol has 119 parties that together account for 44.39% of global emissions. The United States withdrew from the Kyoto pact in 2001; therefore it can enter into force only if ratified by the Russian Federation with its 17% of the 1990 carbon emissions. Recently, a number of remarks by a top Kremlin official, Andrei Illarionov,[3] have led many to believe that the Russian government decided not to ratify the Kyoto Protocol, thereby effectively sending it to its death. These remarks were disclaimed by several government representatives including most recently (on January 17, 2004) the Russian Government representative to the UNFCCC, Alexander I. Bedritsky, who confirmed Russia's commitment to ratify the protocol.[4] However, the issue of when and under what conditions the Russian government will ratify the Kyoto Protocol remains open.

This chapter describes the evolution in the Russian government's position on the Kyoto Protocol, the potential benefits and drawbacks of Russia's participation in the global climate regime, and policy options that seem to be most desirable for different actors in the country.

Russia's Position on Protocol Ratification

Russia's position on the global climate regime has been ambivalent and confusing for many observers (at least for those not familiar with Russia's negotiating behavior and the Kremlin politics of policy-making).

The Russian Federation signed the UN Framework Convention on Climate Change (UNFCCC) at the United Nations Conference on Environment and Development (Rio de Janeiro, 1992) and ratified it in 1994. In 1994, the Russian Government created an Interagency Commission on Climate Change Issues (ICCC) to coordinate efforts of domestic actors in implementing national obligations under the Framework Convention and to elaborate the national stance on more concrete aspects of global climate regime, including the UNFCCC additional protocols.[5] The President of Russia signed the Federal Law *On the Subject of Ratification of the UN Framework Convention on Climate Change* on 4 November 1994, after its acceptance by the Federal Assembly (upper chamber of Russian Parliament. The UNFCCC committed nations to "aim" to stabilize emissions at 1990 levels by 2000. The parties also agreed to compile inventories of their greenhouse gas emissions and submit regular reports (National Communications) on actions they were taking to implement the Convention. Under the Framework Convention the parties should prepare national programs that contain:

> Climate change mitigation measures
> Provisions for developing environment friendly technologies
> Provisions for sustainably managing carbon "sinks"
> Preparations for adapting to climate change
> Plans to engage in climate research, observation, and information exchange
> Plans to promote education, training, and public awareness relating to climate change.

In October 1996 the Russian Government adopted and earmarked funds for the 1997-2000 Federal Target Programme "Prevention of Dangerous Climate Changes and Their Adverse Effects" (FTPC) aimed at the domestic implementation of the UN Framework Convention.[6] It has also earmarked funds for the federal "Russian Energy Efficiency for 1997-2005" program, which has a direct impact on greenhouse gas emissions. However, it is difficult to estimate how many provisions of these programs have been actually implemented while according to the FTPC implementing agency "...funding of the Program can not be considered as satisfactory even for the first steps of activity."[7] The greater part of the Program (inventory of GHG sources, preparation of National Communications, and elaboration of the Climate Change Action Plan) was actually financed by the US Department of Energy and EPA. Apparently, the Federal Target Program was not renewed after 2001.

While the achievements of the domestic climate policy during the last decade in Russia have been modest (to say the least), the Russian delegation to the Conference of Parties that worked to incorporate specific national obligations into the legally binding Kyoto Protocol succeeded in negotiating very beneficial conditions under the global climate regime. In particular, negotiating parties established 1990 as the baseline and set Russia's greenhouse gas (GHG) emission reduction target for the period of 2008-2012 at 100% of the 1990 level.[8] The Protocol also introduced mechanisms of carbon emission trading (with quite lenient regulations) and Joint Implementation (JI) that make it possible for Russia to benefit financially and technologically from other countries' obligations to cut their emissions. Negotiating parties also gave Russia and other Annex I countries more sink allowances thus effectively relaxing the emission constraints negotiated in Kyoto and, in addition, allowed some flexibility for economies in transition in implementing their obligations.

Russia signed the Kyoto Protocol on February 11, 1999 as an interstate agreement that would become binding upon its ratification. Under Russia's ratification procedure, the protocol has to get majority approval in the lower house, the State Duma, and the Federation Council (upper house) before going to the President for signature.

The formal process of ratification was launched in 2001 when the Ecology Committee of the Duma organized the hearings on the Kyoto Protocol. Participants of the hearings supported ratification of the Kyoto protocol (though with several conditions) and recommended that the government draft specific legislation dealing with quota allocation among different actors, creation of national monitoring system, and introduction of incentives to ensure GHG emission reduction.

In April 2002 the government requested the Ministry of Economic Development to submit a National Report on the Climate Change Issues.[9] The report, produced by the Ministry in August 2002 presented an analysis of available data on global climate change and how it would particularly affect Russia. The National Report also included forecasts of national carbon emissions under various scenarios of Russia's economic development along with an assessment of potential advantages and drawbacks of Russia's participation in the Kyoto Protocol. In that respect the authors of the National Report specified that the acceptance of the Protocol should depend upon a "proper" estimate of Russia's exact contribution to the global balance of carbon emissions/absorptions as well as assessment of its current and prospective energy demands. Among many other facts and assumptions, the Report introduced the concept of Russia being a "global environmental donor", whose ecosystems absorb anthropogenic emissions of other developed countries in addition to its national emissions. The authors of the Report questioned the correctness of methods used by science experts of the Intergovernmental Panel on Climate Change (IPCC) to estimate carbon assimilation by different sources and posited that better inventory of the assimilation processes would radically change Russia's position in the whole structure of the Kyoto's Protocol.[10] This statement, indeed, has been supported by recent findings of the Forestry Project run by the prestigious International Institute for Applied Systems Analysis (IIASA) in Laxenburg (Austria) that illustrated the complexity of obtaining a comprehensive accounting of carbon sources and sinks at the national level and demonstrated that including biological sinks among the mechanisms to lower greenhouse emissions renders the Kyoto Protocol unfeasible.[11] This Project and IIASA's related work on carbon accounting confirmed that meaningful implementation of the Kyoto Protocol requires a transparent, consistent, accurate and verifiable system that accounts for all sources and sinks of carbon – a system currently non existent and quite challenging to create.[12]

However, despite these findings and objections, the National Commission posited firstly that change in assessment methods is not feasible at the current stage of the Kyoto process and secondly, that it matters less as long as the economic decline in Russia during the 1990s was so significant that total decrease in emissions for the period of 1990-1999 exceeded 10 billion tons of carbon.[13] In general, the authors of the Report recommended ratification of the Protocol on the grounds that:

> It would provide Russia with substantial political benefits not only as the global environmental leader, but as the "rescuer" of the Kyoto process,
>
> It would create a precedent for the development of market-based global environmental policy, which is beneficial for Russia as a "global environmental donor," and
>
> It would provide economic benefits to the struggling economy of the country through the Joint Implementation (JI) projects and carbon emission trading.[14]

For some time it seemed that Russia was heading towards ratification. Russian Prime Minister Mikhail Kasyanov announced at the 2002 Earth Summit in Johannesburg, South Africa, that "Russia is preparing to ratify the Protocol, which hopefully will take place in the nearest future."[15] In June 2003 the working group of the Presidium of the State Council of the Russian Federation — personally chaired by President Vladimir Putin — recommended ratification of the Kyoto protocol. In August 2003 the Deputy Minister of Economic Development and Trade and Co-Chairman of the Russian Interagency Committee on Climate Change, Mukhamed Tsikanov, expressed his opinion that Russia could ratify before the end of 2003, which was confirmed by representatives of the Ecology Committee in Duma, the lower chamber of Russia's Parliament. On September 19, 2003, the Russian Interagency Committee on Climate Change approved a draft of the legal basis for incorporating the Kyoto Protocol into the national legislation.

Since then, however, the issue has dragged with the Russian government, and Moscow's position on the Protocol has become murkier. In August 2003, Vladimir Potapov, the Deputy Secretary of Russia's Security Council, published an article in the government newspaper *Russian Gazette,* alluding that the Kyoto ratification would put Russia in comparative disadvantage with its main economic competitors, China and India, and that more sophisticated political and economic interests in the West were trying to take advantage of Russia. A month later, on September 30, 2003, President Vladimir Putin announced at the Climate Conference in Moscow that Russia would take its time to evaluate the consequences of ratification and set the national policy around Russia's interests,[16] though later in an interview he said: "I believe ratification and making the Kyoto Protocol effective would be a step in the right direction." Two weeks later, at the Asia-Pacific Economic Cooperation Summit in Bangkok, he stated that the Russian Federation was in favor of ratifying the Kyoto Protocol. In a subsequent interview he posited that it would be difficult for Russian authorities to persuade the State Duma to ratify the document.[17] Indeed, a delegation of members of the European Parliament reported after their visit to the Russian Duma in September 2003 that Russian parliamentarians were quite explicit that they needed more support from the EU before Russia would ratify the Kyoto Protocol, notably more support with regard to Russia's entrance into the WTO or a preferential trade agreement.[18] Just a week after the Bangkok meeting, President Putin, renowned for his figurative language, proclaimed that "...Russia will not turn into a milk cow at whose expense other countries will solve their problems."[19]

The end of 2003 brought even more surprises. In late October 2003, the Ministry of Natural Resources initiated a pilot stage for Kyoto mechanisms on regulation and trade of carbon emissions in three regions of the Russian Federation. Meanwhile the relevant ministries submitted all required documents to the government, which was then supposed to send the Kyoto Protocol for ratification to the State Duma. But Andrei Illarionov, President Putin's then economic advisor and long-time opponent of Kyoto, pitched in his voice, purportedly speaking for the President: "In its current form, this protocol cannot be ratified."[20] Illarionov cited several reasons for such a decision:

> The Kyoto Accord is not based on sound scientific grounds. Moreover, there is not enough scientific evidence that the phenomenon of global warming even exists;[21]
>
> Kyoto's implementation will hinder Russia's prospects for economic development, in particular implementation of emission limits for 2008-2012 could constrain the Russian government's current overarching goal of doubling Russia's GDP – "...which means dooming the country to poverty, backwardness and weakness." [22]
>
> Kyoto may also damage Russia's international competitiveness: with the United States out of the treaty, there will be no buyers for Russia's "hot air" while exemptions granted by the treaty to India and China (increasingly, Russia's industrial competitors) give their manufacturers an unmerited advantage; and
>
> Russia's ratification of the protocol and therefore its coming into force would not be "fair" to President Putin's "good friend" – U.S. President George W. Bush who might find his administration under additional international and domestic pressure as environmental foot-dragger.

Although M. Tsikanov, Deputy Minister for Economic Development and Trade in charge of Interagency Committee on Climate Change, a few other government officials, and eventually Prime Minister Kasyanov disclaimed Illarionov's statement, it provoked a lot of speculations by experts and media on what Russia's real intentions were. Was this the death of Kyoto or one more attempt (following many)[23] to use a diplomatic trump card to extract even more concessions from Europe, Japan and possibly other actors interested in the success of the new regime? Officially, only the Russian parliament can reject the Protocol with authority, after the President decides when and whether to submit it for approval. Therefore, Illarionov's position (though supported by some prominent officials and politicians) does not necessarily mean that Russia is officially out of Kyoto. However, factors brought to light by Mr. Illarionov, his supporters, and opponents are definitely in the background of Russia's climate policy and worth exploration.

Mr. Illarionov's rhetoric presents a simplistic version of polemics going on in Russia's political and academic circles on several main issues:

Will Russia benefit from forestalling future climate change?

What will be the impact of the Kyoto regulations for the Russian economy?

What are the costs of implementation and its feasibility?

What are the net benefits to Russia of participation in a global climate regime and what are the side effects for Russia's participation in other global regimes?

What are the political ramifications of Russia enacting the Kyoto pact for its relations with the US administration, which is fiercely opposed to it? And, vice a versa, what is the price of Russia's rejection of Kyoto for its relations with European countries, Japan and Canada as well as developing countries who will be losing substantial investment opportunities in this case?

And, finally, following Mr. Illarionov arguments and ever-present concern in Russian politics, the "conspiracy theory":

Is there an American trail? Is Russia's behavior a reflection of new US-Russian bilateral climate policy? Or, vice a versa, is there a European trail? Is the European bureaucracy trying to use Russia's naiveté in matters of environmental geopolitics to "glue" Russia to Europe as a raw materials appendage?

Should Russians be Interested in Forestalling Future Climate Change?

The process of global climate change reportedly affects the territory of the Russian Federation. The mean temperature rise for Russia over the past hundred years has been 0.05 degrees Celsius per decade.[24] Over the last century, average temperatures increased by 3.5 degrees in eastern Siberia and the Far East. The mean annual anomalies within the Russian Federation have shown positive temperature values since 1987, exceeding the norm both throughout Russia and in most of its regions.[25] As President Putin so eloquently described these trends at the World Climate Conference, "If it were two or three degrees warmer, that would be no big deal. Maybe it would even be a good thing; we would spend less money on fur coats and grain harvests would increase."[26]

A stunning misperception of the complex global environmental process is probably understandable in the case of the former KGB officer-turned President, especially if some of his leading scientists have conflicting positions on this issue. For years there have been speculations that Russia could actually benefit from global warming processes. Yuri A. Israel, Vice-Chairman of the Intergovernmental Panel on Climate Change — the body that oversees the entire Kyoto process — often went on record suggesting that Russia might benefit if global warming makes its colder regions more productive. Kirill Kondratyev, another influential global climate expert with the Russian Academy of Sciences, has challenged the accuracy of computer models used to predict global climate change.[27] Alexander I. Bedritsky, the Head of the Federal Service of Russia for Hydrometeorology and Environmental Monitoring (Roshydromet), twice elected Vice-President of the UNFCCC Conference of the Parties, argues that the Protocol's advocates have failed to provide evidence that greenhouse gas emissions are a key force behind global warming.[28]

The Third Communication of the Russian Federation to the UNFCCC (signed by A.Bedritsky) specified that the anticipated rise of CO_2 concentrations may largely be positive and a contributing factor for increasing Russian agricultural productivity.[29] Russian experts expect global warming to be generally favorable for development of cereal crop production and livestock fodder base in the country.[30] They also expect that anticipated warming would have a positive effect on forest productivity and carbon accumulation in Northern Siberia and the Far East.[31]

This position, however, does not prevail in Russia. Most environmentalists accept the scientific reality of climate change and argue that the problem is not in warming *per se*, but in the increasing imbalance of the global climate system, which has negative consequences for Russia as well as other countries of the world. Examples of growing severity and frequency of extreme weather events are numerous. In 2002 there were unprecedented floods in southern Europe and the Caucasus as well as drought and record number of forest fires in central Russia, while the summer of 2003 was the warmest in European records, with large-scale economic losses and death of thousands of people. For the first time in recorded history, the Lena River in Siberia became so shallow that river boats (the only means to provide the local population with food and other basic supplies) were barely able to navigate it.[32] Annual climatic abnormalities account for increased frequency and intensity of floods, hurricanes, whirlwinds, heavy showers, and snowstorms that have occurred frequently in the last decade in Russia. Similarly, landslides, mudflows, and avalanches have been more frequent. Such climate-change related natural calamities have dire ecological consequences: flooding of land tracts, destruction of forest land, and pollution of surface waters.

Apparently, the southern and northern regions of Russia are the worst affected. Southern Russia should expect more droughts, soil erosion, desertification, and spread of tropical diseases.[33] Russian environmentalists are most concerned about the northern front: the consequences of potential permafrost melt. Permafrost currently covers 67% of the territory of the country.[34] According to the estimates, in the next 20-25 years in western Siberia the border of insular permafrost will be shifted northward by 200-450 km,[35] which may cause serious negative consequences for settlements, transport communications, pipelines, and other elements of infrastructure constructed over layers of permafrost.[36] The northern areas of Siberia and the Far East may turn into impassable swamplands unsuitable for human habitation and economic use.

Direct consequences of climate change include casualties from natural disasters (tsunami waves in the Far East, floods, storms and hurricanes). Additional studies by Russian epidemiologists demonstrate a high "human price" paid by the population for global warming: increased incidence of cardiovascular symptoms on hot summer days; intoxication from petrochemical smog in big cities; increased incidence of infectious diseases following breakdown of sewer systems, especially in permafrost regions; sharp rise in mosquito-transmitted diseases, such as malaria, Dengue fever, Crimean and Omsk hemorrhagic fevers; and increased areals of tick-transmitted infections, such as tick encephalitis, Lime disease, and Q fever.[37] Epidemiologists are particularly concerned about the northward spread of noxious insects and associated diseases; for instance, the incidence of malaria over the last decade has increased in Russia by a factor of 6 with higher than ever spread of the disease over the territory of the country.[38] Experts from Russia's prestigious Higher School of Economics estimate that Russia's participation in the Kyoto Protocol could help to prevent risk of morbidity associated with atmospheric pollution (mainly respiratory, cardiovascular diseases, and cancer) by 40,000 cases a year as soon as 2012.[39]

Long-term consequences of global climate change are not well understood. Though the authors of National Communication III (mainly climatologists) did not expect the effect of climate change to be significant over the European territory of Russia for the next 50 years, environmentalists, on the other hand, foresee such events as radical change in biota, decrease in soil fertility, invasion of alien species, and unprecedented increase in the population of forest and agricultural pests much sooner[40]

A 2003 report produced jointly by scientists from Kassel University in Germany, Moscow State University, and the Center for Ecology and Forest Production of the Russian Academy of Sciences, supports concerns of Russian environmentalists and contradicts the "climate establishment" view. According to this report, official calculations of the benefits of global warming for Russia's agriculture fail to take into account regional variations. Only 15 of the 89 administrative regions of Russia provide the rest of the country with much of its food and it is exactly these 15 areas in the south and west that will suffer summer heat and droughts. The estimated number of people affected by these droughts by the 2020s is 77 million. The report concludes: "Our findings challenge the belief that climate change will generally benefit Russian agriculture and water resources. Instead they point out how extreme events such as droughts may become more frequent in key areas of Russia and may pose a threat to the food and water security of its people." [41]

In general, it seems that there are enough well-documented reasons why Russia as a country should be interested in forestalling climate change. Any potential benefits of temperature increase for agricultural input in some regions of Russia would be far outweighed by the negative consequences of the increased climate instability for all sectors of the economy and human life. Russia could respond to this threat through a combination of domestic mitigation and adaptation policies and/or by actively participating in the global climate regime by ratifying the Kyoto protocol or joining some alternative international scheme.

Potential Impact of Kyoto Regulations for the Russian Economy

In the UNFCCC, Russia is classified as an Annex I country – an industrialized country, which has made a commitment to reduce its emissions. It is also a country, which is "undergoing the process of transition to market economy" and therefore is allowed some flexibility in meeting its obligations.

Under the Kyoto Protocol, Russia is an Annex B country, which has made a commitment to reduce or stabilize emissions. In particular, Russia's CO_2 emissions in the period of 2008-2012 should not exceed a baseline level of emissions in 1990. This agreement actually allocates to Russia extremely generous emissions allowances based on its 1990 industrial performance - the last year Soviet factories chugged along at full speed before communism collapsed and the same year Russia was the second most energy-intensive economy in the world (after Ukraine). Countries that have limits set above their current production can sell off their credits to other countries that do not meet their targets. Countries also receive credits through various shared "clean energy" programs and "carbon sinks", i.e., forests and other systems that remove carbon from the atmosphere.

The Kyoto Protocol defines three flexibility mechanisms to promote the reduction of GHG emissions and facilitate the achievement of national targets:

> *Joint Implementation (JI)*: under Article 6, the Protocol provides a Mechanism for Annex I Parties to implement projects that reduce emissions, or remove carbon from the atmosphere, in return for Emission Reduction Units (ERUs),
>
> *Clean Development Mechanism (CDM)*: under Article 12, the Protocol provides a mechanism for Annex I Parties to implement projects that reduce emission in non-Annex I Parties, in return for Certified Emission Reductions (CERs) and sustainable development in the host countries, and
>
> *International Emissions Trading (IET)*: under Article 17, the Protocol provides a mechanism for the Purchase/Sale of GHG emission allowances allocated to Annex I Parties as assigned amounts. Annex I Parties can allocate their assigned amounts to emitting facilities.

The main attraction of the Kyoto pact to Russia is emission trading. Russia has a potential for sale of "hot air" – the difference between the huge amount of carbon dioxide that Russia produced in 1990 (at which level Russia's quota is set) and the meager amount it has been producing since the collapse of the Soviet Union and the economic downturn. In the 1990s, Russia's industrial output and greenhouse gas production have dropped by about one-third. Total anthropogenic emissions of greenhouse gases from the territory of Russia in 1999 (in CO_2 equivalent) amounted to 61.5% of the 1990 level.[42] Many observers still maintain that under the Kyoto Protocol, Russia and Ukraine might be the only major sellers of allowances and therefore might enjoy substantial market power. Economic models developed at the Massachusetts Institute of Technology (USA) suggest that under the system of international emissions trading limited to Annex B countries, Russia might be the major net beneficiary.[43] According to Greenpeace, Russia could make $20 billion annually from selling quotas, about a quarter of 2003 budget revenues.[44] It seems that Russia has everything to gain by ratifying the Protocol and selling its carbon credits to the countries that need it. These estimates might become true under several conditions:

> If Russia will still have enough "hot air" to sell during the 2008-2012 budget period;

If it will be eligible for emissions trade by meeting requirements of the UNFCCC,

If other countries will be willing to buy Russian "hot air" permits at the appropriate (for Russia) price, and

If Russia will have institutional capacity to process carbon purchase transactions.

Several research teams tried to answer the first question: will Russia still have emission quotas for trade in the "first budget period" of 2008-2012? In 1998, the International Institute for Applied Systems Analysis (IIASA) published its assessment of the GEG emission dynamics in the former Soviet Union. Using six alternative scenarios of economic growth and technological change the IIASA experts estimated Russia's carbon bubble (surplus) between minimum 9 Mt/C (scenario of high economic growth and carbon-intensive technologies) and 877 Mt/C (the most likely outcome). According to IIASA, under all scenarios of economic development Russia will have an emission surplus, which it could sell if Kyoto comes into force. These estimates put the Russian bubble's likely value from 2008 through 2012 between $4 billion and $34 billion a year. [45]

The Study of Russian National Strategy of GEG Emission Reduction, prepared in 1999 by the State Committee for Environmental Protection of the Russian Federation, the Bureau of Economic Analysis of the Russian Government and the World Bank, used a different input-output model and concluded that under business-as-usual development (that is, in the absence of technological innovations and specific policy instruments) but with relatively rapid GDP growth of 4.5% a year, Russia will have no significant volume of quotas for trade in 2008-2012 and will hardly be able to meet its national commitments. However, introduction of new technologies and carbon tax might bring Russia's trading potential up to 2.7 Bt/C. [46]

The 2002 Third National Communication, prepared largely by experts from the Roshydromet, presented three possible scenarios of CO_2 emissions in the period of 2001-2012 and further. Depending on the dynamics of such basic factors as GDP, energy intensity of GDP and carbon intensity of energy consumption, GHG emissions in Russia are projected to increase from 0.8% per year to 2.5% per year (the latter under the most probable scenario of 4.5% GDP increase and 2% decrease of GDP energy intensity). In the last scenario, the 2012 GHG emissions (in carbon equivalent) have been projected at 93.4% of the 1990 baseline. [47] Even in light of the new economic development goals for doubling the Russian GDP in ten years proposed by President Putin, experts are saying that in 2013-2017 total emissions of the country "with a reasonable margin of certainty will not exceed their 1990 levels because of implementation of energy efficiency and energy-saving measures." According to the Ministry of Economic Development and Trade, the Russian GHG emission quota surplus during the period 2008-2012 could be of the order of 1.5 to 3 Bt/C, or 300-600 Mt annually available to trade.[48]

With all the differences in the above-cited assessments, it seems that practically under all scenarios developed by international and domestic experts with the use of different methodologies, Russia will not lack emission allowances during the first budget period (unless it deliberately chooses to combine economic growth with remaining the most energy-intensive economy in the world). Even given the very last trends (higher than anticipated rates of the economic revival during 2000-2003 accompanied by increase in GHG emissions) the latter scenario is unlikely. By the estimates of the Russian Institute of Economic Studies of Transition, Russia's GDP has been growing by 6.1% per year since the start of economic recovery, accompanied by a 2.5% annual increase in GHG emissions.[49] The decline of carbon intensity of the economy demonstrated during the last three years can be attributed to its structural changes and does not yet include major investments to improve energy efficiency anticipated by the Federal Energy Strategy of Russia.[50]

Therefore, Russia will most probably have emission quotas for trade in the "first budget period" of 2008-2012 even if its economic growth surpasses current predictions. Will it be able to sell these quotas at net profit? Less than three years ago, when the US was still a likely member of the Kyoto pact while China and India were not, the answer to this question would be an unequivocal YES!!! However, the decision of the US to stay out of Kyoto with China and India joining the agreement has created a very different economic picture.

Projections of the carbon market (its volume and prices) differ substantially.[51] According to various estimates, most European countries and Japan will experience a shortage in emission quotas.[52] The fact that the costs of lowering carbon emissions in Russia varies between $12 and $27 per ton (versus $50-180 per a ton in most developed countries) seemingly makes this country the most logical source of emissions trading. However, there is large uncertainty about the "price" for traded allowances and over total volume of emissions trading market. The economic model developed at the Massachusetts Institute of Technology (MIT) suggests that under the Kyoto Protocol with trading limited to Annex B countries (including the United States), in 2010 the former Soviet Union would sell allowances for 345Mt carbon at $127 per ton for revenues of $43.77 billion and probably enjoy a net gain from trade of $33.2 billion.[53] Interpolating the MIT model to the situation when the US demand for 106 Mt is removed from the market brings the price of allowance down, probably to $60 per ton. The fact that China and India ratified the Kyoto agreement has changed Russia's prospects for a financial windfall from emission trading even more drastically. China is expected to flood the market with low-cost allowances, which in the absence of the US demand would tremendously affect Russia's potential to sell its credits at net profit. In addition, the supply of emission credits available from central and eastern European countries which have already ratified the Kyoto Protocol and are also a part of the emerging EU Emissions Trading Scheme, effectively limits Russia's access to the European market as well. Though the EU, Japan and Canada persistently push Russia toward ratification, the Russian Government has not been able to extract firm commitments to purchase it emissions credits from any of these countries. Therefore, the "invisible market hand" in reality might well destroy (or at least drastically diminish) the billion-dollars-worth windfall expected by Russia from "hot air" sales.

The "Joint Implementation" (JI) mechanism seems to be a better bet for Russian business. JI would Implementation will allow partners from developed countries to invest in projects in Russia and earn credits from the emission reductions achieved. Most of the JI opportunities in Russia are in the energy sector and relate to energy efficiency improvements, renewable energy, and fuel switch. Through JI projects, Russian companies would have access to new technologies and management systems. Russian analysts envisage benefits at the regional and local levels coming from carbon-related investments: "For industry, fuel and energy complex, housing and communal services, it means renovation of capital assets and reduction of costs for fuel-energy resources; for agriculture it means conversion to new methods of land management and increasing crop capacity of fields; for forestry it means both enhancing of forest lands and opportunities of timber industry development."[54]

The Dutch, Japanese, Canadians, and the World Bank have all expressed interest in investing in Russia's energy-efficiency projects, but without ratification there will be no reduction credits for these investors and therefore no incentive. The EU intends to allow the conversion of joint implementation credits into emission allowances under its internal emission trading scheme which will further increase this incentive. Several EU Member States are planning to set up public funds to promote joint implementation projects. The number of joint implementation projects will depend upon the market as well as the framework conditions for JI in Russia.

Russia has a very high JI potential (comparatively low costs for emission reductions). There are plenty of efficiency gains to be made in projects such as patching leaky oil pipelines, switching to natural gas and phasing out dirty coal firing plants. However, there are two other criteria that define the real prospects of JI development in transition countries: JI institutional capacity and JI investment climate.[55] Institutional capacity for participation in the flexible mechanisms is often defined as a combination of the host country's political will, its ability to meet eligibility criteria and, to some extent, the prior experience gained through the JI Pilot Phase in the late 1990s. It implies the existence of strong institutional frameworks, as well as appropriate levels of governance and transparency. Russia's institutional capacity for implementing the flexible mechanisms is being ranked among the lowest among transition countries. This assessment is being supported by the analysis of the pilot phase of JI in Russia, which demonstrated that undeveloped institutions (absence of relevant legislation and guidelines on credit allocation, confusion about functions of different institutions, etc.) severely undermined the efficiency of the implemented projects.[56] In addition, a recent study conducted by the EBRD ranked Russia among the least attractive countries for JI investments.[57] Carbon investors face higher risks than traditional investors – they are more susceptible to political and regulatory risks. The prevalence of corruption, absence of clear mechanisms for dispute resolution, financial instability, delayed institutional reforms, and questionable degree of structural change – are factors that in combination may prevent potential investors from tapping into Russia's high JI potential. In all fairness, one has to admit that during the last three years Russia's progress in institutional modernization has been substantial (institutional modernization included tax reform, further deregulation of economy, development of market-friendly legislation, etc.). However, a current combination of low JI institutional capacity and still unattractive JI investment climate makes significant private investments in Russia less likely.

Under these circumstances, the GHG market in Russia is likely to be driven by public investment from the OECD countries and will probably follow political and economic priorities of the donors. The future of the carbon market depends on the level of commitment of the Russian Government to institutional modernization of the country and the creation of more favorable investment climate.

What are the Obligations of the Russian Federation under the Framework Convention and its Kyoto Protocol?

The 1997 Kyoto Protocol introduced legally binding emissions targets for industrialized countries for the period of 2008-2012. Later, the Bonn and Marrakech Accords clarified Protocol rules in terms of implementation, accounting and compliance. Countries that ratify the Kyoto Protocol are under obligations to "play the game."

Opponents of Kyoto ratification maintain that the costs of Russia's participation in the pact are prohibitively high. Also, in their perspective the benefits of the Kyoto pact for Russia are questionable because as long as the US remains absent, Russia will not be able to sell its quotas while in future the country would have to buy additional quotas for its emissions from somebody else and pay high penalties for noncompliance.[58]

Proponents of the Kyoto ratification, and in particular an odd alliance of the Russian Regional Environmental Center, National Carbon Union, and Russia-WWF,[59] maintain that Russia will not face difficulties meeting its obligations during the first compliance period and that it will have to meet most of these obligations anyway as a member of the Convention.[60]

Russia's obligations under the Kyoto Protocol include the following:

> To limit/stabilize its emissions of greenhouse gases at their 1990 level over the first commitment period of 2008-2012 (Annex B); before the commitment period begins, the country must file a report providing emissions data for its base year.
>
> To implement climate change policies and measures that have a mitigating effect on climate change and may include (upon the decision of the national government) enhancing energy efficiency, promoting renewable energy, favoring sustainable agriculture, recovering methane emissions through waste management, removing subsidies and other market distortions, protecting and enhancing greenhouse gas sinks, and reducing transport sector emissions (Article 2).
>
> To put in place a national system for estimating its anthropogenic greenhouse gas emissions and removals no later than in 2007 (Article 5); the national inventory should be prepared in accordance with the methodology and reporting requirements specified by the Protocol.
>
> To inform the FCCC Secretariat of the creation of a National Coordinating Center on Joint Implementation projects as well as main principles and procedures used to select such projects (Article 6).
>
> To prepare a national inventory of anthropogenic sources of GHG emissions and their sinks and to submit relevant information to the UNFCCC on a regular basis (Article 7).
>
> To create a National Registry to track and record transactions in emission reduction units (ERUs) under Joint Implementation projects, assigned amount units (AAUs) under emissions trading and removal units (RMUs) generated through sink activities in the LULUCF sector (Article 17). It has to submit a description of these facilities to expert review teams.
>
> To cooperate in scientific research and development of databases to decrease uncertainty.

The Kyoto Protocol and the Marrakech Accord set strict accounting, reporting and review procedures, which Russia has to follow if it ratifies

the agreement. In particular, as well as other Annex I Parties, Russia has to include in its National Communication:
- details of its national system and national registry,
- how use of the mechanisms is supplemental to domestic action,
- details of policies and measures implemented to meet emission targets.

It also has to include in the greenhouse gas inventories prepared under the convention:
- any data specific to land use, land-use change, and forestry (LULUCF) sector,
- any changes in national systems or national registries,
- transfers and acquisitions of emission credits, and
- actions to minimize adverse impacts on developing countries.

The Protocol sets out detailed procedures for potential noncompliance, which may include officially announcing countries in noncompliance, suspending their eligibility to "sell" credits under emissions trading, and – the extreme measure – expelling a country from the Protocol.

Russia's capacity to implement its commitments under the UNFCCC and the Kyoto's Protocol in the nearest future is questionable. Ironically, the main reason is not even financial constraints. According to the Ministry of Economic Development and Trade, Russia already has in place most elements of the system of emission monitoring required by the Kyoto Protocol but needs to update and modernize. Costs of annual inventory and maintaining a National Registry are estimated at 25 million rubles (for the period of 2003-2005) and 20 million rubles annually starting in 2006.[61] The Ministry estimates that these costs will not exceed 1% of all environmental costs, which in turn comprise less that 0.03% of the Russian budget expenditures. To date, Russia lacks critical activity data and uses methodology incompatible with IPCC source categories. However, a recent Russian study suggests that establishment of a national GHG registry is still feasible even in the absence of significant financial resources, but would require a clear institutional division of responsibilities.[62] The main factor that hinders Russia's capacity to implement its commitments is institutional: persisting confusion about division of functions and responsibilities among different actors and their struggle for influence (and in perspective, for benefits) in shaping domestic climate policy.

The challenge faced by the Russian Government in respect to its obligation under the Kyoto protocol is considerable: elaboration of new legislation (on property and management rights on emissions, emissions trade, carbon investments, dispute resolution relating to carbon exchange, etc.) and designation of a particular agency (or creation of a new one) that would be able to effectively implement this legislation in practice. However, as a member of the UNFCCC Russia has to meet institutional requirements anyway and the sooner it starts working on them, the more chances Russian business has to reap at least some benefits from the process. Such institutional changes might also help in promoting more efficient functioning of domestic economy.

Is It Politics?

In a different political setting the government's ambivalence on the ratification of the climate pact might be explained by the opposition of a large business lobby or juxtaposition of various actors with relatively strong power but opposing interests. In Russia, the composition of stakeholders and their positions is quite unusual.

The *Russian population* in general is not well aware of the problem and environmental concerns are not among their top priorities in any event. There are only a handful of experts and social activists who have an understanding of the complexity of the issue and its ramifications for Russia.

Epistemic Community

Transnational networks of experts, who share "a set of symbols and references," common values, and approaches to problems, are believed to be instrumental in the formation of international regimes. Formation of an epistemic community in Russia strains imagination. Several highly-placed officials in the Russian Academy of Science (Vice-President Laverov and Acting Member of the Academy, Izrael) are strictly against Kyoto, while many experts with the same Academy (V.Danilov-Danielian) and some other scientific institutions strongly advise its promptest ratification. Both groups collaborate with foreign colleagues and have some international and domestic support for their opposing views and suggested policies; however, neither group is strong and influential enough to change a position of the Russian Government whether in one direction or the other. In addition, some academic and popular journals publicize discussions among ecologists-turned-politicians and former professors of political economy-turned-ardent environmentalists of the concept of "natural rent," according to which Russia as "environmental donor" is entitled to payments from the world community for numerous environmental services it provides to the world. The concept of "entitlement" for some extra benefits because of Russia's environment, history, current state of economy, and other factors is quite widespread among scholars, politicians, and populace.

Several *NGOs* (Socio-Ecological Union, the Center for Energy Efficiency and the Center for Energy Policy) include climate issues in their activities. Regretfully they have no or very little political or lobbying power as well as no direct access to the process of decision-making.

Business

Russian industries, whose interest in the issue and relative bargaining power are stronger than those of other stakeholders, are split over the Protocol. A relatively small number of industrial giants produce the major bulk of industrial CO_2 and methane emissions in Russia. Unexpectedly, major national polluters–from large energy companies (GAZPROM and United Energy Systems) to Russian aluminum, key steel mills and metal smelters–have been lobbying for ratification. These companies are even offering to help pay to administer the program once it is adopted.[63] The major attraction for these actors is potential foreign investment and emission credit purchases that could help to modernize or replace their aging infrastructure. Major Russian greenhouse gas producers had already begun discussions few years ago with international companies on transferring emission credits.

The gas sector produces over 10% of the national carbon emissions. Over 50% of the equipment in this sector is in urgent need of reconstruction or replacement and its investment needs for the period upto 2020 are estimated in the range of $170-200 billion.[64] To meet increasing demand of the domestic market and its foreign obligations the national gas sector urgently needs to develop new fields, construct pipelines, and rehabilitate corroding pipelines and other aging infrastructure. The Kyoto mechanism provides unprecedented opportunities for this sector. GAZPROM (natural gas state-run monopoly entity in control of over 90% of Russia's natural gas production) has reported that the company has significant potential under Kyoto to reduce leakage of natural gas and an independent program is already under way. For the last few years the company has been involved in several mitigation-related projects.

The Russian utility sector is a natural monopoly dominated by Unified Energy System, or UES (known in Russia as RAO EES). It is a joint-stock company, wherein the Russian government owns 52 percent shares and appoints its president, but which is operated by a board of directors with remaining shares distributed between foreign investors and employees. UES, which controls nearly all the large thermal power and hydropower plants and approximately 70% the country's distribution system, is the worst polluter in the country. Like the rest of the energy sector its equipment is largely obsolete by Western standards and according to UES CEO Anatoly Chubais, $55 billion in investment would be needed over the next 10 years for maintenance and modernization efforts. The company is interested in Russia's promptest ratification of the Kyoto's Protocol. It has developed one of the best inventories of greenhouse gases in the country and submitted several project proposals to European partners (these proposals can be implemented only if Russia joins Kyoto). According to Mr. Chubais, the company now has "a larger volume of certified quotas than leading European countries".

The state-owned MINATOM (nuclear ministry) also supports Kyoto as well as a prominent business organization — The Union of Entrepreneurs and Industrialists. Several pro-Kyoto major businesses have formed the National Carbon Union to push the virtues of the pact in Moscow, to coordinate activities of Russian businesses in attracting investments through Kyoto mechanisms, and to participate in designing the national emission market.

On the other side, Russia's biggest export sector, the oil industry, worries that the Kyoto Protocol would curb oil consumption, thereby depressing oil prices. This industry would not like to see renewable energy sources cutting demand for its key export. The oil sector has tremendous financial clout and like the domestic coal industry apparently has been putting significant pressure on President Putin.

Regions

It is reported that 80% of Russian regional leaders support ratification of the Kyoto Protocol and are competing to bring European investors to their regions. Regions and cities are important stakeholders in climate policy: on one side, they account for a significant share of GHG emissions and have strong incentives to participate in project-based trading.[66] On the other side, their reluctance to participate in mitigation projects may well result in project failure. In addition, effective participation of regions is omnipotent for data collection and in developing a national GHG inventory. The National Carbon Sequestration Organization that was established in Moscow in 2001 has been working on disseminating information on Kyoto-related issues at the regional level and on encouraging regional authorities to elaborate and implement region-wide plans for GHG emissions management and reduction.[67] Some Russian regions have indeed been actively preparing their involvement in the Kyoto-related projects and are in the process of negotiations with potential foreign partners.

Bureaucracy

Many government officials within Russia support Kyoto's ratification, however, the exact functions and political weight of different agencies in the formation of national climate policy are not too clear.

The Interagency Commission on Climate Change Issues (ICCCI) created in 1994 lists twenty-nine ministries and state committees among its members, including major producers and consumers of fuel and energy.[68] The status of the Commission limits its role to compiling materials and coordinating work among participating agencies without binding authority. In practice, four federal agencies (Ministry of Economy, Ministry of Fuel and Energy, the State Committee for Environmental Protection (later dissolved), and State Committee for Hydrometeorology and Environmental Monitoring i.e. Roshydromet) were key players in elaborating domestic climate policy; they also sent representatives to international negotiations. Initially, the ICCCI was headed by Hydromet, a governmental body without the status of a ministry and with low influence over other members of the Commission. In 2000 the leadership of the ICCCI was shifted to the Ministry of Economic Development and Trade, a much more powerful actor in the domestic bureaucratic hierarchy. In 2002 the Russian Government placed responsibility for Kyoto's preparation activities with five agencies (Ministry of Economic Development and Trade, Ministry of Energy, Ministry of Natural Resources, Ministry of Foreign Affairs, and Roshydromet).[69] However, their responsibilities and functions are still not clearly defined. Anecdotal evidence suggests that leadership and personnel of these agencies are not overly enthusiastic about Kyotos ratification. The anti-Kyoto position of Roshydromet and its Head A. Bedritsky was noted above. The Ministry of Economic Development and in particular its Minister Gref, have been constantly accused by Russian environmentalists of deliberately holding the process back to avoid being overloaded by the difficult task of quota selling.[70] Meantime, Deputy Minister of Natural Resources Kirill Yankov went on record saying that though his Ministry is ready to organize a system for controlling GHG emissions, in his opinion, ratification of the Kyoto pact would breed yet another layer of bureaucracy tasked with issuing emission permits; instead, Russia should pursue an independent national climate program shaped along the US experience.[71] Meantime, while expressing their discomfort with the Kyoto conditions, these agencies have been simultaneously and without much coordination working on the mechanisms of its practical implementation, such as JI projects and monitoring. They also have different visions of the main issues of Kyoto implementation, in particular which domestic actors would be allowed to participate in the Kyoto mechanisms, how allowances would be distributed, who would regulate the process of implementation and monitor the compliance. To respond to these questions, these agencies designed different institutional structures (or their elements) that are to be created in Russia should the Kyoto regime come into force.

Legislature

Russian legislature, the Federal Assembly, consists of the Federation Council (members appointed by the top executive and legislative officials in each of the 89 federal administrative units) and the State Duma (half the seats are distributed by proportional representation from party lists, and half the seats allocated to popularly elected representatives from single-member constituencies). The Federation Council has not been actively involved in the climate policy but has demonstrated no opposition to the FCCC or Kyoto and representing the regions may well be in support of it. The State Duma, for its part, ratified the Framework Convention and is supposed to ratify the Kyoto protocol. The Russian Democratic Party YABLOKO has been the strongest supporter of the Kyoto pact in the Duma. It intends to submit draft documents required for ratification of the Kyoto Protocol to the Russian parliament in autumn 2003. According to YABLOKO's Chairman Grigory Yavlinsky, Russia would benefit from ratification of the Kyoto Protocol both economically and environmentally. In particular, he anticipates that ratification of the Protocol would encourage Russian industries to introduce new industrial technologies.[72] However, in the December 2003 elections YABLOKO and other liberals were virtually eclipsed from the Duma,[73] where two-thirds of the seats now belong to Putin's main backer, the United Russia faction (not known for proenvironmental concerns). Representatives of the United Russia also chair all the Duma's Committees. Such composition of power in the new Russian legislature virtually means that Duma's stance on Kyoto would heavily depend upon what the President wants to hear.

External actors

If the Russian government were to ratify the Kyoto agreement now, it would affect President Bush, who is running for reelection in November 2004 and who has been challenged on his anti-Kyoto stance by numerous domestic and international groups. The United States is interested in Russia's rejection or at least delay of Protocol ratification and apparently several US delegations to Moscow have been discussing development of a separate US-Russia climate policy. Creation of the Chicago Climate Exchange indicates that even with no US participation in Kyoto, it would still be closely involved in emissions trading and therefore might become the main market for Russia's emissions. It is also clear to Russia that some of its important energy projects (including export of Russian oil to the American market) depend on US goodwill. In distinction from Europe, the US has been providing Russia with trade incentives (including with regard to the WTO) and has some other levels of leverage, which could be used for retaliation should Russia go against the US interests in global climate policy.

On the other hand, the European Community is strongly interested in Russia's support and has been ardently working on developing the EU-Russian Energy Dialogue with the eventual objective it creating a "Eurasian Climate Change Area". Thus far, the European Commission has promised roughly two million Euros in technical aid as soon as Russia ratifies. However, the Russian Government considers this sum extremely low and requests cash guarantees rather than verbal promises before signing up to the pact which will fail without its backing. M.Tsikanov went on record recently saying that no decision on Kyoto's ratification has yet been taken due to the "passive position of our partners - the

European Union and Japan."[74] Given the emerging EU Emissions Trading Scheme, which will apparently favor EU-accession states over Russia, attempts by the Russian Government to obtain assurance of guaranteed purchases of Russian natural gas and emissions credits at acceptable prices from individual European state and Japan are quite understandable.

Such a combination of domestic and external power and interests puts Kyoto ratification into the basket of political issues on which Russia's government prefers to procrastinate and to extract more benefits.

Conclusion

- Forestalling climate change is in the long-term interests of the Russian Federation and the Russian Government should undertake effective domestic policies to mitigate the effects of global climate change and to adapt to their negative consequences. Ratification of the Kyoto Protocol is ultimately in the interests of the Russian Federation but will require substantial changes in its climate policy and institutional structure.

- Though recent remarks of highly placed officials and President Putin himself reveal poor understanding of the climate change problem and signal Russia's intent to abstain from the Kyoto Protocol, these comments should probably not be taken at face value but viewed rather as a strategic attempt to provoke a practical response from Europe, Japan and other actors in the field of global climate policy.

- Despite confusing remarks and strange jokes, there is no evidence that the Russian Government is taking the issue of global climate policy lightly (especially given its relatedness to the energy market and the international ramifications of decision).

- Russia has legitimate reasons for concern about potential benefits and costs of Kyoto membership for its economy and international relations. Economic aspects of the Kyoto arrangements have changed substantially during the last few years and Russia's anticipated windfall from implementation of flexible mechanisms (international emissions trading and Joint Implementation) will neither be as high as expected in the later 1990s nor guaranteed. Under these circumstances, Russia will press the European Commission and individual countries for additional action (guaranteed access to market, guaranteed prices, guaranteed investment etc.) while continuing to develop collaboration with the United States on a bilateral climate policy.

- The final decision on Kyoto ratification will not be made before Russia's presidential elections in March 2004 and will eventually depend upon a combination of external and domestic economic factors at play then rather than upon environmental concerns.

· Nothing happens fast in Russia. Delaying the decision much longer may result in overarching economic and political losses for the country.[75] Even when/if the decision to ratify the Kyoto Protocal is finally taken, its implementation will take more time than anticipated. To reap benefits from the agreement, national authorities and other interested stakeholders should start working on a strong national climate policy and implementation program immediately.

Footnotes

1 As of 2001, the area of land under forest was 794.3 M ha or about 46.5% of the territory of the country. Annual CO_2 sequestration by live phytomass of Russian forests in 1990-1999 was estimated at 300-600 Mt CO_2 per annum. See: Ministry of Economic Development 2002. *National Report on the Climate Change Issues*, Moscow, p. 21 and p. 24; Inter-Agency Commission of the Russian Federation on Climate Change 2002. *Third National Communication of the Russian Federation*. Moscow, p. 12.

2 Ministry of Environment Protection (1995) *Natural Environment in Russia*. Moscow, ECOS, p.4.

3 Illarionov, 2003. cited by Sergei Blagov. *Cybercase News Service*, September 9, 2003; Illarionov's interview during the World Climate Convention in Moscow (October, 2003) reported by *Inter-Tass*; Additional remarks cited by *Associate Press*, November 30, 2003. Illarionov explained his position in more detail at a seminar at the Carnegie Center in Moscow on December 16, 2003. *BBC Monitoring International Reports, December 16, 2003.*

4 Reported by WWF Russia Press Officer on January 17, 2004, see: http://www.wwf.ru/resources/news/article/eng/718

5 Resolution of the Government of the Russian Federation N 34, 01/22/1994.

6 Ordinance of the Government of the Russian Federation N 1242, 10.19.1996.

7 Hydromet 2000. *Climate Change Action Plan Report*. Executive Summary. Moscow.

8 1990 was the last year of Soviet-era levels of industrial activity. In 1991 heavy industry practically collapsed after the Soviet Union broke apart.

9 The text of the National Report (in Russian) can be found at http://rusrec.ru.

10 National Report, op.cit., p.23.

11 Nilsson, Sten at al 2000. *Full Carbon Account for Russia*. Interim Report IR-00-021, Inst. Appl. Syst. Anal. available at: http://www.iissa.ac.at/Admin/PUB/Documents/IR-00-021.pdf.

12 IAASA documents related to carbon accounting can be found at: http://www.iissa.ac.at/Publications/Catalog/PUB_Project_FOR.html.

13 This sum is close to the whole increase in the emissions by all developed countries during the same period. See: National Report, op.cit., p.24.

14 Ibid., p. 28-29.

15 Cited in: Russian Regional Environmental Center in cooperation with the World Wide Fund for Nature and National Carbon Union 2003. *Kyoto Protocol: Politics, Economics, Environment*. Moscow, p.2.

16 Sergei Leskov, *Global News Wire*, October 4, 2003.

17 Ministry of Foreign Affairs of the Russian Federation, *Daily News Bulletin*. 10.20.2003.

18 Reported by Alexander de Roo, MEP, Head of the EP Delegation to the Russian Duma, cited by Louise van Schaik at http://ceps01.link.be/Article

19 Putin: Russia Not to Be Milk Cow, In: The *Russian Journal Daily*, October 20, 2003, available at: http://www.russiajournal.com/news/cnews-article.shtml?nd=40912.

20 Cited by *Associate Press*, November 30, 2003. Illarionov explained his position in more details at a seminar at the Carnegie Center in Moscow on December 16, 2003, *BBC Monitoring International Reports, December 16, 2003.*

21 Illarionov 2003. Cited by Sergei Blagov, *Cybercase News Service*, September 9, 2003.

22 Illarionov's interview during the World Climate Convention in Moscow (October, 2003).

23 For a description of the dynamics and specific features of the Russian negotiating position on global climate change issues see Moe, Arild, and Kristian Tangen 2000. *The Kyoto Mechanism and Russian Climate Politics*. The Royal Institute of International Affairs, London.

24 Ministry of Environment 1995. *Natural Environment in Russia*. Ecos, Moscow, p.30.

25 Ministry of Environment Protection and Natural Resources of the Russian Federation 1995. *Natural Environment in Russia*, Ecos, Moscow, p. 30.

26 Reported by Sergei Lesov, *Global News Wire*, October 2003.

27 *Kyoto Protocol News*, October 26, 2003.

28 Reported by Milan, Vanessa Houlder 2003. *Russian doubts and emissions concerns cloud Kyoto pact*. In Financial Times, December 15, 2003.

29 Inter agency Commission of the Russian Federation on Climate Change (2002) *Third National Communication of the Russian Federation*. Moscow, p. 75.

30 Ibid. p. 76.

31 Ibid. p.78.

32 Russian Regional Environmental Center in Cooperation with the World Wide Fund for Nature and National Carbon Union 2003. *Kyoto Protocol: Politics, Economics, Environment*. Moscow, p.11.

33 Danilov-Danielyan, Victor 2003. *Kyotskii Protokol: Kritika Kritiki* (in Russian). Available at the website of Russian Regional Environmental Center at http://rusrec.ru/kyoto/articles/art_climate_critics.htm;

34 Interagency Commission of the Russian Federation on the Climate Change Issues 2002. *Third National Communication of the Russian Federation submitted in accordance with Articles 4 and 12 of the United Nations Framework Convention on Climate Change*, Moscow, p. 19.

35 Interagency Commission of the Russian Federation on the Climate Change Issues 1995. *First National Communication*. Moscow, p.17.

36 Danilov-Danielyan, Victor 2003 *Kyotskii Protokol: Kritika Kritiki* (in Russian). Available at the website of Russian Regional Environmental Center at http://rusrec.ru/kyoto/articles/art_climate_critics.htm. Anisimov, O.A. and Belolutskaya M.A. 2002. Otsenka vliianiia izmeniia klimata I degradatsii vechnoui merzloty na infrastructury v severnykh regionakh Rossii [Assessment of the Affect of Climate and Permafrost Degradation upon the Infrastructure in the Northern Parts of Russia]. In: *Meteorologiia and Gydrologiia*, 6: 15 – 22.

37 Ministry of Health 2001. *State Report on Sanitary-Epidemiological Status in the Russian Federation in 2000*. Moscow; Revich, B. And Maleev V. 2003. Potepleniie Klimata – Vozmozhnyie Posledstiviia dlia Zdoroviia [Global Warming – Possible Consequences for the Health]. Publ. Russian Regional Environmental Center available at: http://rusrec.ru/kyoto/articles/art_climate_health.htm

38 Revich. Boris et al 2003. Novaya Ugroza [New Threat]. Russian Regional Environmental Center, Mascow, pp. 6-7.

39 Russia's Regional Ecological Center 2003 Nezavisimaya Otsenka Posledstvui Prisoiedineniia Rossii k Kiotskomu Protokolu [Independent Assessment of Consequences of Russia's Accession to the Kyoto Protocol], Moscow, p.24-25.

40 Danilov-Danielyan, Victor 2003. Stoit Li Nam Radovat'sa Potepleniiu Klimata? [Should We Be Happy about Global Warming?] (in Russian). Available at the website of Russian Regional Environmental Center http://rusrec.ru/kyoto/articles/.

41 Reported by Paul Brown, environment correspondent to *The Guardian (UK)* February 26, 2003.

42 Interagency Commission of the Russian Federation on the Climate Change Issues 2002. *Third National Communication of the Russian Federation submitted in accordance with Articles 4 and 12 of the United Nations Framework Convention on Climate Change.* Moscow, p. 8.

43 Dellerman, A.D. et al 1998. *The Effects on Developing Countries of the Kyoto Protocol and CO$_2$ Emissions Trading.* Report 41, MIT Joint Program on the Science and Policy of Global Change, December 1, 1998.

44 Natalya Olefirenko, head of the climate project at Greenpeace in Moscow.

45 Victor, David et al 1998. *The Kyoto Protocol Carbon Bubble: Implications for Russia, Ukraine and Emission Trading.* Int. Inst. Appl. Sys. Anal. Laxenburg. Mt/C – million tons in CO$_2$ equivalent.

46 Golub, A. et al (1999) *National Strategy Studies: Study on Russian National Strategy of Greenhouse Gas Emissions Reduction,* The World Bank, Moscow. Bt/C – billion tons in CO$_2$ equivalent.

47 Interagency Commission of the Russian Federation on the Climate Change Issues 2002. *Third National Communication of the Russian Federation submitted in accordance with Articles 4 and 12 of the United Nations Framework Convention on Climate Change.* Moscow, p. 73.

48 Ministry of Economic Development and Trade 2002. *National Report on Climate Change Issues* (in Russian). Moscow, p. 29.

49 Cited by Sergei Leskov, 1994. Ratify without Selling. In: *Izvestiia Nauki,* January 2004, available at: http://www.inauka.ru/english/article38094.html.

50 The Federal Energy Strategy of Russia to 2020 envisages the 1.14-1.36-fold increase in domestic consumption of primary energy in the period 2001-2020 with a corresponding 1.7-2.1-fold decrease in the energy intensity of the GDP. See: Fundamental Provision of Energy Strategy of Russia for the Period up to 2020, approved by the Government of the Russian Federation, protocol # 39, November 23, 2000. Moscow (in Russian).

51 William Nordhaus, for instance, estimated that under the Kyoto protocol with US participation, the price of an allowance in 2010 would be about $55 per ton of carbon, while without US participation it would drop to about $15. Projections of the allowance price without US participation in Kyoto range from $0 to $65, depending on Russia's market power, choices to bank allowances, and other factors. *See:* Nordhaus, W. 2001. Global warming economics. *Science* 294, November 9, 2001: pp.1283-1294.Loschel, A. and Zhong Xiang Zhang (2002) *The Economic and Environmental Implications of the U.S. Repudiation of the Kyoto Protocol and the Subsequent Deals in Bonn and Marrakech.* FEEM Working Paper 23, 2002.

52 Various experts estimate that the EU emissions will exceed its obligations by

the Protocol by 150 Mt CO_2 equivalent, while Japan, Canada, New Zealand and Norway will exceed their obligations by 300-400 Mt.

53 Dellerman, A.D. et al 1998 *The Effects on Developing Countries of the Kyoto Protocol and CO2 Emissions Trading*, Report 41, op.cit.

54 Starikov, I. 2003 Regional Approach to State Regulation and Management of Carbon Investments, *Business*, special issue on the Kyoto Protocol, p. 40.

55 Gassan-Zade, Olga. 2003 *Economies in Transition: At the Crossroads of Development.* Int. Inst. Sustainable Development, Winnipeg Canada.

56 Evans, M. et al 2001 *The Climate for Joint Implementation: Case Studies from Russia, Ukraine and Poland.* Working Paper. Pacific Northwest National Laboratory,

57 Frankhauser, S. and Lavric, L. 2003 *The Investment Climate for Climate Investments: Joint Implementation in Transition Countries.* EBRD

58 Illarionov, A. 2003 Remarks at a seminar at the Carnegie Center, Moscow on December 16, 2003, *BBC Monitoring International Reports, December 16, 2003.*

59 Russian Regional Environmental Center (RREC) was established by the European Commission and the Academy for Civil Service under the auspices of the President of the Russian Federation, formerly the Academy of the Communist Party of the Soviet Union. National Carbon Union is the noncommercial partnership of the largest Russian corporations, responsible for a significant part of industrial greenhouse gas emissions in the country and looking for opportunities to use the Kyoto mechanisms to attract foreign investments. WWF-Russia is a Russian branch of World Wildlife Fund - one of the most influential international conservation organizations.

60 Russian Regional Environmental Center in cooperation with the World Wide Fund for Nature and National Carbon Union 2003. *Kyoto Protocol: Politics, Economics, Environment.* Moscow, pp.11-12.

61 Ministry of Economic Development 2002. *National Report on the Climate Change Issues.* Moscow, p.27.

62 Berdin, Vladimir 2003. Development of the Registry System and GHG Reductions in a Pilot Project of Russia. Presentation at the Workshop "Implementing Kyoto in Russia and CIS: Moving from Theory to Practice", Moscow, Higher School of Economics, April 9-10, 2003. available at: http://www.climate-strategies.org/russiaworkshop.org

63 Murphy, Kim 2003. Polluters Join Call for Russia to Ratify Kyoto Pact, *Los Angeles Times*, September 5, 2003.

64 Osnovnyie Polozheniia Energeticheskoy Strategii Rossii na Period do 2020 goda [Main Points of the Energy Strategy of Russia until 2020]. Moscow, 2003.

65 Cited in *The Russia Journal* / RBC 29 Sep 2003

66 Lots of emissions are generated by district heating systems, which serve nearly 80% of the Russian population. These systems are energy-intensive, inefficient and in Russian climate have to run for at least seven months a year. Most systems are owned (at least partially) by municipal governments, who therefore acquire a major stake in project-based crediting.

67 Detailed information on the activities of the National Carbon Sequestration Organization can be found at their website: http://www.natcarbon.ru

68 Government Statute of the Russian Federation # 346, 04.22.1994 (in Russian).

69 Press Release #580 April 11, 2002, Part 2. *On the Preparation of the Ratification of the Kyoto Protocol to the UN Framework Convention on Climate Change* (in Russian).

70 WWF Russia 2003. *Russia' Ratification of the Kyoto Protocol: Questions and Answers from WWF Russia,* May 8, 2003.

71 *News from MNR Russia*, August 2003 (in Russian).
72 Yavlinsky, Grigory, 2003. Speaking at a meeting of the Round Table attended by ecologists and parliamentarians, held in Moscow in June 2003. YABLOKO Press-Release on 05.06.2003.
73 YABLOKO currently has only 4 seats in the 450-seat chamber versus more than 300 seats that belong to United Russia.
74 M.Tsikanov, cited in *Pravda (Moscow)*, 12.03.2003.
75 In late 2003, because of Russia still being out of Kyoto two Russian companies – Amur Thermal Power Plant and Kotlas Pulp and Paper Mill – lost about 15 million Euros which could have been invested in Russian economy under the Dutch ERUPT program. *See*: Russian Regional Environmental Center in Cooperation with the World Wide Fund for Nature and National Carbon Union, 2003. *The Kyoto Protocol*, op.cit. p. 15.

CHAPTER 18

Sri Lanka: Its Industry and Challenges in the Face of Climate Change

Ajith de Alwis
University of Moratuwa, Dept. Chemical and Process Engineering, Sri Lanka

Sri Lanka: from an agrarian economy to an industry economy

Sri Lanka was once known as the "granary of the east". The agrarian heritage is still visible and many understand and yearn for our future to follow in the same direction. In the present situation the country is unable to feed its citizens in a satisfactory manner and hence we are dependent on imports for many of our food needs. The population is increasing, though the rate of increase is much below those observed in neighboring countries (i.e., 1.2% annual population growth). These challenges of sustaining a growing population plus their needs are intended to be met through industrialization of the economy. Various governments have focused and still do on developing the manufacturing as well as the service sectors, though the emphasis has changed from the former to the latter. *An economic strategy based on agro-exports, free trade zone, cheap labor, and sun-and-sand tourism is a strategy for staying poor* is a concept which could be supported with examples but perhaps would still not be fully understood by the decision-makers (www.competitiveness.lk). This fact was reiterated to the UK government by UK's Engineering Council in the 1990s when drastic declines in manufacturing capabilities were noted in the country. Sri Lanka epitomizes this statement as the present economy is based on labor and traditional exports etc. Basically, economic development of a country is achieved via success of Industrial, Agricultural and Service sectors. The Sri Lankan industrial sector is based on a number of "Labor-Intensive" and "Resource-Based" industries. As a result Sri Lanka is not in a strong position to face the growing global competition. Sri Lanka shifted to a market-oriented economy (even without a developed infrastructure) in 1977, and the national economy has shown a steady growth with GDP increasing from US $4.1 billion in 1977 to US $15.8

billion in 1998. Due to continuous adoptions of policies to create an open market, Sri Lanka is now reputed to be the most open economy in South Asia. Industry in Sri Lanka means factory industries and plantation industries (i.e., tea, rubber factories). The manufacturing subsector of the "Industrial sector" in GDP dropped from 23.1% to 16.5% from 1977 to 1998. (www.worldbank.org). The manufacturing sector, however, remains heavily dependent on a few labour-intensive industries. Over two-thirds of the manufacturing value added is derived from textile, apparel and leather and the food, beverage, and tobacco subsectors. Textile and apparel is the only sector which has increased percentage of value addition in industry from 14% to 37% since 1970-1998. With increasing liberalization of the economy and stimulating foreign direct investments, Sri Lanka hoped to be an emerging economy by the year 2000. However, due to absence of a clear strategy and an understanding of the implications of the goal, this failed to materialize. A manufacturing share in the industrial sector at the level of 20% can be taken as one prerequisite for NIC status. The share of manufacturing in the industrial sector was 16.5% in 1998 for Sri Lanka. At the end of 1999, Sri Lanka's status was "lower middle income" with US $810 per capita. These changes were brought about by income to the country from the Middle East workforce as well as from the service sector. For example, Sri Lankan migrant workers remitted a sum of Rs 87.6 billion to the government coffers in 2000. Though the GNP values indicate improvements the wealth divide is increasing. Thus Sri Lanka is in dire need of real development and only meaningful developments in the industry sector can bring about the change required.

Brief history of Industrialization in Sri Lanka

Sri Lanka's efforts at developing an industrial base can be traced back to the 1930s, when the colonial government, under strong pressure from local legislators, prepared plans for state-sponsored factories, because the private sector was considered unwilling to take on the responsibility for this type of activity. The British colonial government as a matter of policy had discouraged industrial development in the tropical colonies and though a textile mill was set up in 1883 and a brewery in 1884, significant industries were never established (de Silva, 1991). At that time, Sri Lanka's industries were mainly limited to the small-scale factories processing agricultural crops such as tea, rubber and coconut and manufacturing simple products such as matches and carbonated drinks. The onset of World War II hastened the government's plans for intervention, when some factories were set up to meet the war-time shortages. The first policy paper for an independent Sri Lanka was the booklet on the subject of industrialization by the Department of Commerce and Industries released February 4th 1948 (the day of independence) titled, *It Can Be Done*. This clarion call for action still awaits realization, however.

These state factories proved uneconomical when trade channels were reopened after the war; the newly independent government was keen on rehabilitating them under state ownership. This attempt proved a failure and for a brief period in the mid-1950s, the government attempted to transfer the factories to the private sector. The government backed out from that policy path due to public opposition, and this in turn led to the progressive strengthening of the role of the Government in the field of industry. This period of state intervention and controls lasted for nearly two decades. Certain industries enjoyed state monopoly and some others opted for joint state-private ownership. Heavy tariffs and import bans were imposed to transfer a whole range of import-substitution industries. Licenses were required not only to start an industry, but also to import necessary raw materials. The public sector ventured into larger scale industry with the support of some foreign countries in such areas as petroleum refining, steel and cement. These protectionist policies did foster a cluster of import-substitution industries, most of which were mainly processing or assembling raw materials or components, which were imported under lower import duties.

With passage of time, the limitations of this strategy became evident. The licenses were issued on the basis of some assessment of the likely domestic demand, which gave certain monopolistic power to industrialists, who had a captive market. Product quality was poor and there was little incentive for industrialists to improve quality. Moreover, the domestic market was so small, and exportability negligible as a result of poor quality, the firms could not achieve economies of scale. The state industries slowly became afflicted with problems of overstaffing, inadequate management, and lack of incentives. The government was only too willing to subsidize in cases of losses on the grounds of welfare. Generally, the investments made by the state in industry yielded a very low return, invariably lower than the cost of borrowing by the Government.

Another popular reaction against state regulations by way of change of governments, led to liberalization of the economy and elimination of most controls from 1977. The post-1977 era essentially promoted export oriented industrialization within a framework of a liberalized trade regime. However, the state industrial sector had been slow in joining this process for several reasons. One reason was privileged position as a result of state patronage. The heavily protected private industries were also reluctant to let competition erode their power. Thus, although most trade restrictions were replaced by tariffs as far back as November 1977, the reduction of high protective tariffs to reasonable levels proved to be a slow process. The licensing of local industries was disbanded only in 1989, while "Privatization" or divesting commenced only after the concept of "Peoplization" was accepted as an important aspect of government policy in 1989. However, the disbanding of import quotas had mixed effects on local industry after 1977. This enabled some industries to carry lower levels of stocks, enhance capacity utilization and improve production in general. Increased competition, although under high tariffs, had some beneficial impacts for those who rose to the occasion and took up the challenge. However, such changes were not evident in the state sector where monopolistic privileges continued. In 1989, a new industrialization strategy was unveiled by the government which led to the Industrial Promotion Act No. 48 of 1990, which provides for the identification and registration of industries carried on or to be commenced. This also established regional industrial cells. Today Sri Lanka is said to offer many benefits to investors and is termed an investor friendly country. However, this friendly policy environment has not resulted in a substantial flow of FDI (foreign direct investment).

The economic changes are visible on looking at the GDP contributions. In the early 1960s, the share of agriculture in the GDP (%) was 40 with manufacturing having 14%. For the first time 1996 saw manufacturing overtaking agriculture with contributions of 21 and 18% respectively. With the bulk of the manufacturing contribution coming from the textile and apparel sector and with poor value addition, the industry sector in Sri Lanka is yet to achieve a healthy diversity, which is critically required.

The industrial composition of Sri Lanka is given in Table 18.1 (Dept of Census and Statistics, 1997) and clearly depicts the composition and employment level. Textile and food processing dominates the sector.

Table 18.1: Industrial Composition of Sri Lanka

Industrial Category	% of national total
Mining	2.9
Food processing	16.9
Textiles	45.8
Wood products	3.1
Paper production and printing	3.4
Chemical & petroleum	9.1
Nonmetallic minerals	5.5
Metal products	0.3
Machinery & equipment	5.2
Other manufacturing	3.3
Electricity and gas	3.0
Water	1.5

Wignaraja (1977) argues that the lack of a coherent, proactive industrial strategy in Sri Lanka has led to the development of a fragile, low-skill export base a base—that has few linkages to the domestic economy and one which effectively 'traps' Sri Lanka into producing low-skill exports. It is perhaps fair to conclude that we have not followed the industrialization with a clear vision and by not taking the right step at the right time the expected developments have not materialized. The present industry is not creating much wealth for the nation and the outlook in many quarters is to set up industries primarily for employment creation.

Today the government is spending no public money in developing industries and the private sector due to high cost of capital (18%-20%) is not investing in industrial ventures either. This situation has brought the industrial sector development to a near halt.

Energy in Sri Lanka

Energy availability is crucial to any activity and if industry is to make any significant headway and contribute to the economy, energy availability is a key criterion. The "Energy" factor also requires an explanation. The energy sector development in the post-Independence era of Sri Lanka has been slow and lethargic. Today energy consumption appears to be directly related to the level of living of the populace and the degree of industrialization of the country.

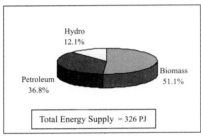
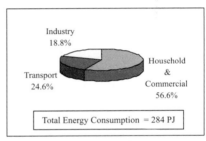

Fig. 18.1: Sri Lankan evergy distribution and consumption.

Figure 18.1 indicates the Sri Lankan energy distribution and consumption pattern as exists today (Perera and Sugathapala, 2002). The energy sector in Sri Lanka is dominated by bioenergy, and especially by fuel wood. The total annual primary energy supplied in Sri Lanka during 1998 was about 326 PJ (or 7,786 thousand toe). This comprises of 51.1% biomass, 12.1% hydroelectricity and 36.8% petroleum. The final energy consumption, was amounted to about 284 PJ (or 6,778 thousand toe), 18.8% of which was consumed by the industrial sector, 24.6% by the transport sector and 56.6% by household and commercial sectors. Consumption, once dominated by the manufacturing sector, has changed to one in which households and commercial buildings have become the major consumers.

The national power generation system has an installed capacity of about 1,835 MW consisting of about 1,150 MW hydro, which contributed 3,915 GWh energy in 2000. The thermal component is 685 MW, which provided 3,486 GWh in the same year. The per capita power consumption stood at 247 kWh in 1998. Demand for electricity is rising at the rate of 8-

10% per annum. Sri Lanka has to import all the fossil fuel requirements and the country has exploited most of the sites available for hydro. Thus the future demand has to be met through thermal power. The generating pattern is changing rapidly. The fuel mix in the electricity generating system has changed from a predominantly hydroelectric system to a mix of hydro and thermal. The distribution efficiency has worsened with network losses in Sri Lanka having increased from 18.8% in 1996 to 21.5% in 1999. CEB's planning process envisions an addition of 1,800 MW of coal-fired plants within the next 15 years.

Short-term supply crises have faced the Sri Lankan economy starting from 1979, in 1983, 1987, and 1992. In 1994 these short-term crises were much more frequent. These have a resulted from delays in providing additional supply capability to the national grid. The year 1996 witnessed a peak in this supply-and-demand mismatch when there were 8-hour blackouts coupled with additional electricity use restrictions. The year 2001/2002 is also witnessing mandatory load-shedding programs due to supply shortfalls. Today hydropower systems are under the purview of the CEB. CEB owns and operates oil-fired power plants (gas turbine and diesel systems); independent thermal power plants exist as well. There are also many emergency power plants operating on diesel or furnace oil. The renewable source contribution to the Sri Lankan grid is minimal. As a result of a World Bank/GEF supported program the grid has witnessed the addition of a 3 MW wind farm as well as electricity from 16 minihydro plants (with an output of approx. 30 MW). The reason for these power shortages is the inability of the CEB to build the larger power plants as specified in the long-term generation plan. The reasons for this are many and one significant factor has been environmental opposition to thermal power stations based on coal. The coal power station scheduled was shifted from Trincomalee to Mawella and then to Norochcholai, a process taking place in the early 1980s. Much work has been done and studies published; environmental clearance has also been given but the plant has failed to materialize.

How this energy crisis affects the industry stems from the cost and availability of energy. In recent years industries such as apparel and garment sector have always had captive power generation. Industries tend to have their own generating units running in diesel. CEB also hires generators. Usually in times of power cuts and limitations industries are forced to operate their stand-by generators. Though a reliable supply is available from the customer's point of view, the operating costs are prohibitive. One significant factor is that unlike in other countries the industrial tariff is more expensive than the tariff structure with domestic consumers who enjoy a 30% subsidy and a 50% subsidy of the first 30kWh bracket per month based on the average price. In the region perhaps only Bangladesh has a tariff situation similar to Sri Lanka.

In addition, Sri Lanka imports all the fossil fuels and the country has only one refinery (50,000 bpd). This growing trend of thermal power generation has resulted in importation of refined products. In 1998, Sri Lanka imported 2,824 ('000 tons) of petroleum products. Of this, 28% were refined products. This situation is worsening by the day and the import bill is also rising, reaching US $100 million in 2000.

These uncertainties have resulted in further growth problems in the economy as well as in the industry sector. However, at this moment one certainty is that all shortfalls, as well as future needs will be mostly met through thermal power with increasing consumption of fossil fuels.

Climate and the Environment: The impact on Sri Lanka

Sri Lanka's coastline stretches over 1,585 km of beaches, lagoons, estuaries, etc. The land area is approx 65,610 km^2. The country is a small island nation by normal economic concepts and we have usually considered ourselves a small island nation. However, from a climatic change concept Sri Lanka falls in the category of "big island nations". With the advent of climate change issues, the United Nations defined small island nations as islands with less than 10,000 km^2 land mass and fewer than 500,000 inhabitants (UNESCO, 1994). For Pacific Island Developing Countries (PDIC), comprising 22 political entities, the impacts of climate change are projected to be quite severe if all the IPCC predictions come true. In this context Sri Lanka is no longer a small nation. Thus the country has a different platform to develop itself as globally countries are aligning to follow different development patterns. The cheap fuel era plus the singular development based on "consumption and growth" that fuelled much of the developed countries is not available to us. It must also not be forgotten that it is this unrestricted use which has led to the present ecological crisis.

The rules of development globally have changed with an understanding of energy environment linkages. Once the global warming and ozone depletion concepts were forwarded and by and large accepted (the latter more strongly than the former), the causative agents were traced back to the industrial revolution and its outcomes. The power sector and the industry today operate within a different framework. Prior to this understanding the linear concept of economic development via industrial development and cheaper and abundant energy leading to unbridled growth were accepted norms. The developed countries have been able to implement industrial development partially due to abundant energy at a low price. Not being a party to that growth scenario, Sri Lanka attempts the same "development" in a different environment. What is clear is that options being limited, different strategies are required. Table 18.2 presents some energy-environmental and economic indicators for Sri Lanka.

Table 18.2: Some energy-environmental and economic indicators for Sri Lanka (IEA, 1999)

Energy production	4.25 Mtoe
Net imports	3.09 Mtoe
Electricity consumption	4.25 TWh
CO_2 emission	8.48 Mt of CO_2
Per capita CO_2 emission	0.46 tCO_2/capita
Population millions, (1999)	19.0
GNI per capita, $(1999)	830
GDP per unit of energy use (1998)	
PPP$/ kg oil equivalent	8.0
CO_2 emission per capita (1997) Mt	0.4
Share of electricity generated by coal	0%
Population density, rural people/sq. km arable land	1664
Annual deforestation % change (1990-2000)	1.6
Nationally protected areas, % land	13.5
Freshwater resource per capita, m^3	2626
Access to improved water sources, % of total population	83

The issue of environment has added a different dimension to Sri Lankan project planning and execution. One of the last major hydropower projects – the Upper Kotmale Hydro Project (UKHP)—has faced much opposition due to the conflict between need for development and preservation of unique natural assets (i.e. depletion of 07 waterfalls has been predicted if the project goes ahead). On the other hand, the coal power stations in all proposed sites have faced opposition from perceived problems such as thermal discharges, acid rain. It is not sure, however, whether pragmatic environmentalism is being practiced in the country. Though the project-based environmental impact assessment process is well developed and practiced in the country, the ground level reality has been different. For example the Upper Kotmale project was not given approval after the EIA process whereas the coal project received approval. Today with another power crisis on, it appears that the Upper Kotmale Project will go ahead.

Temperature Studies

The mean annual temperature in the coastal areas below 150m in elevation ranges from 26°C to 28°C while in the hill country above 1500 m, the range is 15°C to19°C. Most of Sri Lankan weather monitoring has been carried out to support agriculture. An analysis of more than 100 years of temperature and rainfall data at 14 meteorological stations in the country was carried out by Fernando and Chandrapala (1992). This analysis indicated that the annual mean air temperature, particularly over the period 1961-1990, has been increasing at all the 14 stations by varying amounts, the highest rate of increase being 0.036° C per annum at

Anuradhapura. Such data gathering is carried out by the Department of Metereology. The department today also supports the Center for Climate Change Studies where the mandate has been broadened. The center maintains a database on climate studies and related issues carried out in Sri Lanka. There are about 600 entries and much is on weather and plantation-related activities.

Sri Lanka and the Climate Change Convention

Sri Lanka ratified the Climate Change Convention on Nov 23, 1993. The requirements under the Kyoto National Action Plan (as required under Article 12) and the preparation of a National Inventory of Greenhouse gas emissions (as required under Article 4) have been completed. The initial national communication under the United Nations Framework Convention climate change was presented on Oct 27, 2000. The Ministry of Environment is the focal point in implementing the convention and has a coordinating committee in place.

Sri Lanka has responded well to international environmental treatises. It is stated that Sri Lanka's progress in meeting goals set under the Montreal Protocol is excellent. This is action with respect to a global problem. However, in meeting these targets one should not ignore the local aspirations. It is important to have the local population well informed and in step with activities. However, this will be difficult if the perceived understanding is that Sri Lanka is meeting global obligations but at the cost of local needs. If the burden is shifted to the local populace the net effect is not satisfactory. It is with this in mind that the US action withdrawl from Kyoto agreements is seen: local needs first and then global requirements. The faith and commitment of the significant contributors (US being the no. 1 CO_2 contributor) are important indicators to the public. In Sri Lanka, under the present conditions, people may not see the value of planning to reduce GHG emissions and steps taken in that direction if they do not alleviate local conditions.

The region of South Asia where Sri Lanka lies is poised to become a major contributor to GHG emissions, largely due to the growth of India. Per capita emissions in the region are currently quite low, with total production of CO_2 representing only about 5% of global emissions. However, the rate of increase is about 7% per annum, which is twice the global average.

It is obvious that if we are to improve the living standards of Sri Lankans and raise the economic well-being of the country, energy consumption must be greatly expanded. It may be wise for Sri Lanka to accept the spirit of the situation and follow a common sense policy based on the objectives, which lead to protection of the environment and ensure sustainable development. These principles are:

- limit anthropogenic emissions of the GHGs by carrying out processes efficiently;
- develop techniques which will improve the quality of life with low environmental footprints;
- understand the importance of efficient resource management and principles of sustainable consumption;
- protect and enhance sinks and reservoirs.

Underlying these commitments remain the challenges to society.

It is obligatory under the convention to report national inventories of anthropogenic GHG emissions and removals. Fernando (1999) carried out the activity for the period 1993-1995 and the national GHG inventory, which is in the initial national communication, is presented in Table 18.3.

Table 18.3: National greenhouse gas inventory

Greenhouse gas categories	CO_2 (Gg)	CH_4 (Gg)	N_2O (Gg)
Total (net) national emission (Gg per year)	33,630.22	1,098.375	162.8657
1. All energy	5,447.668	927.785	156.0257
Fuel combustion			
Energy and transformation industries	482.358	0.021	0.0041
Industry:			
contribution from fossil fuel	801.08	1.92	0.07
Transport:			
national navigation, rail transport and domestic aviation	2,946.31	263	116.12
International bunkers (not included in national total)	1,540.82	74	21.6
Commercial and institutional	32.49	21.57	1.29
Residential: contribution from fossil fuel	834.18	640.17	38.51
Other: agriculture, forestry, and fisheries sector	351.25	0.024	0.0106
Biomass burned for energy		0.98	0.021
Fugitive emissions			
Oil refining, transport and storage		0.1	
2. Industrial processes	300.55		
3. Agriculture		156.86	3.42
Enteric fermentation		135.86	
Rice cultivation		21	
Manure management		47	3.34
Burning of Field crops			0.08
4. Land-use change and forestry	27,882		
Changes in forest and other woody biomass	14,773		
Forest and grassland conversion	3,748		
Abandonment of management lands			
Forest soils	9,361		
5. Other sources		13.73	
Domestic wastewater management		0.59	
Commercial wastewater management		13.14	

In 1995, a total of 34,114 Gg of GHGs were emitted in Sri Lanka with 95% of it coming from CO_2 while 2.61% and 2.16% are attributable to CO and CH_4. The largest contribution to GHG emissions in Sri Lanka is through the change in forest and woody biomass stocks, forest grassland conversion, liming, and organically amended soils. Industry contribution is a small fraction of the total. The greenhouse gas emission can be expected to increase. In the power sector during the period 1972-1990 almost 90% of power generation was hydroelectricity. The situation has changed today to about 60% hydro. As the hydro opportunities are almost over, the future expected demand is expected to be met through thermal which means that GHG emissions can increase significantly. No update to this inventory is available. It can also be seen that the Sri Lankan inventory has not calculated the sink potential.

As a non-Annexe I country, Sri Lanka is not expected to reduce its GHG emission. Mechanisms such as the CDM give us opportunities to realize some development via participation in activities which will be climate friendly but funded by developed countries. Sri Lanka has developed a national policy on the CDM mechanism to explore possibilities in benefiting from the CDM provisions in the Kyoto Protocol. The Clean Development Mechanism (CDM) is a flexibility mechanism introduced by the Kyoto Protocol, in order to control the global greenhouse gas emissions. Being a developing country, which emits a relatively low amount of GHG emissions, Sri Lanka has the potential to participate in CDM projects. Any CDM project should be implemented with a clear and long-sighted policy and the Ministry of Environment has set up two CDM study centers in universities to actively engage in national efforts. Table 18.4 presents the draft CDM policy of Sri Lanka.

Table 18.4: Draft National Policy on CDM of GOSL (Batagoda, 2002)

1. Sri Lanka has ratified the Kyoto Protocol, and decided to participate in CDM to assist developed country parties in meeting their GHG emission reduction targets and achieving sustainable development objectives of the country.
2. Sri Lanka has decided to provide maximum contribution to the global effort of long-term stabilization of GHG to address climate change even though its contribution to the problem is negligible.
3. The following sectors will be considered high priority sectors for the implementation of the CDM projects:
    Energy, including industry and transport
    Forestry
    Agriculture, including plantations
    Waste
4. The Secretary/ Environment or a committee approved by the cabinet (National Expert Committee) should make the final decision with regard to the eligibility of a project under CDM.
5. A National Expert Committee on CDM will be established and will be responsible for CDM project evaluation and approval. Until the CDM process is in place at

the international level, any interim CDM projects should be approved by the National Expert Committee. Special approval should be taken on a case-by-case basis.

6. Guidelines for the National Expert Committee should be developed and the committee should have the authority to negotiate on behalf of the government and it should help the Secretary/ Environment in making decisions in approving suitable projects.

7. Until real Carbon Trading is in place, the total emission from all the interim projects approved by the National Expert Committee should not exceed 200,000 tons of carbon.

8. The interim CDM projects should be treated as pilot projects, and they should be monitored by the National Expert Committee, and should be used to develop a comprehensive national CDM policy. If an interim project is evaluated as appropriate, the letters by the Secretary/ Environment issued on such interim projects will only mention that "the project is recognized as a CDM project".

9. A private sector organization will also be accredited as an independent entity for project validation and verification.

10. In addition to the criteria for CDM project development decided by the National Committee, the following should be addressed in the CDM project development:
    i. CDM project should address poverty alleviation and employment generation effort of the government, which is fundamental to sustainable development in the country.
    ii. The project should have a tangible national impact and it should meet the sustainable development agenda of the country.
    iii. Preference should be given to new technologies
    iv. CDM projects that address local environment issues while addressing the global issues should be given priority

11. The initial project contract period should be 10 years. The project proponents should have the ability to renegotiate at the end of the contract.

12. The project developers should follow the guidelines of the Intergovernmental Panel on Climate Change (IPCC) and the UNFCCC. The project proponents should be responsible for the avoiding or sequestration of carbon emissions. Project monitoring should be included in the project design itself.

13. The eligible local parties for submitting CDM proposals are as follows:
    i. The National Expert committee on CDM should be satisfied with the expertise, capacity and experience of the proponents in dealing with the subject area of the project
    ii. The proponent should be a public or registered private sector entity
    iii. The proponent should be a national body.
    iv. If the proponent is a foreign organization, it should have a local partner in order to submit a proposal for a CDM project.

14. Small CDM projects to be considered need to be in accordance with the Bonn Agreement adopted at the COP-6, which is described below:
    i. Renewable energy project activities with a maximum output capacity equivalent of up to 15 megawatts (or an appropriate equivalent).
    ii. Energy efficiency improvement project activities which reduce energy consumption, on the supply and/ or demand side, by up to the equivalent of 15 GWh per year; or
    iii. Other project activities that both reduce anthropogenic emissions by sources and directly emit less than 1 5kilotons of carbon dioxide equivalent annually.

15. The ownership of CDM credits earned through any CDM project lies with the government, and any CDM proceeds should be directly transferred to the CDM Fund established by the government. Any percentage of the revenue decided by the committee will be credited to the government.

16. The National committee should decide the percentage of CDM proceeds that should be paid to the project developer. However, it should not exceed 50% of the total proceeds.
17. Administration of the CDM Fund will be carried out by an Executive Board under the advice of the National Expert Committee on CDM. The National Committee can allocate some funds for promoting new and relevant technologies.
18. Carbon sink or sequestration CDM projects include afforestation and reforestation projects that will be selected based on the decisions made in the ongoing international negotiations.
19. It is recommended that the carbon component should be addressed under the Environment Impact Assessment (EIA) process.

Challenges for Sri Lankan Policy-makers and Industry

It is important that both parties understand the implications and the new paradigms for development. It is also imperative to understand the social development needs. We did not follow a suitable strategic path consistently during the post-independence era and another moment has arrived where we have to understand how to proceed.

Much work has been done in identifying threats from climate change effects to Sri Lanka. The studies have looked at issues in a sectoral manner. Some issues of concern raised for the industry sector as impacts from climate change are (National Action Plan on Climate Change, 1999; MOFE, 2000):

> ➢ The industries in Sri Lanka are mainly located in the coastal belt. An accelerated sea-level rise as well as other related atmospheric disturbances are considered cause for concern in industrial siting, sea-water intrusion on water bodies causing fresh ecosystems to be more saline and preventing water use, increasing costs in wastewater treatment in trying to achieve standard water quality criteria, damage and increased vulnerability to industrial support activities such as roads etc.
> ➢ Solid waste disposal practices currently being carried out along the coastal belt could create undesirable environmental problems in future – damaging water resources etc.
> ➢ Loss of productivity and increasing energy bills in response to warmer environments

Apart from impacts on energy these are the assessed direct impacts on industry. Except in siting decisions and in solid waste management it appears that neither the country nor its industry sector can contribute much in combating the identified impacts. For Sri Lanka the issue is in adaptation though in the industry sector we really could mitigate emissions as well but with a view toward improving benefits via improving efficiency.

The following options are available to industry in carbon dioxide mitigation:

• Energy conservation

- Change of industry to 'soft industry' options - industries with lower energy intensities wherever possible.
- Changes in energy infrastructure
- Fuel mix reformulation
- Renewable energy technologies
- Creation of CO_2 sinks through reforestation
- Removal of CO_2 from combustion gases

Except for the last, the other options could easily be followed in industry. However, the driving force should be to realize industrial efficiency and the GHG reduction could be accounted for as the by-product. The primary driver need not be the GHG reduction as the national contribution is a small value. Policy-makers should understand that markets alone will not drive this movement. The reasons for this statement are many and range from economic characteristics to stitutional problems to human behavior. Thus the policy makers should try to create a conducive climate.

Sri Lankan industries should learn lessons from the past even from the era when the problem was only an energy crisis. This is due to the fact that we have not learned the lessons industry in other countries learned during the first oil crisis. Japan provides us with a good example. Japan in the 1970s followed a policy of energy efficiency via industry restructuring and its improvements in energy efficiency. The industrial sector showed the greatest success in Japan's energy conservation policy. Between the fiscal years of 1973 and 1987 Japan's GNP increased by 72.7% in real terms while energy demand (total primary energy requirement) showed only a 4.3% increase. In other words, energy intensity, or energy consumption per unit of GNP activity improved by 36.2% in this period. The industry did not simply follow a path of switching off unwanted lights, but followed a rigorous strategy involving good housekeeping, equipment and process upgrading and prudent investment. Though some concepts were transferred to the Sri Lankan industry by the Japan Energy Efficiency Office during the 1990s, neither widespread adoption nor dissemination of practices is evident.

Efficiency of utilization is a serious issue for national concern though it is not receiving the attention it deserves. The peak power demand today is around 1,400 MWe. The loss in transmission and delivery ignoring the end-use efficiency is around 21% and has been steadily growing in the last few years. This implies a loss of around 370 MWe, which is a serious loss of generated energy. The issue should be addressed as a priority since for all these losses the answer today is adding more thermal power-generating units on an ad-hoc basis. Again, in terms of economics, the GDP growth rates are in the range of 3-5% whereas the electricity demand is growing at the rate of 8-10%. This really indicates that the whole economic process is relatively energy inefficient with more being needed for a unit output. In comparison, the developed economies have a 1:1 relationship in these two indicators.

It is unlikely that conservation oriented strategies such as carbon taxes even under "effect neutral" conditions would be effective as the SL industry is already burdened under many types of taxes. As per changes in fuel mix via the use of natural gas, much mooted by a segment of the Sri Lankan planning community, it is not clear at present that this option would be viable in Sri Lanka. However, financial and banking institutions should change their present ways of project assessments if industry is to move ahead in investing in some capital intensive industry projects.

Some of the options available are briefly reviewed below:

Industrial Cooperation

It is necessary to develop and foster cooperation within the industry. This is poor at present and there is unequal access to information as well as unequal commitment to supply information. This state of affairs prevents sectoral planning and hampers real development. Cooperation would enable benchmarking to be carried out and everyone to benefit. Lack of reliable data is a significant barrier in realistic planning in the country.

Develop cogeneration systems

Most biomass waste could be processed to supply heat and electricity to industry. The brick and tile, tea and rubber sector generates and uses biomass in a very much suboptimal manner. Processed wastewater treatment systems as well as solid waste management could be channeled to develop cogeneration schemes. It is important to have proper process planning as well as industrial location strategies to obtain benefits of this nature. Recent developments toward having industrial estates and zones with cleaner technologies in place facilitate these developments. Most of the thermal power plants also operate only to produce electricity. Significant quantities of thermal energy is lost. The first combined cycle power plant is also two years away from being coming into operation.

Institutionalizing Energy Audit

From a policy point of view, Japan established an Energy Conservation Law in 1979. This demanded designated factories to allocate resources to manage and optimize energy use. A similar law has now come into force in India as well. This is a valuable learning point for Sri Lanka to put into operation.

Energy audit is a tool which must be employed in an active manner by the industry. Energy audit helps in energy cost optimization and suggests methods to improve the operating and maintenance practices of the system. It is helpful in coping with the situation of variation in energy cost availability, reliability of energy supply, decision on appropriate energy mix, decision on using improved energy conservation equipment, instrumentation, and technology. The concept has been practiced and developed in Sri Lanka by creating a nongovernmental organization, the Sri Lanka Energy Managers Association. However, the benefits have not been widely visible. Leelerathne (1998) showed that though many audits have been carried out and potential demonstrated, the number of industries that have actually carried out measures to improve efficiency is low.

Improvements in Processes and Equipment

Industry still uses less efficient technologies. Replacement of capital assets after their economic lifetime is not common. The usual explanation is that the capital required to implement such replacements is simply not available. The developments taking place elsewhere could easily be utilized by the industry.

There are significant opportunities in the following energy utilization systems employed in industry:

> Efficient lighting schemes and devices
> Energy-efficient motors
> Compressed air systems
> Power factor correction
> Improved refrigeration
> Improved chiller systems
> Factory building design

Energy audits would effectively indicate the possibilities. Of the options stated, the power factor correction has been the most frequent action taken by many factories.

Renewable Energy Utilization

Sri Lanka should really develop its renewable energy utilization at the industry and service level. As an example, in the hotel industry (CEB places hotels in the industry category) much of the electric power used in water heating could be replaced with solar water heating, now a mature technology. Yet solar water heating is not a common practice. With plenty of rain and sun, biomass gasifiers could be utilised for energy delivery. At present, one integrated textile mill uses a biomass boiler (3 MWth) for its thermal needs but at reduced efficiency and potential. In some industrial and service buildings PV roofs could be integrated. This should not be viewed as a novelty but as a normal operation. Solar assisted drying – simple drying to heat pump drying—can be utilized in rubber and tea industries. An active program of solar cooling developments should be pursued as an R and D exercise. This would help in removing the changes to the daytime load curve seen recently due to the burgeoning number of air conditioners. The use of anaerobic digestion technologies as an energy supply mechanism in industry should be actively pursued (Kottner, 2001).

Research and development in the area of solar cooling technologies could be carried out. Demonstration plants using ground-based heat pump systems should be implemented and evaluated. Though studies have been forwarded in the area of tea and copra drying for the coconut sector none has been implemented.

One challenge to the industry is to develop an industry catering to these needs—an environmental and renewable energy service provider sector, which is absent in the Sri Lankan industry today.

Industrial Case Study

An interesting option for industrial development, GHG abatement, economic development, technology development and assimilation exists with the Sri Lankan sugar industry. Sri Lankan sugar industry can learn lessons from Brazil, Taiwan, and Mauritius industries (Deepchand, 2001). Sri Lankan sugar industry is an industry going downhill with a few early factories now in a state of decay after closure. Two major factories are still operating but with much difficulty and lower returns. From an economic basis the country spends vast amounts of foreign exchange in importing sugar. Current estimates show that nearly Rs 10 billion is spent on importing 0.5 mt of sugar annually. Thus there are good grounds for expansion as well as for optimization. Two of the operating mills are in a province where electrification is at its lowest, Uva province. The sugar industry though it utilizes bagasse does so in an inefficient manner and wastage of bagasse (Perera et al., 1997). In one instance an adjacent distillery brings in furnace oil to run its generators to generate electricity.

The Pelwatte sugar factory is the largest sugar factory in Sri Lanka as well as the most important industrial activity in that part of the country. It is also in a district considered to have the highest poverty level in Sri Lanka as well as the lowest electricity penetration. Its boilers are the largest bagasse (crushed cane remnants) boilers in Sri Lanka (45t h^{-1}/two boilers). However furnace oil is at hand for frequently needed boosting. Also furnace oil is the single source of primary energy in the nearby distillery, which produces ENA (extraneutral alcohol) by utilizing molasses from the sugar factory. One reason for needing furnace oil has been that when the bagasse has high moisture content it fails as a good solid fuel. Currently the moisture content of bagasse is around 51% and no attempt has been made to dry it using flue gases etc. Perera et al., (1997) demonstrated that with improvements to bagasse combustion the system could be made to generate 3.1 MWe for the national grid.

The estimate for the mill is that it produces a surplus bagasse quantity of 4.5 tons for every 100 tons of surplus bagasse. This usually means a 1,600 tons surplus bagasse from a typical daily load. The electricity conversion possible is 450 kWh for every 1 ton of mill-run bagasse.

Also, an anaerobic digestion system exists to treat mill wastewater where the gas generated is burned off instead of being put to use.

An ecofriendly system could be developed around these options. Development of a gasifier-based cogeneration scheme and ethanol-based transport are options for the region. Additional energy via the anaerobic digestion option is possible from the treatment of filter mud and wastewater.

Implementation of such a strategy requires the commitment of public policy and public-private partnership. This perhaps is an option for a CDM project.

Conclusion

Business-as-usual is no longer applicable to the Sri Lankan industry. This is not because the globe is becoming warmer, but simply because the sector is highly inefficient. And pollution is simply a reflection of the inefficiency.

Industrial sector is in need of significant upliftment in Sri Lanka and as it stands today is not well attuned to support Sri Lankan economy. The different climate in which it has to operate and develop also calls for different strategies to be adopted. Considering the fact that the country as a whole and the sector in particular is not a significant GHG contributor, the strategy should be to adopt policies which will improve efficiencies and embrace cleaner technology options. Work toward improving efficiency and the pollution contribution so often identified with the industrial sector of Sri Lanka will be an issue of the past. The sector thus could be a model of responsible behavior and bottom line benefits would accrue from the improved competitiveness brought in by being an efficient sector in the national economy.

Considering that there are still uncertainties with respect to climate change – in the area of magnitude and timing of potential impacts – it is not necessary to have a policy dilemma but to follow a prudent policy which would benefit the nation but not compromise other interests. In the industrial sector CO_2 is not the only GHG of interest. Processes contribute other GHG gases as well. CFC which is also a significant greenhouse gas is being effectively phased out under the ozone action program and Sri Lanka has utilized funds from the multilateral funds available for CFC phase-out action.

Amid statements such as "positive response to global warming are regarded as investments in the future of the nations and the planet", the backing out by leading economies simply demonstrates that local national interests precede global responsibilities. Sri Lankans can get mixed information when looking at global developments. The Pacific Island of Vanuatu (a smaller island nation) has outlined its vision for 100% hydrogen economy by 2010 based on renewable energy (Dunn, 2001). On the other hand, the new Bush administration in the USA has proposed various cuts in renewable study schemes (i.e. 8% cut in the hydrogen research budget) and approvals have been given for oil drilling in protected environments. The former with its 100% commitment will only be symbolic in its impact. The latter, already the no. 1 polluter, is backing down but the impacts there will not be symbolic.

The option for Sri Lanka appears to head for efficiency and utilize external options in a prudent manner as it fulfills all criteria of serving the nation as well as reducing global concerns.

References

Batagoda, B.M.S. 2002. Sri Lanka Policy Perspectives on the CDM, Energy for Sustainable Development 6 (1): 21-29.

De Silva, A. 1991. Industrialisation in Sri Lanka: Policies and Options. LOGOS, vol. 30, no. 1

Deepchand, K. 2001, Commercial scale cogeneration of bagasse energy in Mauritius, Energy for Sustainable Development 5 (1): 15-22.

Department of Census and Statistics, Sri Lanka 1997. Annual Survey of Industries.

Dunn, S. 2001. Routes to a hydrogen economy. Renewable Energy World, July/Aug 2001, pp 18-29.

Fernando, T.K. and Chandrapala L. 1992. Global warming and rainfall variability –the Sri Lankan situation. Proc. 5th Inter. Meeting on Statistical Climatology, pp. 123-126. Toronto, Canada, 22-26 June 1992.

Fernando W.J.N. 1999. Greenhouse gas inventory of Sri Lanka 1993-1995, Ministry of Forestry and Environment, Sri Lanka.

IEA, 1999. International Energy Agency, Key World Energy Statistics.

Kottner, M. 2001. Biogas in agriculture and industry potentials: Present use and perspectives. Renewable Energy World, July/Aug 2001, pp 132-143.

Leelerathne, M.W. 1998. Some non-technical factors affecting energy conservation in Sri Lankan industry, SLEMA Journal, 8 (1) pp 25-32.

MOFE (2000).

National Action Plan on Climate Change, Final Draft. Ministry of Environment, Sri Lanka. 1999.

NAPCC (1999).

Perera, K.K.C.K., de Silva, M.P., Wijesiri DPN and Deshapriya M.P.M.C. 1997. How can bagasse be utilised to overcome national power crisis, OUR Engineering Technology, Vol. 3, pp 21-24.

Perera, K.K.C.K. and Sugathapala, A.G.T. 2002. Fuelwood cookstoves in Sri Lanka and related issues. Energy for Sustainable Development, vol 6 (1): 85-94.

UNESCO, 1994. Small Islands. 1994. UNESCO Courier, pp. 23

Wignaraja, G. 1977. Trade Policy, Technology and Manufactured Exports: Sri Lanka's Liberalization Experience. MacMillan Press, Basingstoke, UK.

www.competitiveness.lk

www.worldbank.org

Environmental and Socioeconomic Impacts Associated with Climate Changes in Brazil

Marcia Marques[1] and William Hogland[2]
[1]Rio de Janeiro State University UERJ, Dept. Sanitary and Environmental Engineering, PESAGRO-RIO, Rio de Janeiro, Brazil
e-mail: marcia@marques.pro.br
[2]Kalmar University, Dept. Technology, Kalmar, Sweden,
e-mail: william.hogland@hik.se

Introduction

Although much progress has been made and considerable effort put into developing hydrological models for estimating the effects of climate change (Compagnucci et al., 2001), considerable uncertainty still exists in predicting regional climate changes, future conditions in the absence of climate change, or to what degree climate will become more variable (Grid-Arendal, 2001). Understanding of some key ecological processes associated to climate changes is still insufficient even though on regional scales there is clear evidence of changes in some climate variability indicators, and an increase in proportion of rainfall from extreme events in some regions. The economic damages from weather-related disasters have also increased dramatically (Grid-Arendal, 2001; IPCC, 2001a) although much of these disasters may be attributed to a greater number of people living in vulnerable areas. In many regions in the world, both increase and decrease trends in stream flow have been observed (IPCC, 2001b). However, these trends cannot all be attributed to changes in regional temperature or precipitation. Increasing demand and pressures on the water resources are perhaps the most important causes of the observed hydrological changes. However, widespread accelerated glacier retreat and shifts in stream flow timing in many areas from spring to winter are more likely to be associated with climate change (Compagnucci et al., 2001). In Brazil, climate changes including temperature, precipitation and stream flow have been registered during the last century and a number of environmental and socioeconomic impacts associated to them have been recently described. These issues and their impacts both observed and foreseen are discussed herein.

Brazil: General features

Brazil is the fifth largest country in the world in terms of population, with 163 million inhabitants (IBGE, 2000), as well as land area (8,574,761 km^2). Brazil occupies almost half of South America's land area and is the economic leader of the subcontinent, with the ninth largest economy in the world. The state of Amazonas has the largest area (1.5 million km^2) and the most populous state is São Paulo (about 35 million inhabitants). From the Amazon basin in the north and west to the Brazilian Highlands in the southeast, Brazil's topography is quite diverse. Extensive uplands lie in the southeast and drop off quickly at the Atlantic Coast. Much of the coast is composed of the Great Escarpment which looks like a wall from the ocean. The fact that 92% of Brazil's land mass lie between the tropics, together with its relatively low topography, accounts for the predominantly hot climate, with annual average temperatures above 20°C. Almost all of Brazil is humid as well with either a tropical or subtropical climate. Brazil's rainy season occurs during the summer months (ANA, 2002). Northeast Brazil, a semiarid region, suffers from regular drought,

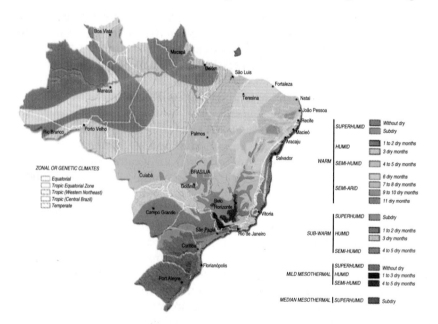

Fig. 19.1: Climate zones in Brazil (after ANA, 2002).

forming the "drought polygon". The climate varies due to geographical and topographical factors, the continental dimensions of the country and the dynamics of air movement, directly influencing temperatures and rainfall (ANA, 2002, Figure 19.1).

Brazil is divided into five regions (Figure 19.2), politically and geographically distinct, which nevertheless share certain physical, human, economic and cultural characteristics.

Each of the five regions (North, Northeast, Southeast, South and Middle West in Figure 19.2) is limited by the borders belonging to its states (Figure 19.2). There are considerable economic disparities among regions, hampering the country's sound economic and social development. When the percentage of Brazil's GDP is divided by region (Table 19.1), it is clear that the Southeast region produces more than half of Brazil's GDP. Additionally, a littoralization trend during many decades resulted in concentration of the population in the coastal cities, in south, southeast and northeast Brazil.

Fig. 19.2: 5 Brazilian regions (colored areas), 9 hydrographical basins (black line boundaries), 26 states and 1 federal district DF (white line boundaries).

Table 19.1: Percentage of the total GDP by region (IBGE, 2000).

Region in Brazil	% of GDP
North	4.4
Northeast	13.8
Southeast	59.4
Central West	5.3
South	17.1
Total	100

Overview of Hydrographic Regions of Brazil

Estimates show that approximately 10% of the world's fresh water is found in Brazil, placing it among the richest countries in freshwater volume terms (ANA, 2002). However, its distribution varies greatly over the year and among the different regions of the country.

The Amazon River system carries more water to the ocean than any other river system in the world. The basin is home to the most rapidly depleting rain forest in the world, losing about 20,000 km^2 annually (ANA, 2002). The basin, occupying more than 60% of the entire country, receives more than 2,000 mm of rain per year.

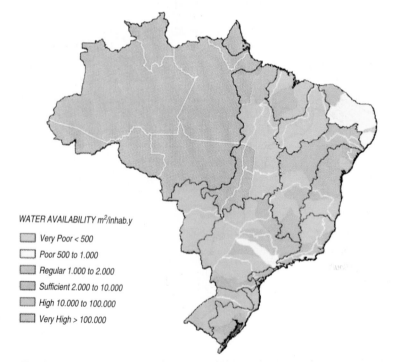

WATER AVAILABILITY m^2/inhab.y

☐ Very Poor < 500
☐ Poor 500 to 1.000
☐ Regular 1.000 to 2.000
☐ Sufficient 2.000 to 10.000
☐ High 10.000 to 100.000
☐ Very High > 100.000

Fig. 19.3: Availability of water according to the water resources and demands per capita in each subbasin in Brazil (ANA, 2002).

Considering the diversity of climates, topography, water resources versus demands, socioeconomic and cultural conditions, there are regions with an abundance of good quality water, semiarid regions subject to long periods of water scarcity and urban areas experiencing serious floods (ANA, 2002). A water availability map of Brazil is shown in Figure 19.3. Based on available information, different regions of Brazil are differently impacted by global changes as briefly illustrated below. For some regions, such as south and southeast Brazil, there is more information substantiating the impacts than for the other regions, such as north, northwest (Amazon forest) and northeast (semiarid) regions.

Particularly in Brazil, climate change can be seen as one of many pressures on the water resources. Many alterations in the hydrological systems have been described as a consequence of land-use and land-management practices, which often lead to deterioration in the resource baseline (Marques, 2001; Marques, 2002; Marques et al., 2002). For rivers in semiarid lands in Brazil, significant negative trends of river flow have been detected over time, but these variations seem to be related to consumption by agriculture and damming, rather than climate-induced changes (INRENA, 1994; Marengo, 1995; Marengo et al., 1997; Marengo et al., 1998; Marques et al., 2002).

Deforestation and changes in the hydrological cycle

The controversy raging over the role played by the forest on the regional hydrological cycle (Calder, 1999), notwithstanding, if the extent of deforestation were to expand to substantially larger areas, rainfall is expected to be reduced in the central-south, and southern regions of Brazil as a consequence of reduction in evapotranspiration (Lean et al., 1996). The Amazon Basin is shared with Brazil, Peru, Ecuador, Bolivia, Colombia, Venezuela and Guyana. More than half of this basin is located in Brazilian territory, but the headwaters located in the Andean zone belong to Bolivia, Peru, Ecuador and Colombia. The drainage area of the Amazon Basin measures 6,869,000 km^2. The headwaters are close to the Pacific Ocean, about 100 km, and run more than 6,000 km until they reach the Atlantic Ocean. About 15 tributaries measure more than 1,000 km in length. The Amazon discharge is about 220,800 m^3.s^{-1}, representing 10-20% of the total river discharge by rivers of the world. Amazon River is by far the world's largest river in terms of stream flow. The Amazon and its tributaries play an important role in the water cycle and water balance of much of South America. Model studies and field experiments show that about 50% of the rainfall in the region originates as water recycled in the forest (Lean et al., 1996). In the Amazon region, even small changes in evapotranspiration affect water vapor fluxes. Therefore, deforestation is likely to reduce precipitation because of a decrease in evapotranspiration, leading to important runoff losses in areas within and beyond this basin. Any reduction in rainfall is likely to affect not only the Amazon region but also Brazil's central-south region, where most of the country's agriculture and timber production are located. Nevertheless, an increase of stream flow since the 1970s has been detected in northwestern Amazon (Marengo et al., 1998) where deforestation is less conspicuous.

The deforestation process in the Amazon has been associated to a large extent with fire and expansion of the livestock and agriculture frontier, forming the "fire arch" (Fig. 19.4). With the extent of deforestation expanding to substantially larger areas (the annual deforestation rate during the period 1977-1998 varied from 0.31 to 0.81% per year, and was 0.48% during 1997-1998, according to INPE, 2000), reduced evapotranspiration will lead to less rainfall during dry periods in Amazon (high confidence level) and rainfall will be reduced in the central-west, central-south, and southern regions of Brazil (medium confidence level), according to Lean et al. (1996).

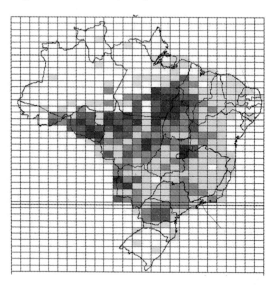

None
1-15 fire sites
16-32 fire sites
33-93 fire sites
94-444 fire sites

Fig. 19.4: Example of satellite monitoring of fires in Brazil. Period: 22-28 August, 2000. Total fire sites within the period: 10,606. A "fire arch" with 33-444 fire sites/ grid embracing the Amazon region from the southeast towards the forest is observed. The "arch" is coincident with the expansion of the livestock and agriculture frontiers (After Embrapa, 2000. Original satellite image: NOAA).

Although the annual rainfall total in Amazon would decrease by only 7% from conversion to pasture, based on simulations with the Hadley Centre model, in August (dry season) the average rainfall would decrease from 2.2 mm day^{-1} with forest to 1.5 mm day^{-1} with pasture-a 32% decrease (Lean et al., 1996). According to the same principle, large-scale deforestation of the Atlantic Rainforest constituting today only 7.3% (94,000 km^2) of the original forest (SOS Mata Atlântica and INPE, 2002) has probably affected the hydrological cycle, resulting in increased surface temperatures, decreased evapotranspiration and reduced precipitation during the last century. To confirm this hypothesis, more detailed investigation should be required, although in Brazil, stream flow records are usually short and very often data sets come from catchments with a long process of human intervention, particularly in the Atlantic Rainforest.

El Niño/La Niña

El Niño and La Niña are warming or cooling, respectively, phenomena occurring in a particular region of the equatorial Pacific Ocean over a three-month period. According to the newest U.S. National Oceanic and Atmospheric Administration-NOAA's new definition (NOAA, 2003), El Niño is a phenomenon in the equatorial Pacific Ocean characterized by a positive sea surface temperature departure from normal (for the 1971-2000 base period) in the Niño 3.4 region (120W-170W, 5N-5S) greater than or equal in magnitude to 0.5°C, averaged over three consecutive months and La Niña is characterized by a negative sea surface temperature departure from normal greater than or equal in magnitude to 0.5°C, averaged over three consecutive months. The definitions are based on an index of El Niño/La Niña water temperature extremes for the El Niño/Southern Oscillation (ENSO) cycle. El Niño/La Niña phenomena disrupt some of the "normal" climatic conditions.

During El Niño, the polar frontal systems are blocked and diverted eastward out to the Atlantic and trade winds are barred at the front (IPCC, 2001a). The blockage system extends from southern Peru to southern Brazil and its position oscillates in relation to the degree of enhancement of the subtropical jet stream and the intensity of the El Niño phenomenon. This situation provokes an anomalously high rainfall in the blocking zone and drought northward, as well as modification of the wind patterns, and consequently of the wind-driven littoral dynamics. ENSO leads to floods in southern South America (part of Argentina, Uruguay and southern Brazil), provoking the reverse effect (droughts) in the rest of the Brazilian coastal regions (Hastenrath and Heller, 1977). In other words, in northeast of Brazil, El Niño is associated with dry conditions, while the southern portion of Brazil (Rio Grande do Sul, Santa Catarina and Paraná states) exhibits anomalously wet conditions during

El Niño (IPCC, 2001a). Under the effect of La Niña, rainfall is significantly reduced in southern Brazil, while in the northern region rainfall increases. The northern Amazon follows the pattern of the northeast coast. The effects of El Niño and La Niña in Brazilian territory are presented in Table 19.2.

During the period 1982-1983, El Niño caused severe drought in northern Brazil and flooding in the southern Brazil. Droughts that led to forest fires were recorded during the very strong ENSO events of 1911-1912, 1925-1926, 1982-1983, and 1997-1998.

In the subtropical areas of Brazil, precipitation exhibits a long-term change, with a sharp increase recorded in the period 1956-1990 after a dry period 1921-1955 (Castañeda and Barros, 1994).

Since the 1960s an increase in stream flow has been recorded in southeastern South America, particularly in rivers less affected by anthropogenic pressures (Genta et al, 1998).

Although the annual rainfall total in Amazon would decrease by only 7% from conversion to pasture, based on simulations with the Hadley

Table 19.2: Variability and impacts of El Niño and La Niña in different Brazilian regions.

	Climatic/ Hydrological Variable	Region	Reference(s)	Observation Period
El Niño[a]	Severe droughts	Northeast Brazil	Silva Dias and Marengo (1999)	1901-1997
	Decrease in precipitation during rainy season	Northern Amazon Northeast Brazil	Aceituno (1988); Richey et al. (1989); Marengo (1992);	1931-1998
	Negative large anomalies of rainfall during rainy season	Northeast Brazil	Silva Dias and Marengo (1999); Hastenrath and Greischar (1993)	1930-1998 1912-1989 1849-1984
	High precipitation and flooding	Southern Brazil	Rebello (1997)	1982-1983
La Niña[b]	Increase in precipitation, higher runoff	Northern Amazon Northeast Brazil	Marengo et al. (1998) Meggers (1994)	1970-1997 Paleoclimate
	Severe droughts	Southern Brazil	Grimm et al. (1996 and 2000)	1956-1992

[a] Extremes of the Southern Oscillation (SO) are responsible in part for a large portion of climate variability at interannual scales in Latin America. El Niño (ENSO) events represent the negative (low) phase of the SO.
[b] La Niña is the positive (high) phase of the SO.

Centre model, in August (dry season) the average rainfall would decrease from 2.2 mm day^{-1} with forest to 1.5 mm day^{-1} with pasture-a 32% decrease (Lean et al., 1996). According to the same principle, large-scale deforestation of the Atlantic Rainforest constituting today only 7.3% (94,000 km^2) of the original forest (SOS Mata Atlântica and INPE, 2002) has probably affected the hydrological cycle, resulting in increased surface temperatures, decreased evapotranspiration and reduced precipitation during the last century. To confirm this hypothesis, more detailed investigation should be required, although in Brazil, stream flow records are usually short and very often data sets come from catchments with a long process of human intervention, particularly in the Atlantic Rainforest.

Impacts on ecosystems and fish stocks

On decadal to centennial timescales, changes in precipitation and runoff may have significant impacts on mangrove forest communities (IPCC, 2001a). There are indications that major declines of sardine stocks previously attributed exclusively to overexploitation in some years (e.g. 1975 and 1987) were mainly due to oceanographic anomalies, such as less intense intrusions of the northward flowing colder, nutrient-rich South Atlantic Central Waters (SACW) onto the inner shelf and coastal regions. The process is controlled by atmospheric/oceanic conditions of regional scale. The relative contribution of both climate changes and overexploitation of fisheries to the declines in sardine stocks in south/southeast Brazilian coast is not fully–understood (Rossi-Wongtschowski et al., 1996).

Garcia and Vieira (2001) analyzed the species composition and species diversity in the Patos Lagoon estuary (southern Brazil, Rio Grande do Sul state) before, during, and after the 1997-1998 El Niño. A total of 20 hauls were made monthly at four beach stations from August 1996 to August 2000, using beach seine hauls. Species were grouped as follows: (i) estuarine resident, (ii) estuarine dependent, (iii) marine vagrant, and (iv) freshwater vagrant. Species diversity was evaluated by the H' index, species richness by the rarefaction method and evenness by the Evar index (see Garcia and Vieira, 2001). Confidence intervals were obtained by the bootstrap method. Since El Niño phenomenon causes higher than average rainfall in southern Brazil and directly affects river discharge, the salinity in estuaries changes. During the event, rainfall exceeded the average and salinity was lower than average (Garcia and Vieira, 2001). Fish species diversity was higher in Patos Lagoon estuary during the EN due mainly to an increment in the number of freshwater species, and somewhat to an increase in species evenness.

Socioeconomic impacts associated with El Niño/La Niña

In southern Brazil the economic impact due to flooding caused by El Niño is severe and affects many sectors; however, these impacts are sporadic and associated to the events per se. The sectors most affected by flooding are: services, urban/housing, transport, industry, agriculture and fishing. Civil engineering works for flood prevention and recovery of affected areas represent considerable costs to the local, state and eventually, federal government, and also contribute to the economic impacts. Plantation forestry is a major land use in Brazil and is expected to expand substantially over coming decades (Fearnside, 1999). Climatic changes can be expected to reduce silvicultural yields to the extent that the climate becomes drier in major plantation states such as Minas Gerais and Espírito Santo (eastern Brazil) among other regions, as a result of global warming and/or reduced water vapor transport from the Amazon (Eagleson, 1986). Dry-season changes can be expected to have the greatest impact on forest yields. UKMO model results (Gates et al., 1992) indicate that annual rainfall changes would cause yields to decrease by 8% in southern Brazil and increase by 4% in the northeastern Brazil. During the June-August period, yields would decrease 14% in southern Brazil, and 21% in the northeast (Fearnside, 1999). Periodic occurrences of severe droughts associated with El Niño in the northeast agricultural region have resulted in occasional famines (Kiladis and Díaz, 1986). Severe food shortages occurred in this region in 1988 and 1998 (Kovats et al., 1998).

During the last 2 decades, the strongest effects from El Niño in the Brazil Current were experienced in 1982-1983, 1986-1987 and 1997-1998 (Rebello, 1997). Southern Brazil suffered the most severe socioeconomic impacts (Rebello, 1997). Three states were badly impacted: Rio Grande do Sul, Santa Catarina and Paraná. Of the total area of the state of Santa Catarina (95,000 km^2), 79% were affected by flooding, including 135 cities, and evacuation of 300,000 inhabitants. Taking only the city of Blumenau, the magnitude of the flooding event is illustrated by the following indicators (Defesa Civil de Blumenau, 1997 in Rebello, 1997): (i) 31 days of duration-from the 6th July to 5th August, 1983, (ii) July, 9th 1983, the Itajaí river reached 15.34 m above the mean value, (iii) previous experience of flooding at this level of severity was recorded in 1911, (iv) 50,000 inhabitants moved out from their homes and 38,000 households affected, (v) recovery costs of USD 8.7 billion (estimated in 1997).

The harvesting losses in 1983 due to the flooding in different states were (Rebello, 1997): (i) State of Rio Grande do Sul with 693,777 ton; (ii) State of Santa Catarina with 1,626,298 ton; (iii) State of Paraná with 1,568,700 ton.

The flood in 1983 was considered a result of synergy between El Niño and stream flow changes due to and erosion and silting up (Rebello, 1997).

Greenhouse effect

It has been foreseen a temperature increase in Brazil of 1-6°C, according to the emissions scenario (Nobre, 2001; de Siqueira et al., 1994, 1999). Agriculture is impacted by temperature increase in different ways: (i) it reduces crop yields by shortening the crop cycle (e.g. maize, wheat, barley, grapes) and (ii) it reduces fishing and forestry productivity.

Lack of consistency in the various Global Spectral Model precipitation scenarios (Roads et al. 1999) renders a precise assessment difficult of crop production under climate change, even when the relationships between precipitation and crop yields are well known. Some of the relatively weak cold surges may exhibit unusual intensity, causing frosts and low temperatures in coffee-growing areas of southeastern Brazil, resulting in heavy damage and losses in coffee production (Marengo et al., 1997). Global warming and sea-level rise are likely to influence the hydrological cycle, agricultural yield and threaten human health and property security (IPCC, 2001a). However, complex climatic patterns, which result in part from interactions of atmospheric flow with topography, intermingled with land-use and land-cover change, make it difficult to identify common patterns of vulnerability to climate change in a given region (IPCC, 2001a).

Global warming and regional climate change resulting from land-cover modification may be acting synergistically to exacerbate stress over the region's tropical ecosystems (IPCC, 2001a). In several cities in south/ southeast Brazil, studies on long-term time series for temperature, from the beginning of the 20th century, indicate warming tendencies (Sansigolo et al., 1992). This could be attributable to urbanization effects or to systematic warming observed in the South Atlantic since the beginning of the 1950s (Venegas et al., 1996, 1998).

Temperature anomalies related to the El Niño event and coral bleaching has long been observed along the Brazilian coast. In Abrolhos archipelago (an environmental protected area in the littoral of Bahia state, northeast Brazil) two episodes of coral bleaching have been recorded as related to temperature rise of the sea surface (Leão, 1999). The first episode was during a sea surface temperature anomaly in the summer of 1994 when 51-88% of colonies of the genus *Mussismilia* were affected (Castro and Pires, 1999 in Leão, 1999). The second one was related to the strong El-Niño Southern event that began by the end of 1997 in the Pacific Ocean, and also caused a rise of the sea surface temperature on the eastern coast of Brazil. The most affected species were *Porites branneri* and *Mussismilia hispida*, both with more than 80% of their colonies totally bleached, *M. harttii* with an average of 75% of its colonies affected, and *Porites asteroides* with all colonies showing some signs of bleaching (Kikichi, 1999 in Leão, 1999). According to the authors, although the specie *Agaricia agaricites* did not show a totally bleached colony, more than 90% of them were pale in color.

One methodological approach recently applied for assessing economic impacts due to global changes is to estimate the aggregate monetized impact, based on current economic conditions and populations, for a 1.5-2.5 °C temperature increase (Tol et al., 2000 in Beg et al., 2002). According to such assessment, developing countries such as Brazil have greater economic vulnerability to climate change. At lower levels of climate change, damages might be mixed across regions; for example, poorer countries are likely to be net losers and richer countries might gain from moderate warming.

Changes in the sea level

Worldwide impoundment of water has reduced sea levels; moreover, the concentration of water in reservoirs at high latitudes has increased, albeit minutely, the speed of the earth's rotation and changed the planet's axis (Chao, 1991, 1995). Even though, according to IPCC (2001a), a slightly rising rate of global warming and sea levels has been observed in the world, modeling indicates that there will be an increase in frequency and intensity of atmospheric fronts, leading to growing problems of coastal erosion because of so-called spring tides, a phenomenon which is most evident in the southern part of South America, including southern Brazil. In southern Brazil, many engineering constructions along the coast have been destroyed during the last decades. The sea level along the Brazilian coast has been subjected to changes between 18-40 cm per century (IPCC, 2001a) and there has been a sea level rise in the middle section of the Brazilian coast, which is found to be within the normal worldwide variation and changes in ocean streams. Studies of vulnerability to sea-level rise (Perdomo et al., 1996) have suggested that southern Brazil and Uruguay could suffer adverse impacts, leading to losses of coastal land and biodiversity, saltwater intrusion, and infrastructure damage. Impacts likely would be multiple and complex, with major economic implications. Rising sea-level may eliminate mangrove habitat at an approximate rate of 1% per year (IPCC, 2001a). Decline in some of the region's fisheries at a similar rate would be observed because most commercial shellfish use mangroves as nurseries or refuges. Coastal inundation stemming from sea-level rise or flatland flooding resulting from climate change may seriously affect mangrove ecology and associated human economy (IPCC, 2001a).

Economic activities such as tourism and fishing, settlements and structures are particularly vulnerable to physical changes associated with sea-level rise. Tourism—one of the most prominent industries in Brazil, particularly in the coastal zone—is one economic sector to be severely impacted by sea-level rise. Protection, replenishment, and stabilization of existing beaches would represent a relevant contribution to the economic impacts. However, it is difficult to separate the impact of climate-induced sea-level rise from erosion resulting from changes in the sediment transport dynamics due to damming associated with the continuous effect of the sea on the coast.

Climate changes and health impacts in Brazil

The magnitude of the impacts of climate change on health primarily depends on the size, density, location, and wealth of the population. It is likely that extreme weather events increase death and morbidity rates (injuries, infectious diseases, social problems, and damage to sanitary infrastructure) particularly in developing countries (IPCC, 2001a). There is evidence that the geographical distribution of vector-borne diseases in Brazil (e.g. malaria, dengue) change when temperature and precipitation increase. El Niño and La Niña are expected to cause changes in vector populations and the incidence of waterborne diseases in Brazil on long time scales (IPCC, 2001a). The exact distribution of these diseases, however, is not clear. Additionally, floods represent risks mainly to the population that live in risky areas, not only from safety but also the health viewpoint, because of waterborne diseases. Areas of high risk include all poor settlements and *favelas* on hill slopes and river banks in metropolitan areas of the south, southeast and east Brazil. Flood and drought periods are related to migrations or relocations, giving rise to social and community impacts, which affect a significant part of population. The severity of the impact is high during events but the duration short and not frequent, since episodes of population displacement from areas affected are usually followed by population return to the original area. According to IPCC (2001a), under climate change conditions, subsistence farming might be severely impacted in areas such as the Brazilian northeast. In many coastal societies, cultural values are associated with the use of a wide range of natural products from the coastal wetlands and surrounding waters (Field, 1997). Sea level change and erosion promote changes in the coastal environment.

Other global changes in Brazil

In Brazil, there is no clear evidence of increasing effects of UV-b radiation on marine or freshwater organisms as an effect of ozone layer destruction. However, the destruction of ozone layer would probably increase the effects of UV-b radiation and more research is necessary to assess possible impacts due to the ozone layer destruction. Additionally, changes in CO_2 source/sink function of aquatic systems have not been properly assessed in Brazil, although on a global level, modifications in the CO_2 exchange between the atmosphere, the land and the sea have been investigated as well as the theory that global changes interfere in ocean's capacity to release/capture CO_2 (IPCC, 2001b).

Discussion

In a worldwide perspective, general worsening on environmental as well as socioeconomic impacts is foreseen due to global changes (Nobre, 2001). The following comments refer to the expected impacts in a long-term perspective, as has been discussed for Latin America (IPCC, 2001a; Nobre, 2001).

Among the global climate changes foreseen for the next 100 years, in a "business as usual" scenario regarding greenhouse effect, the most significant ones for Brazil are: temperature and sea level rise, changes in the rainfall regime and alterations in the distribution of extreme climate events, such as drought and flooding (Nobre, 2001). However, insufficient information about the other global change issues makes impossible construction of an overview picture of the impacts global changes are causing in Brazil.

According to Beg et al. (2002), although climate change does not feature prominently within the environmental or economic policy agenda of developing countries, evidence shows that some of the most adverse effects of climate change are expected to occur in developing countries, where populations are most vulnerable and least likely to easily adapt to climate change. Climate change could exacerbate current inequalities due to uneven distribution of damage costs, in addition to the cost of mitigation and adaptation efforts (Beg et al., 2002).

Although climate change may bring benefits for certain regions of Latin America (e.g. rainfall intensification in the Brazilian semiarid region during La Niña), increasing environmental deterioration, combined with changes in water availability and agricultural lands, may reduce these benefits to a negligible level (IPCC, 2001a). The adaptive capacity of socioeconomic systems in Latin America is very low, particularly with respect to extreme climate events, and vulnerability is high. Some economic and health problems could be exacerbated in critical areas, fostering migrations from rural and small urban settlements into major cities and giving rise to additional stress at the national level and, at times, adversely affecting international relations between neighboring countries (IPCC, 2001a). Therefore, under climate change conditions, the risks for human health may increase.

Some synergies already exist between climate change policies and the new sustainable development agenda in some developing countries, such as energy efficiency, renewable energy, transport and sustainable land-use policies (La Rovere, 2002). In Brazil, renewable energy production and efficiency improvements in energy use in the 1980s have made a significant contribution to reducing GHG emissions. The program of energy efficiency improvements in the use of electricity (PROCEL) alone, has led to significant GHG emission mitigation (La Rovere, 2002). Rovere predicts that changes in rainfall patterns and ENSO-induced climate change may further affect the already limited availability of water resources and aggravate the risk of famines due to the disruption of agricultural and cattle-raising activities. Understandably, vulnerability to the adverse impacts of climate change is one of the most crucial concerns of developing countries engaged in climate policy discussions, including Brazil. It is also a critical element in planning any long term climate and development strategy. According to IPCC (2001a), climate change does not in itself stimulate development of new adaptive strategies, but it encourages a more adaptive, incremental, risk-based approach to water management.

Conclusion

Precipitation changes in Latin America do not follow a consistent trend. In the Amazon region, the most important finding is the presence of periods with relatively wetter or drier conditions that are more relevant than any unidirectional trend (Nobre, 2001). Rainfall in north/northeast Brazil exhibits a weak positive trend. Precipitation in southern Brazil increased abruptly in the 1956-1990 period after a dry period from 1921 to 1955. In the Pampas, there was a positive trend in precipitation over the 1890-1984 period. The Southern Oscillation is responsible for a large part of the climate variability at interannual scales in Latin America. El Niño is associated with dry conditions in northeast Brazil and northern Amazon region, whereas southern Brazil exhibit anomalously wet conditions. On the other hand, drought occurs in southern Brazil during the positive phase of the Southern Oscillation. Additionally, although no conclusive evidence can be given, aquatic productivity has been altered as a result of global phenomena such as ENSO events. If El Niño or La Niña were to increase, Brazil, as well as Latin America would be exposed to these conditions more often. In Brazil, dramatic changes in land use, particularly in forest coverage, erosion rate, damming and sediments silting/transport dynamics make identification of the relative contribution of global changes to the observed impacts difficult, particularly for trends on stream flow. The regional temperature rise due to deforestation might act synergistically with the global greenhouse effect, reducing precipitation and leading to fires, particularly in the Amazon Rainforest. Since some of the most adverse effects of climate change will be in developing countries, such as Brazil, due to the environmental and population vulnerability, strategic assessment to predict the impacts of these changes should be one of the most relevant topics in the environmental policy agenda of these countries during the coming years.

References

Aceituno, P. 1988 On the functioning of the Southern Oscillation in the South American sector, Part I: Surface climate. Monthly Weekly Review, 116: 505-524.

ANA. 2002 *Overview of Hydrographical Regions in Brazil*. National Agency of Water ANA Report. Brasília.

Beg, N., Morlot, J.C., Davidson, O., Afrane-Okesse, Y., Tyani, L., Denton, F., Solona, Y., Thomas, J.P., La Rovere, E.L., Parikh, J.K., Parikh, K. and Rahman, A.A. 2002 Linkages between climate changes and sustainable development. Climate Policy 2: 129-144.

Calder, I.R. 1999 *The Blue Revolution: Land Use and Integrated Water Resources Management*. ISBN 1-85383-634-6, 192 pp.

Castañeda, M.E. and Barros, V.R. 1994 Las tendencias de la precipitación en el Cono Sur de América al este de los Andes. *Meteorológica*, 19, 23-32.

Chao, B.F. 1991 Man, water and global sea level. *EOS, Trans. Brazil Amer. Geophys. Union* 72: 492.

Chao, B.F. 1995 Anthropogenic impact on global geodynamics due to reservoir water impoundment. *Geophys. Res. Lett.* 22: 3529-3532.

Compagnucci, R., da Cunha, L., Hanaki, K., Howe, C., Mailu, G., Shiklomanov, I. and Stakhiv, E. 2001 Chapter 4. Hydrology and Water Resources. In: *Climate Change 2001: Impacts, Adaptation and Vulnerability*. UNEP, WMO. Cambridge.

de Siqueira, O.J.F., Farias, J.R.B. and Sans, L.M.A. 1994 Potential effects of global climate change for Brazilian agriculture: Applied simulation studies for wheat, maize and soybeans. In: Implications of Climate Change for International Agriculture: Crop Modelling Study. C. Rosenzweig and A. Iglesias (eds.). EPA 230-B-94-003, U.S. Environmental Protection Agency, Washington, DC, USA.

de Siqueira, O., Salles, L. and Fernandes, J. 1999 Efeitos potenciais de mudanças climáticas na agricultura brasileira e estratégias adaptativas para algumas culturas. In: *Memórias do Workshop de Mudanças Climáticas Globais e a Agropecuária Brasileira, 1-17 de Junho 1999, Campinas, SPO, Brasil*. Brasil, pp. 18-19. Ministério do Transporte. 2002. Avaliação da contribuição da navegação do SF ao incremento da competitividade da agricultura na bacia. Projeto de Gerenciamento Integrado das Atividades em terra na bacia do Rio São Francisco (ANA, GEF, PNUMA,OEA) Departamento de Hidrovias, MT.

Eagleson, P.S. 1986 The emergence of global-scale hydrology. *Water Resources Res.* 22 (9): 6-14.

Embrapa. 2000 Monitoramento por Satélite: Monitoramento Orbital de Queimadas – Brasil 22-28 de Agosto de 2000. Brasília.

Fearnside, P.M. 1999 Plantation forestry in Brazil: the potential impacts of climatic change. *Biomass and Bioenergy*, 16(2): 91-102.

Field, C.D. 1997 *La restauración de ecosistemas de manglar*. Sociedad Internacional para Ecosistemas de Manglar. Okinawa, Japan.

Garcia, A.M. and Vieira, J.P. 2001 O aumento da diversidade de peixes no estuário da Lagoa dos Patos durante o episódio El Niño 1997-1998. *ATLÂNTICA*, Rio Grande, 23: 133-152.

Gates, W.L., Mitchell, J.F.B., Boer, G.J. , Cubasch, U. and Meleshko, V.P. 1992. Climate modelling, climate prediction and model validation. In: *Climate Change 1992: The Supplementary Report to the IPCC Scientific Assessment*. J.T. Houghton, B.A.

Callander and S.K. Varney (eds.). Cambridge University Press, Cambridge, UK, pp. 97-134.

Genta, J.L., Perez-Iribarren, G. and Mechoso, C.R. 1998 A recent increasing trend in the streamflow of rivers in southeastern South America. *Journal of Climate*, 11, 2858–2862.

Grid-Arendal. 2001 *Vital Climate Graphics.* United Nations Environmental Program UNEP. ISBN 8277010095. 65p.

Grimm, A., Teleginsky, S.E. and Freitas, E.E.D. 1996 Anomalias de precipitação no Sul do Brasil em eventos de El Niño. In: *Congresso Brasileiro de Meteorologia*, 1996, Campos de Jordão. Anais, Sociedade Brasileira de Meteorologia, Brasil.

Grimm, A.M., Barros, V.R. and Doyle, M.E. 2000 Climate variability in southern South America associated with El Niño and La Niña events. *J. Climate*, 13, 35-58.

Hastenrath, S. and Heller, L. 1977 Dynamics of climatic hazards in north-east Brazil. *Quart. J. Roy. Meteo. Soc.*, 103: 77-92.

Hastenrath, S. and Greischar, L. 1993 Further work on northeast Brazil rainfall anomalies. *J. Climate*, 6: 743-758.

IBGE. 2000 Brazil in Figures. *Instituto Brasileiro de Geografia e Estatística* IBGE. 367 pp. ISSN 0103-9288. Rio de Janeiro.

IPCC. 2001a Chapter 14: Latin America. In: *Climate Change 2001: Impacts, Adaptation and Vulnerability.* Inter-Governmental Climate Change Panel IPCC. UNEP-WMO. Cambridge University Press. Cambridge.

IPCC. 2001b *Climate Change 2001: The Scientific Basis.* Inter-Governmental Climate Change Panel IPCC. UNEP-WMO. Cambridge University Press. Cambridge.

INRENA. 1994 Descargas de los Rios y Almacenamiento de Reservorios y represas de la Coasta Peruana, 88pp. *Instituto Nacional de Recursos Naturales* IRENA. Minist. de Agricultura, Lima.

Kiladis, G.N. and Díaz, H.F. 1986 An analysis of the 1877-78 ENSO episode and comparison with 1982-83. *Monthly Weather Review*, 114: 1035-1047.

Kovats, S., Patz, J.A. and Dobbins, D. 1998 Global climate change and environmental health. *Int. J. Occup. Environ. Health* 4(1): 41-51.

La Rovere, E.L. 2002 Climate change and sustainable development strategies: a Brazilian perspective, Paper commissioned by the OECD. Federal University of Rio de Janeiro, http://www.oecd.org/env/cc.

Lean, J., Bunton, C.B. and Rowntree, P.R. 1996 The simulated impact of Amazonian deforestation on climate using measured ABRACOS vegetation characteristics. In: *Amazonian Deforestation and Climate* J.H.C. Gash, C.A. Nobre, J.M. Roberts and R.L. Victoria (eds.). John Wiley Sons, Chichester, UK. pp. 549-576.

Leão, Z.M.A.N. 1999 Abrolhos-The South Atlantic largest coral reef complex. In: C. Schobbenhaus, D.A. Campos, E.T. Queiroz, M. Winge and M. Berbert-Born (eds.) *Sítios Geológicos e Paleontológicos do Brasil.* CPRM, DRM. Rio de Janeiro.

Marengo, J. 1995 Variations and change in South American streamflow. *Climate Change*, 31: 99-117.

Marengo, J., Nobre, C.A. and Sampaio, G. 1997 On the associations between hydrometeorological conditions in Amazonia and the extremes of the Southern Oscillation. In: *Consecuencias Climáticas e Hidrológicas del Evento El Niño a Escala Regional y Local.* Extended Abstracts of Memorias Técnicas–Seminario Internacional. 26-29 Nov 1997, Quito. pp. 257-266.

Marengo, J.A., Tomasella, J. and Uvo, C.B. 1998 Trends in streamflow and rainfall in tropical South America: Amazonia, eastern Brazil and northwestern Peru. *J. Geophysical Res.*, 103 (D2):1775-1783.

Marengo, J.A., Tomasella, J. and Uvo, C.B. 1998 Trends in streamflow and rainfall in tropical South America: Amazonia, eastern Brazil and northwestern Peru. *J. Geophy. Res.* 103 (D2): 1775-1783.

Marques, M. 2001 *Scaling and Scoping Report Sub-region 39 Brazil Current.* Global International Waters Assessment GIWA UNEP/GEF, 185pp.

Marques, M. 2002 Causal chain analysis applied to transboundary waters: The pollution concern in Paraíba do Sul River Basin, Brazil. In: *First Int. Symp. Transboundary Waters Management.* Monterrey. Proc. ISTWM, v. I, p. 623-630. Assoc. Mex. Hidrologia, Monterrey, Mexico.

Marques, M., Knoppers, B., Machmann Oliveira, A. 2002 Transboundary basin of São Francisco, Brazil: Environmental impacts and causal chain analysis. In: *First Int. Symp. Transboundary Waters Management.* Monterrey. Proc. ISTWM, v I, p. 487-494.Assoc. Mex. Hidrologia, Monterrey, Mexico.

Meggers, B.J., 1994 Archaeological evidence for the impact of mega-Niño events on Amazon during the past two millennia. *Climatic Change,* 28(1-2): 321-338.

Nobre, C. 2001 Mudanças climáticas globais: possíveis impactos nos ecossistemas do país. *Parcerias Estratégicas,* 12: 239-258. Sep 2001.

NOAA. 2003 *NOAA Gests U.S. consensus for El Nino/La Nina Index definitions.* National Oceanic and Atmospheric Administration. September 30, 2003.

Perdomo, M., Olivo, M.L. M.L., Bonduki, Y., and Mata, L.J. 1996 Vulnerability and adaptation assessments for Venezuela. In: *Vulnerability and Adaptation to Climate Change.* J.B. Smith, S. Huq, L.J. Lenhart, L.J. Mata, I. Nemesova and S. Toure (eds.). Kluwer Acad. Publ., Dordrecht, Netherlands, pp. 347-366.

Rebello, E. 1997 Anomalias climáticas e seus impactos no Brasil durante o evento "El Nino" de 1982-83 e previsão para o evento El Niño de 1997-1998. In: *Consecuencias Climáticas e Hidrológicas del Evento El Niño a Escala Regional y Local.* Extended Abstracts of Memorias Técnicas–Seminario Internacional. 26-29 Nov 1997, Quito.

Richey, J., Nobre, C. and Deser, C. 1989 Amazon River discharge and climate variability: 1903 to 1985. *Science,* 246: 101-103.

Roads, J., S. Chen, M. Kanamitsu, H. Juang, 1999 Surface water characteristics in NCEP global spectral model reanalysis. *J. Geophys. Res.,* 104, 19307-19327.

Rossi-Wongtschowski, C. L. D. B., Saccardo, S. A. and Cergole, M. C. 1996 Are fluctuations in Brazilian sardine catches related to global-scale climate changes? *An. Acad. Bras. Ci.* 68 (supl. 1): 239-250.

Sansigolo, C., Rodriguez, , R. and Etchichury, P. 1992 Tendências nas temperaturas médias do Brasil. *Anais do VII Congresso Brasileiro de Meteorologia,* 1: 367-371.

SOS Mata Atlântica and INPE. 2002 *Atlas dos Remanescentes Florestais da Mata Atlântica 1995-2000. Situação original (Domínio da Mata Atlântica do Decreto Federal 750/93) e Situação atual.* SOS Mata Atlântica e Instituto National de Pesquisas Espaciais, INPE. São Paulo.

Silva Dias, P. and Marengo, J. 1999 Águas atmosféricas. In: *Águas Doces no Brasil-capital Ecológico Usos Múltiplos, Exploração Racional e Conservação.* A. Cunha Rebouças, B. Braga Jr. and J.G. Tundizi (eds.). IEA/USP, p. 65-116.

Venegas, S., L., Mysak, L., and Straub, N. 1996 Evidence for interannual and interdecadal climate variability in the south Atlantic. *Geophy. Res. Let.* 23: 2673-2676.

Venegas, S., Mysak, L. and Straub, N. 1998 Atmosphere-ocean coupled variability in the south Atlantic. *J. Climate,* 10: 2904-2920.

Section v
BEYOND KYOTO

The Future of Global Climate Change Policy: Developing Country Priorities After Kyoto

Adil Najam

Fletcher School of Law and Diplomacy, 160 Packard Avenue, Medford, MA 02215, USA; Fax: 617 627 3005; Phone : 617 627 2706 email: adil.najam@tufts.edu

Introduction

The Sixth and Seventh Conferences of the Parties (COP-6 and COP-7) to the United National Framework Convention on Climate Change (UNFCCC) managed to resuscitate what had seemed to be a sinking Kyoto Protocol, despite the US decision to abandon the agreement. However, the subsequent COPs (COP-8 and COP-9) have demonstrated that these changes still leave the Protocol riddled with all the many problems that had dogged the original agreement while further diluting its content significantly.

While the survival of the Protocol may be something to celebrate, from the perspective of the developing countries of the South, the Protocol — which had been imperfect to begin with — is now all the more imperfect (Agarwal et al., 2001; Najam and Page, 1998; Najam, 2001). The longer the coming into force is delayed, the less there is to celebrate.

So, where do we stand at the fifth anniversary of the Kyoto Protocol? Is the world's climate any better off than it was five years ago? Are the developing countries any better off? Unfortunately the answer to the last two questions has to be negative, and resoundingly so. However, while much precious time has been lost, some may still be salvaged. This paper will take stock of the Kyoto Protocol at its fifth anniversary from a developing country perspective.

This paper is based on the author's earlier work, including that done with Youba Sokona, Saleemul Huq, Atiq Rahman and William Moomaw. The author acknowledges his gratitude to all for having helped shape these ideas.

The key argument is that the Kyoto process has been focused—even obsessed—with the short-term need to launch the policy process and get the industrialized countries to agree to some targets, no matter how meager. It is time now to refocus on the long-term objectives of the UNFCCC, particularly on its stated goals regarding sustainable development. In this regard, the developing countries are now confronted by both challenges and opportunities. The challenges emerge from the fact that developing country concerns, which had always been marginal to the thrust of the UNFCCC, have become even more marginalized in recent negotiations, as focus has concentrated on getting the Northern countries (those listed in Annex 1) to accede to the Kyoto Protocol (see Table 20.1). This has happened at the cost of sidestepping, if not outright ignoring, Southern priorities (Najam, 2001; Sokona, 2001). On the other hand, the supposed revival of a sustainable development focus from the Johannesburg Summit gives the developing countries an opportunity to reestablish the link between climate change and sustainable development (Huq and Sokona, 2001). Such a link is enshrined in the text of the UNFCCC but has been systematically ignored in its operational provisions; most especially in the Kyoto Protocol (Najam and Sagar, 1998).

A review of the concerns expressed by the developing countries of the South about the direction the global climate change regime is taking is presented first. Then certain key Southern interests for future negotiations on the subject are identified. We conclude by arguing that there is a need to launch a process that revitalizes the global climate change debate and make the regime more inclusive as well as more efficacious.

Southern Concerns

The original UNFCCC was not exactly viewed as a great victory by the developing countries (Dasgupta, 1994; Hyder, 1994; Rajan, 1997; Sagar and Kandlikar, 1997). Since then, the climate regime has become even less sympathetic to the concerns of the South (Agarwal et al., 2001; Huq and Sokona, 2001; Najam, 2001). This has largely been a case of neglect and inattention, rather than outright assault. For the most part, it has been a direct result of the overwhelming preoccupation by policy-makers, scholars, and activists with getting Annex 1 countries to agree, and then accede, to the Kyoto Protocol. In focusing on this short-term objective, the longer term goals of the UNFCCC – especially those related to sustainable development – have tended to slip. The result has been a systematic marginalization of the core interests of the developing countries.

While developing country governments and scholars have raised a number of specific concerns regarding the direction in which the global climate regime has evolved, these relate generally to three large categories of concerns:

Table 20.1. Kyoto commitments and climate-negotiation-related indicators (for 1990) for selected countries

Kyoto commitment, (percent relative to base period emissions)	Selected countries	Cabon dioxide emissions (000' tons)	GNP per capita ($)	Carbon dioxide emissions per capita (tons)	Carbon dioxide emissions per unit GDP (tons per 000' $)
		Selected Annex-1 Countries			
108	Australia	2,88,965	16,516	16.91	0.98
94	Canada	4,62,643	19,705	17.44	0.81
92	Czech Republic	1,65,792	3,049	16.00	5.25
92	France	3,66,536	20,966	6.49	0.31
92	Germany	10,14,155	21,861	12.76	0.59
92	Ireland	30,719	11,349	8.77	0.68
94	Japan	11,55,000	24,205	9.35	0.39
92	Luxembourg	11,343	34,614	30.41	1.10
92	Netherlands	1,74,000	18,939	11.22	0.61
100	New Zealand	25,476	12,192	7.61	0.59
101	Norway	35,514	26,387	8.37	0.31
94	Poland	4,14,930	1,459	10.87	7.04
92	Portugal	42,148	6,774	4.00	0.63
100	Russian Federation	23,88,720	3,897	16.11	4.13
92	United Kingdom	5,77,012	16,723	10.08	0.59
93	United States	49,57,022	22,046	19.83	0.90
		Selected non-Annex-1 countries			
	Brazil	1,97,905	2,877	1.34	0.41
	China	23,40,635	313	2.06	6.60
	India	6,81,248	347	0.80	2.28
	Indonesia	2,13,422	613	1.20	1.87
	Iran	1,87,986	2,139	3.37	1.56
	Mexico	3,13,826	2,856	3.76	1.27
	Republic of Korea	2,43,434	5,875	5.68	0.96
	South Africa	2,94,107	2,758	7.93	2.76

Annex-1 country data from the UNFCCC (FCCC/CP/1996/12/add.2) and *World Development Indicators 1997* (World Bank, Washington, DC). The German GNP and GDP numbers are for 1991 due to the nonavailability of 1990 figures.
Non-Annex-1 country data from Marland G. and Boden, T.A. *1997. Trends: A Compendium of Data on Global Change*. Carbon Dioxide Information Analysis Center, Oak Ridge National Laboratories, Oak Ridge, TN; and *World Development Indicators 1997* (World Bank, Washington, DC).
The emissions data for the Annex-1 countries and the non-Annex-1 countries may not be exactly comparable due to differences in the methodologies utilized for estimating these sets of data.

> ➤ First, the principle of equity – both inter- and intragenerational – which was so central to the discussions of global climate change until the adoption of UNFCCC, has been sidelined in the discourse since then, especially since the Kyoto agreement.

> ➤ Second, the focus of the regime has become skewed toward minimizing the burden of implementation on polluter industries and countries, instead of giving priority to the vulnerabilities of the communities and countries at greatest risk and disadvantage.
> ➤ Third, the regime has now distinctly become a regime for managing the global carbon trade and has lost sight of its original mandate of stabilizing atmospheric greenhouse gas concentrations.

Issues of **equity and responsibility between and within generations** have been among the central themes in the policy as well as scholarly discussions on global climate change (see Weiss, 1989; Agarwal and Narain, 1991; Jamieson, 1992; Gadgil and Guha, 1995; Shue, 1995; Banuri and Sagar, 1999; Meyer, 1999; Baer et al., 2000; Carraro, 2000; Munasinghe, 2000). The discussion on this issue was particularly heated during the years leading up to the UNFCCC. Although it still figures as a recurrent theme in the scholarly literature, it seems to have lost its salience in the policy discourse. Indeed, equity seemed to be among the first casualties of the Kyoto process, where even the pretence of some form of equity between emission reduction targets was quickly abandoned amid the arbitrariness and global horse-trading on which the agreement was ultimately based (Reiner and Jacoby, 1997; Najam and Sagar, 1998). While intragenerational equity was always deemed a problematic notion by Northern policy-makers, even the lip service that had routinely been paid to intergenerational equity seems to have gone out of fashion.

The abandonment of the equity principle – particularly with respect to the least developed countries, and particularly in the context of the related principle of "common but differentiated responsibility" – is of grave concern to the South. Indeed, the essence of the term equity has been convoluted by the US Congress demanding "equity" between the percentage emission cuts for Annex 1 countries and their developing country counterparts. It is sad that in the very same breath the US is both willing and able to deny any call for equity in emissions themselves. As the desire for efficiency overwhelms both equity and responsibility the distinction between "luxury" and "survival" emission is lost and any discussion of global or generational fairness becomes all but moot (Agarwal and Narain, 1991).

The third assessment report of the Intergovernmental Panel on Climate Change (IPCC; see especially Working Group II report) has made it abundantly clear that even if the Kyoto Protocol were implemented in full, the impacts of global climate change would start being felt within the next few decades and that **the most vulnerable communities and countries are those which are already the poorest and least able to adapt to these changes**. The threat is especially pressing for the least developed countries (LDCs) and the small island developing countries

(SIDs), where any economic development they may be able to achieve in the next few decades is in real danger of literally being swept away due to human induced climate change. In the past, climatic disasters such as floods, cyclones and droughts may have been attributable to nature alone; in future they will definitely have a component that is human induced. More importantly, it is also clear that the contribution of these countries to the climate change problem is miniscule. The result is that those who have been least responsible for creating the crisis are most at risk by its ravages (Rayner and Malone, 1998; Banuri and Sagar, 1999).

The reconvened sixth COP at Bonn did agree to set up a number of funds including a Climate Change Fund (to capacity building and transfer technology) and the LDC Fund (to assist LDCs in climate change adaptation). While the intent of these funds is noble, it is difficult to place too much confidence in their potential, because: (a) they are voluntary, (b) they are to be managed via the still-controversial Global Environmental Facility (GEF), and (c) even after a few years they remain poorly funded (Huq and Sokona, 2001). Similarly, the solution proposed in the Kyoto Protocol— participation in carbon trade via the Clean Development Mechanism (CDM)—is unlikely to benefit the poorest countries, which are unlikely to attract private sector funding in any case. It is more than likely that the CDM will follow the path of foreign direct investment; the much trumpeted benefits will accrue to a handful of the larger developing countries, leaving the bulk of the South on the sidelines of the global carbon market.

Flowing directly from the above is the concern that the so-called "flexibility mechanisms" of the Kyoto Protocol have turned it into **a global carbon trade regime that has lost sight of the original mandate of the UNFCCC,** i.e., stabilization of atmospheric greenhouse concentrations. Significant problems with the Kyoto regime—including the issue of "low hanging fruit", trades in "hot air", the exclusion of poorer countries and marginal groups, and the sheer inadequacy of the Kyoto targets (Malakoff, 1997; Najam and Page, 1998; Sokona et al., 1998; Agarwal et al., 1999; Banuri and Sagar, 1999; Meyer, 1999; Banuri and Gupta, 2000) —have long been known and highlighted. These lingering concerns were tempered by the belief that despite all the holes in it, the Protocol was a step in the right direction. However, it was and remains quite clear that the problems inherent in the Protocol will need to be somehow addressed, and soon. Moreover, the concessions made in recent COPs (especially on the issue of sinks) and absence of the world's largest carbon emitter from the regime, have made an already inadequate agreement all the more inadequate (Najam, 2001).

Most importantly, there is a danger that Kyoto has now become so much of a mechanism for managing global carbon trade that the issue of real emission cuts has been marginalized. With actual and meaningful emission cuts by the world's largest polluters, stabilization of atmospheric concentrations will not only be more difficult, but unlikely (Malakoff, 1997). This concern is most pronounced for the most vulnerable coastal countries for whom the delay in actual emissions cuts could have dire consequences, especially if the much touted flexibility mechanisms of the Kyoto Protocol fail to deliver the expected benefits of carbon trading. For the emitter countries of Annex 1, it makes full sense to pin their hopes on a successful global market in carbon trade; for low-lying LDCs most vulnerable to climate change, the possibility of failure, no matter how remote, is both unacceptable and unimaginable.

Southern Interests

While the South's concerns about the climate regime have evolved as the Kyoto Protocol has taken shape, the longer term interests of the developing countries have remained relatively unchanged over the last decade or longer. While specific (and generally shorter term) interests of particular countries and regions vary, the key interests of the developing world as a whole can be characterized in three categories:

> ➢ Creation of a predictable, implementable, and equitable architecture for combating global climate change that can stabilize atmospheric concentrations of greenhouse gases within a reasonable period of time, while giving all nations a clear indication of their current and future obligations based on their current and future emissions.
> ➢ Enhancing the capacities of communities and countries to combat and respond to global climate change, with particular attention on adaptive capacity that enhances the resilience of the poorest and most vulnerable communities.
> ➢ Efforts to combat global climate change and the pursuit of sustainable development are two sides of the same coin. For either process to work, each must reinforce the other. To be at all meaningful, any global climate regime must have sustainable development as a central goal – at the declaratory as well as operational level.

Most environmental issues are long-term issues. Climate change is particularly so. The test of any climate regime is not simply what it will or will not do in the next few years, but also what it is likely to achieve over the next many decades, even centuries. Any policy architecture put into place today is likely to remain with us for a very long time (Jacoby et al., 1998). It is therefore very important that **the policy architecture we construct be robust enough to withstand the political as well as climatic tests of time**. The Kyoto Protocol, even though it is a step in the right direction, leaves much to be desired in terms of its implications for long-term policy, all the more troublesome since it is also unlikely to produce many short-term benefits (Cooper, 1998). Moreover, the arbitrariness of the Kyoto targets and the lack of any objective basis for their selection leaves the countries of the world – developing as well as industrialized – largely directionless on what might be expected of them in future (Najam and Sagar, 1998).

An alternative, more robust architecture would be one that defines its targets not in terms of symbolic short-term measures, but long-term atmospheric stabilization, which gives all countries a clear signal on what is likely to be expected of them in future; which is based on clear and objective principles derived directly from the UNFCCC, and which is seen to be fair and equitable by all countries, North and South. Developing countries must now seek every opportunity to reinitiate discussion on the larger architecture of the future climate regime. This is not to suggest an abandonment of the Kyoto Protocol; rather, this is to build on the Kyoto promise by returning to UNFCCC basics. In all likelihood this will require moving to per capita emission targets and a "contraction and convergence" policy scenario aimed at atmospheric stabilization in the post-Kyoto phase (Agarwal et al., 1999; Meyer, 1999). Such targets could be applied to all countries, North and South, thereby responding to the US demand for treating all countries equally. Instead of a convoluted system of arbitrary percentage cuts for different countries, having a standard global emissions budget linked directly to atmospheric stabilization would not only be more elegant and equitable, but also more manageable in the long term. Indeed, such a system could be a first step towards a more meaningful clustering of related agreements around a broader regime for all issues related to the atmospheric commons (Najam, 2000).

"Capacity building", much like technology transfer, has been a much abused term in the rhetoric of climate policy. Both North and South reiterate by rote the importance of building capacity, yet neither has shown much willingness to invest meaningfully in it (Banuri and Sagar, 1999). In introducing the twin concepts of "adaptive" and "mitigative" capacity (by working groups II and III respectively), the third assessment of the IPCC (2001) has made a significant contribution to the policy discourse by outlining what types of capacities are required, by whom, and when. The most pressing challenge in this regard is to **strengthen the social, economic and technical resilience of the poorest and most vulnerable against extreme climatic events**. This highlights the need to focus on issues of adaptation, especially in LDCs and SIDs where the threat of climate change is more immediate as well as more intense (Huq and Sokona, 2001). As already mentioned, COP-6 made a rather symbolic gesture in this direction by setting up a set of voluntary funds. It is time now for the world to put its money where its mouth is.

While the developing country interest in capacity enhancement is self-evident, the new element is our growing understanding of *where* capacity needs to be enhanced and *what* capacities need to be supported and strengthened. In short, the capacity to adapt to climatic impacts, i.e., social, economic and technical resilience, is needed most desperately where the vulnerabilities are the most pronounced; namely, at the local and community levels (Bohle et al., 1994; Ribot et al., 1996; Burton, 1997; Rayner and Malone, 1998; Downing and Baker, 1999). However, effective capacity building at this level would require rethinking—both *how* we do capacity building and *who* we do it with. The shift toward strengthening the social, economic and technical resilience of vulnerable local communities would come from working directly with civil society and community organizations. This would be more difficult as well as more expensive. However, the payoff of such an investment would also be higher, both in terms of climate policy and in terms of sustainable development.

Sustainable development remains a pivotal interest not just for the South, but for the entire world. Indeed, as the most recent IPCC assessment (IPCC, 2001) has made clear, the supposed dichotomy between the global climate policy and sustainable development policy is false (also see Munasinghe, 2000). **Combating climate change is vital to the pursuit of sustainable development; equally, the pursuit of sustainable development is integral to lasting climate change mitigation.** The pursuit of sustainable development is a clearly stated goal of both the UNFCCC and the Kyoto Protocol (see, for example, the preamble and Articles 2 and 3 of UNFCCC and Articles 2 and 10 of the Kyoto Protocol). Yet, there has been a clear hesitancy from those operating in the "climate arena" to deal with sustainable development seriously. While the third assessment report of the IPCC has included a chapter linking the two, the linkage is far from integrated into the bulk of the report (IPCC, 2001). Indeed, despite much developing country impetus, the IPCC seems reluctant to pursue the links between sustainable development and climate change at any serious level.

This systematic denial of sustainable development's importance to climate policy may or may not impact the future of sustainable development but will almost certainly adversely impact the future of the global climate regime. Stated more simply, sustainable development is needed because it can provide the conditions in which climate policies can be best implemented (Munasinghe, 2000). It is unfortunate that sustainable development is now being portrayed as being only in the interest of the South. In fact, the so-called Rio compact placed sustainable development quite clearly as a common interest of all countries, developing as well as industrialized; a common interest around which related North-

South bargains could then be built on other issues, including climate change. Unfortunately, this has not yet happened. However, the WSSD and its focus on sustainable development provides the opportunity to forcefully reestablish the link between sustainable development and climate change.

Looking Beyond Kyoto

Let's face it, the global climate change regime, as identified by the Kyoto Protocol, is stuck in a rut. The latest COP (in Milan, Italy) only proved the point. So, where do we go from here?

Conceivably, the nations that have already ratified the Kyoto Protocol could discover uncharacteristic boldness and decide to put the provisions of the Protocol to test; with or without the Russians. The technicality that now gives the Russians the deciding vote was entirely arbitrary to begin with. There is nothing stopping the countries that have ratified the Protocol from launching an accelerated program of implementation without allowing the erstwhile cold warriors - the Americans and Russians - to hold the Protocol hostage.

Even as climate change becomes more urgent with each new study, the momentum for action is fast disappearing. Instead simply rehashing old arguments and battered positions, we need to seriously rethink the climate negotiation process and re-align to the original intent and aspirations of the UNFCCC. This would require the post-Kyoto climate regime to build upon three key realizations.

A Crisis of Imagination. First, the problems of the climate regime are not just political pathologies of expediency but also a crisis of imagination. Ironically, the one global problem that is most clearly long-term has elicited the most short-term policy reactions. Making decadal commitments can lead into blind alleys of short term compliance that preempt the century long strategies needed to actually meet the goal of protecting the climate system and promoting human wellbeing. The climate regime needs to take a bold and imaginative stance to determine not only the costs of action today but also the costs of inaction tomorrow. It is mistaken to believe that a series of well-meaning but poorly thought-out interim measures will ultimately lead us to a safe stabilization of atmospheric greenhouse gases. It would be more prudent to start from a conception of where we eventually want to be and work backwards to what we need to do today to get there.

A Malady of 'Mal-Consumption.' Second, the notion of 'mal-consumption' should be placed at the center. The problem is not just the levels of consumption but also the nature of consumption. The livelihoods and wellbeing of people, and particularly equity concerns, must be central to any policy enterprise. But a right to use sub-optimal technologies when better alternatives are available cannot contribute to wellbeing. For the North this means unshackling itself from the trap of obsolescence: i.e., using nineteenth century fuels (e.g., coal) and mid-twentieth century technologies (e.g., power plants, electric grids). The North's folly is not only that it remains imbedded in outmoded processes, but that it continues to give legal cover to their perpetuation (e.g., US subsidization of its steel behemoths). For the South this means giving up its demand to repeat the mistakes that were made by the North. Paradoxically, because the developing countries are where they are, they have the ability to make better decisions than the North. There is no sustainable development model in the North to emulate; if one is to emerge it will have to be crafted by the South itself.

Making the Global, Local. Finally, stale debates and finger pointing at the inter-state level has to give way to fresh thinking at the people and community level about approaches that can work. The most promising initiatives and the most robust implementation of climate friendly change are happening at the level of enterprise, of communities, institutions and even by individuals. Across Europe and the United States, local governments at the municipality level are creating their own action plans to implement the Kyoto targets. They are doing so without any edict from their national governments and at odds with national policy. Similarly, communities across the developing world are moving to efficient cookers and water heaters, not because they are being told to, but because it makes sense to their livelihoods aspirations. The role of global climate policy must be recast to focus more on unleashing the spirit and potential of civic enterprise and to remove the hurdles to policy innovation at the local levels.

In short, it is time to seriously rethink the climate regime: first, by rooting it firmly in the original long-term vision of the Climate Convention; second, by questioning the very basis of technological decision-making in both North and South; and, third, by facilitating and unleashing the potential of local communities for meaningful climate action.

References

Agarwal, A. and Narain, S. 1991. *Global Warming in an Unequal World*. Centre for Science and Environment, New Delhi.

Agarwal, A., Narain, S., and Sharma, A. (eds.), 1999. *Green Politics: Global Environmental Negotiation-1: Green Politics*. Centre for Science and Environment, New Delhi.

Agarwal, A., Narain, S., Sharma, A. and Imchen, A. (eds.), 2001. *Green Politics: Global Environmental Negotiation-2: Poles Apart*. Centre for Science and Environment, New Delhi.

Baer, P., Harte, J., Haya, B., Herzog, A.V., Holdren, J., Hultman, N.E., Kammen, D.M., Norgaard, R.B., and Raymond, L., 2000. Equity and greenhouse gas responsibility. Science, 289: 2287.

Banuri, T. and Sagar, A. 1999. In fairness to current generations: Lost voices in the climate debate. *Energy Policy.* 27(9) 509-514.

Banuri, T., and Gupta, S., 2000. The Clean Development Mechanism and Sustainable Development: An Economic Analysis. Asian Development Bank, Manila, Philippines.

Bohle, H.G., Downing, T.E., and Watts, M.J., 1994. Climate change and social vulnerability: Toward a sociology and geography of food insecurity, *Global Environmental Change* 4(1): 37-48.

Burton, I. 1997. Vulnerability and Adaptive response in the context of climate and climate change. *Climatic Change 36 (1-2), 185-196.*

Carraro, C. (ed.) 2000. Efficiency and Equity of Climate Change Policy. Kluwer Academic Publishers, Dordrecht, Netherlands.

Cooper, R.N. 1998. Toward a Real Global Warming Treaty. *Foreign Affairs* 77(2): 66-79.

Dasgupta, C., 1994: The climate change negotiations, In: I. Minstzer and J.A. Leonard, (eds.). *Negotiating Climate Change: The Inside Story of the Rio Convention,* pp. 129-48, Cambridge Univ. Press, Cambridge, UK.

Downing, T.E., and Baker, T., 1999. Drought vulnerability: Concepts and theory, in D. A. Wilhite, ed. *Drought.* Routledge, New York.

Gadgil, M. and Guha, R. 1995. *Ecology and Equity: The Use and Abuse of Nature in Contemporary India,* Penguin Books, New Delhi, India.

Huq, S. and Sokona, Y., 2001. Climate Change Negotiations: A View from the South. Opinion: World Summit on Sustainable Development. International Institute for Environment and Development, London.

Hyder, T.O., 1994. Looking back to see forward, In: I. Minstzer and J.A. Leonard, eds., *Negotiating Climate Change: The Inside Story of the Rio Convention,* pp. 201-26. Cambridge Univ. Press, Cambridge, UK.

IPCC. 2001. *Climate Change 2001 – Third Assessment Report of the Intergovernmental Panel on Climate Change.* Including reports of all three working groups. Cambridge Univ. Press, Cambridge, UK.

Jacoby, H.D., Prinn, R., and Schmalensee, R., 1998. Kyoto's unfinished business. *Foreign Affairs* 77: 54-66.

Jamieson, Dale 1992. Ethics, public policy, and global warming. Science, Technology, and Human Values 17(2): 139-153.

Malakoff, D. 1997. Thirty Koyotos needed to control warming. *Science* 278: 2048.

Meyer, A. 1999. 'The Kyoto Protocol and the Emergence of "Contraction and Convergence" as a framework for an international political solution to greenhouse gas emissions abatement'. In: O. Hohmeyer, and K. Rennings, (eds.) *Man-Made Climate Change: Economic Aspects and Policy Options.* Physica-Verlag, Heidelberg.

Munasinghe, M. 2000. Development, equity and sustainability in the context of climate change' In: M. Munasinghe and R. Swart (Eds.) *Proc. IPCC Expert Meeting on Development, Equity and Sustainability, Colombo, Sri Lanka 27-29 April.* IPCC and World Meteorological Organization, Geneva.

Najam, A. 1995. International environmental negotiations: A strategy for the South. *Int. Environmental Affairs* 7(2): 249-287.

Najam, A., 2000. The case for a law of the atmosphere. Atmospheric Environment, 34(23): 4047-4049.

Najam, A., 2001. Deal or no deal. Down to Earth, 10(11): 50-51.

Najam, A. and Page, T. 1998. The climate convention: Deciphering the Kyoto Protocol. *Environmental Conservation.* 25(3): 187-194.

Najam, A. and Sagar, A., 1998. Avoiding a COP-out: Moving towards systematic decision-making under the climate convention. *Climatic Change* 39(4).

Rajan, M.K., 1997. *Global Environmental Politics.* Oxford University Press, New Delhi.

Rayner, S. and Malone, E. (eds.), 1998. Human Choices and Climate Change. Batelle Press, Columbers, OH (USA).

Reiner, D.M. and Jacoby, H.D. 1997. Annex I Differentiation Proposals: Implications for Welfare, Equity and Policy. MIT Joint Program on the Science and Policy of Global Change, Report No. 27. Massachusetts Institute of Technology, Cambridge, MA.

Ribot, J. C., Najam, A., and Watson, G., 1996. Climate variation, vulnerability and sustainable development in the semi-arid tropics. In: J. C. Ribot, A. R. Magalhaes, and S. S. Pangides (eds). *Climate Variability, Climate Change and Social Vulnerability in the Semi-Arid Tropics*. Cambridge Univ. Press, Cambridge, UK.

Sagar, A. and Kandlikar, M. 1997. Knowledge, rhetoric and power: International politics of climate change. *Economic and Political Weekly*, December 6, p. 3140.

Shue, Henry 1995. 'Equity in an international agreement on climate change. In: Richard Odingo, et al. (eds.). *Equity and Social Considerations Related to Climate Change*. ICIPE Science Press, Nairobi, Kenya.

Sokona, Y. 2001. Marrakech and beyond. Down to Earth 10(11): 53.

Sokona, Y. and Denton, F. 2001. Climate change impacts: Can Africa cope with the challenges? Climate Policy 1(1): 117-123.

Sokona, Y., Humphreys, S., and Thomas, J.P., 1998. What prospects for Africa? In: J. Goldemberg (ed.). *Issues and Options: The Clean Development Mechanism*. UNDP, New York, NY.

Weiss, E.B. (1989). In Fairness to Future Generations: International Law, Common Patrimony, and Intergenerational Equity. Transnational Publ., Dobbs Ferry, NY.

CHAPTER 21

Reclaiming the Atmospheric Commons: Beyond Kyoto

John Byrne[a], Leigh Glover[b], Vernese Inniss[c], Jyoti Kulkarni[d], Yu-Mi Mun[e], Noah Toly[f], and Young-Doo Wang[g]
The Center for Energy and Environmental Policy,
University of Delaware, Newark, Delaware, USA, 19716.

Introduction

To date, international negotiations to address climate change have largely focused on the problem of designing market-based tools to encourage efficient adjustments in the carbon intensity of the global economy. The Conference of Parties[1] (COP) process has promoted this profit-focused strategy in the belief that it is essential to facilitate cooperation in the reduction of greenhouse gas (GHG) emissions. Its work is codified in the Kyoto Protocol[2] and the Marrakech Accord.

The COP process may rightly claim success in realizing a treaty of targets and commitments to lower the release of one of the most ubiquitous chemicals associated with human activity. But it is not clear if activity under the auspices of the treaty will, in fact, reduce emissions. Further, there is a reasonable basis for concern that the treaty may shift the burden of action for greenhouse gas (GHG) reductions to countries with little or no responsibility for the problem. The absence of a penalty for withdrawal by the world's largest GHG emitter – the US – likewise raises doubt about the efficacy of the treaty. And the decision to refocus attention on "adaptation" rather than "mitigation" during the 2002 "Delhi round" suggests that confidence in the treaty is waning (Byrne et al 2002).[3]

Our analysis leads us to conclude that the Kyoto Protocol-Marrakech Accord is unlikely to improve climate justice or sustainability. The alliances underlying the COP-negotiated agreements appear to be largely economic in character, not ecological. As a result, its outcomes are better predicted as elements of neoliberal globalization strategy (Byrne and Glover, 2002) than as commitments to ecological principles, values or goals.

If international negotiations are to avoid being coopted as a venue merely for deciding economic advantage, we argue that two principles — equity and sustainability — must guide deliberations. More broadly, we argue that the paradigm of capitalization which has guided nature-society relations in the industrial era needs to be replaced with one that reclaims our climate and atmosphere as elements of a global commons.

Negotiating a Future Climate: An Overview of the COP Process

Throughout the ages, human beings have reflected on the heavens with reverence and fascination. It is troubling to recognize that our era will forever be known for forcing the human sense of awe to compete with a studious interest in the mechanics of climate and the chemistry of the sky. Undeniably, though, the era of atmosphere management is upon us. Its constitutional origin can be traced to the approval of the Kyoto Protocol and the ongoing negotiations that seek to interpret and operationalize it.

The Kyoto Protocol (framed at COP-3) sets binding emission targets for 25 wealthy societies and 13 countries in transition, which are listed in Annex B[4] of the Protocol. Individual Annex B countries were assigned different targets under the principle of "common but differentiated responsibility." Their collective GHG emission reduction target was set at 5% below their aggregate 1990 level. This collective reduction is to be achieved during the Protocol's first budget period (i.e., between the years 2008 and 2012 — see Article 3.1 of the Kyoto Protocol to the United Nations Framework Convention on Climate Change, 1997).

At COP-4 in Buenos Aires and COP-5 in Bonn, great attention was given to a range of market-based policy instruments (called "flexibility mechanisms" in the Kyoto Protocol) that would assist wealthy countries in lowering emissions. Under the flexibility mechanisms, Annex B countries are allowed to purchase emission permits from other Annex B countries that presently release GHGs at a rate below their Kyoto targets, or have lower cost CO_2 reduction options that can be more rapidly realized through emissions trading. Annex B countries may also receive credits toward target reductions through project-based emission reductions or sink expansions in other Annex B countries through Joint Implementation (JI). Finally, Annex B members can earn certified emission reductions (CERs) from project activities in developing countries and apply them in order to comply with GHG reduction targets through the Clean Development Mechanism (CDM).

COP-6 (held in the Hague and Bonn) produced a number of decisions that further shaped national strategies and options under the Kyoto Protocol. The most influential of these was the permission of essentially unrestrained emissions trading. As a result, Annex B participants can take full advantage of available emission permits beyond their borders to meet national commitments (this is a particular problem for efforts to achieve effective emissions reduction, as explained below). One option created with these negotiations is the purchase of emission credits from Russia and other economies in transition whose current releases are well below their 1990 levels. In effect, an Annex B member can assist economies in transition to upgrade technology efficiency and then claim the difference in the resultant GHG emissions at the same time that economies in transition increase their emissions to 1990 levels. This curious option is discussed at length below.

COP-6 also allowed national carbon 'sink' enhancements to offset GHG emissions in national GHG accounting. The methodology for calculating sink improvements (for example, through reforestation) is provided in Article 3.3 of the Kyoto Protocol to the United Nations Framework Convention on Climate Change, 1997. Any claimed activities must have occurred since 1990 and have been the outcome of human activity. COP-6 revisions enabled countries to count changes in all sources of carbon sinks, most notably land use, land-use change, and forestry (LULUCF), but restricted the level of claims against forest sinks.[5] Inclusion of carbon sinks makes the Kyoto Protocol comprehensive, covering all known elements of the carbon cycle immediately affected by human activity,

Despite acquiescence to its demands for unlimited trading and a liberal interpretation of LULUCF opportunities, the US withdrew from the UNFCCC negotiations before continuation of the COP-6 meeting in Bonn (2001). Voicing nearly identical economic concerns to those of the elder President Bush in 1992 (at the United Nations Conference on Environment and Development in Rio de Janeiro, Brazil), the younger President Bush indicated that the US would follow its own "voluntary" GHG reduction policy, setting in motion what has become a unilateralist orientation in international affairs. Australia has now also withdrawn.

In the wake of the US withdrawal, implementation rules for the flexibility mechanisms were articulated in the 2001 Marrakech Accord (COP-7) that reflect significant compromises thought to be necessary to secure ratification. Without US participation, the cooperation of Japan, Russia, and Poland are essential to bring the Protocol into force. Consequently, a host of monitoring, verification, and compliance issues raised by delegates from these countries were addressed. In this regard, securing sufficient participation for ratification took priority over issues concerning the effectiveness of the policy mechanisms adopted and refined between COP-4 and COP-7 in addressing problems of unsustainability and inequity.

COP-7 also continued to polish regulations that govern the Protocol's various flexibility mechanisms and sink allocations. However, many of these decisions have the effect of reducing the level of emissions abatement necessary through domestic measures in developed nations by allowing purchase of foreign emission credits, accreditation for foreign investments that reduce emissions and enhance carbon sinks, and inclusion of an array of domestic carbon sinks as offsets to domestic emissions. For example, through the CDM, Annex B nations can purchase credits from non-Annex B nations for afforestation and reforestation projects, but according to a limit of 1% of a country's target emissions. Emissions trading between Annex B nations can be pursued apart from any supplementarity restriction, and full use can be made of surplus emission credits. Credits earned by any of the aforesaid methods can be used immediately, banked for future use (in the Protocol's second budget period, for example), or sold in the emerging emission permits market.

COP-8 (New Delhi, 2002) offered a glimpse of the future of the treaty negotiation process with the exodus of the US. Sensing that the UNFCCC might have little effect on emissions during the first budget period, several Southern countries, including COP-8's host, led an effort to establish a disaster relief fund (to be financed by the wealthy tier) that would assist the poor to "adapt" to climate variation. While details remain for additional discussion, the principal outcome of COP-8 was the commitment to such a fund. In effect, treaty participants have agreed to negotiate compensation for their expected failure to avert climate change (Byrne et al 2002).

The dawn age of managing the atmosphere has already been witness to triumphs and travails. But as the Parties search for direction and hope amid the numerous difficulties associated with organizing human interactions with the sky, an unapologetic belief in markets and profits steers the treaty to its envisioned destination – an efficiently, and profitably, bargained adjustment of global carbon intensity. Still, there are disquieting questions. Will carbon intensity decline because of the treaty? Or, will the treaty simply create another market where money can be made?

Loopholes in the Kyoto Protocol

Notwithstanding COP-8's "no confidence" vote, the treaty retains at least a formal commitment to try to reduce GHG emissions. Each signatory to the Protocol listed under Annex B has an individual national target for emissions reduction, which amounts to collective reduction of 5.2% below the collective 1990 level of emissions. We converted these national targets into the OECD and FSU/EE groupings and derived the Kyoto Protocol target for each on a per capita basis: 11.71 tCO_2-e for the OECD and 13.23 tCO_2-e for the FSU/EE (Fig. 21.1).[6] While the Kyoto target for the FSU/EE group is below its baseline rate for 1990, it is higher than the actual emissions by the group in 1998. Economic recessions experienced in the

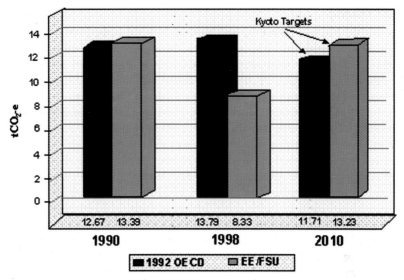

Sources: 1990,1998 Emissions – Marland et al., 2001; 2010 Emissions Targets
– Kyoto Protocol, Population – world Resources Institute, 2001

Fig. 21.1: Actual and target per capita Annex B GHG emissions under the
Kyoto Protocol.

region since the breakup of the Soviet Union are the cause of the dramatic
decline in GHG emissions and the source of an unusual opportunity
under the Protocol (discussed shortly).

Annex B has steadfastly voiced concerns throughout the COP process
that improper policy actions could harm the group's economies and, for
this reason, has been uninterested in high domestic emission reduction
targets. Instead, the group has preferred "practical" "realistic" targets
and market-sensitive options that enable members to tailor their reduction
strategies, including the ability to trade with other nations, in order to
find the most efficient (i.e., cheapest) actions to reduce GHG emissions.

Led by the US, Australia, and Japan, Annex B has promoted the view
that the transition to a low-carbon future is largely an economic and
technological question best handled (with the proper incentives and
enforceable rules) in the global marketplace. This shared belief in markets
as guides to national action on a global environmental problem reflects
Annex B's core commitment to a policy paradigm in which priority is
given to resolutions of environmental conflicts that are least-cost and,
where possible, conducive to economic growth. While pursuing economic
rationality, unfortunately, the Kyoto mechanisms have created significant
loopholes in terms of sustainability, two of which are critically reviewed
below.

Hot-air trading

Economic problems in the FSU and Eastern Europe since 1990 have led to significant decreases in GHG emissions throughout the region. Because national performance in meeting UNFCCC is assessed with reference to 1990 baselines, the FSU/EE bloc will likely not need to undertake any domestic GHG abatement programs. Instead, members are in the interesting position of being able to sell emissions growth to other Annex B nations whose releases are above the 1990 baseline.

The inclusion of the EE/FSU in Annex B has thus produced an opportunity for "virtual reductions" (Byrne et. al., 2001) that may be substituted for actual decreases in GHG emissions. Specifically, it is possible under the Kyoto Protocol for OECD members to assist EE/FSU members of Annex B to "efficiently" increase their GHG emissions, while counting this effort as a deduction to OECD members' national emission accounts. This widely known implication of the trading mechanism permitted under the Kyoto Protocol has created what is now commonly termed "hot air".

Under business-as-usual (BaU) projections by the US Energy Information Administration (EIA, 2000), the OECD countries are expected to increase their emissions by 15.8% over the 1998-2010 period, while total GHG emissions in the EE/FSU nations are anticipated to grow by 13.9% during the same period (Marland et al., 2001). In other words, under a BaU scenario, OECD countries (including the US and Australia) will exceed their collective Kyoto target by approximately 3,400 Mt CO_2-e in 2010, while the EE/FSU bloc will release 1,160 Mt CO_2-e less than their Kyoto target. Thus, "hot air" is estimated to meet 36.5% of the total GHG reduction requirement for the OECD countries.[7] Of course, "hot air" availability will increase further if the EE/FSU emission forecast by EIA happens to be high, which is possible since the prospect of additional economic problems for this bloc is considerable.

It is reasonable to expect that "hot air" will provide the lowest cost emission credits, after sinks, for Annex B traders. Compared to trades with non-Annex B countries, transaction costs and infrastructure incompatibilities are likely to be lower, and commercial relations more extensive and longer lived, for Annex B-to-Annex B trading.

Sinks

In theory, accounting for sinks as an element of the carbon cycle is unimpeachable. Some environmentalists and those seeking to bolster an array of developmental objectives embraced the inclusion of sinks in the UNFCCC as additional support for laudable objectives such as habitat and catchment protection, agroforestry development, rain-forest preservation, prevention of land clearance, and so on. Indeed, COP-6 reiterated that these activities contribute to the conservation of biodiversity and sustainable use of natural resources and therefore should be included as a means for nations to meet Protocol targets. Climate change policy under this provision offers the opportunity to support other environment and development objectives while also being responsive to the need for building a "low-carbon" future. Further, it seemingly offers a way in which climate policy can emphasize domestic action (instead of trading away national responsibility) and at the same time economically meet reduction targets.[8]

A broad array of land-based activities is admissible as sinks and credits for them are largely unrestricted (only sinks resulting from forest management are limited under Appendix Z from COP-6). COP negotiations have only limited sink CDM activities to afforestation and reforestation in this first commitment period (i.e., 2008-2012), and capped available credit by these means to 1 % of a country's target reductions.

Since the principle of crediting carbon storage as a means to meet Kyoto targets has been adopted by the COP, the race has been on to register national sinks and to partner with other nations to expand sink capacities and then take credit for them through JI and CDM. The magnitude of available sink credits through these two mechanisms, is sufficient to enable certain Annex B members to avoid domestic emission reduction entirely.

Efforts to incorporate LULUCF into the Convention have been fraught with basic uncertainties in the measurement of sequestration and fluxes, compounding efforts to construct an effective sinks policy. Production of the national GHG inventories, as required under the UNFCCC, has highlighted how indeterminate the LULUCF component is, even for those nations with the best data and research bases. The IPCC's Special Report on the subject provided a sound description of the current state of knowledge, but further highlighted just how few generalizations could be made about sequestration for any given location (IPCC, 2000).

Even if the aforementioned difficulties with the measurement of these factors were resolved, there are a number of ecological concerns that raise doubts about the efficacy of LULUCF measures. For example, the most effective species for optimizing carbon sequestration are fast-growing trees with short rotations, yet this plantation practice would reduce biodiversity. Reconciling the Kyoto Protocol's intention that LULUCF contribute to broader ecological goals with practices to enhance sequestration could prove difficult.

Climate change policy can only be effective if there are permanent reductions in global GHG emissions. At present, the rules that allow carbon sequestration to offset emissions encourage only a temporary reduction of global emissions. Any number of events, such as fire, disease, or climatic factors, can release sequestered carbon into the atmosphere. In a sense, carbon sinks are simply deferred emissions and are therefore incomparable to actual reductions in GHG emissions, because they fail the test of permanence. Sinks allow GHG emissions to be greater than would otherwise be permitted and pass to future generations an increased burden.

Notwithstanding concerns raised by the IPCC, the COP is proceeding on the basis that quantification, measurement, and verification of sequestration is now possible. This policy appears to be driven less by accurate knowledge than confident expectations of profit.

Assessing the Kyoto Protocol

Together, the "hot air" and "sink" loopholes provide the means for a paradoxical result — compliance with the Kyoto Protocol's target of a 5.2% reduction in GHGs by increasing Annex B emissions. A comparison of current and forecasted Annex B emissions with the magnitudes of "virtual" reductions allowed via these loopholes demonstrates the possibility of this result.

Since signing the UNFCCC in 1992, OECD countries have posted steady annual increases in GHG emissions. Of this group, only a few can claim to be on a path of emissions reduction (arguably Germany, the United Kingdom, and Sweden). Other countries such as Australia, Canada, Greece, Ireland, Portugal, and Spain increased their emissions by more than 10% between 1990 and 1998. Most obvious in its continued emissions growth is the US, which posted a 13.1% increase over the same period. As a group, the OECD bloc has seen substantial economic growth over the decade since the Earth Summit, while the economies of the former Soviet Union and Eastern Europe have languished. This bifurcation in economic paths has its parallel in GHG emissions. While emissions of the OECD group grew by nearly 9% between 1990 and 1998, those of the EE/FSU actually declined by almost 40%.

Expectably, there is substantial disparity in national per capita releases of GHGs by region and income. For example, average annual per capita emissions of OECD countries were 13.79 CO_2-e in 1998, whereas average non-Annex B 1998 per capita emissions were 2.14 CO_2-e, less than one-fifth that of the OECD (Fig. 21.2).

Neither trend is expected to change under "business-as-usual" conditions. Per capita emissions from the OECD are forecast by the US, Energy Information Administration (EIA) in its BaU scenario to continue increasing to 15.97 CO_2-e by 2010 (EIA, 2001). In contrast, the per capita emissions of developing nations (non-Annex B under the Kyoto Protocol) are expected to rise to only 3.13 CO_2-e by 2010 (Fig. 21.2).

But the Kyoto Protocol is presented by its negotiators as altering the BaU conditions of our future. Thus, it seems reasonable to ask if its

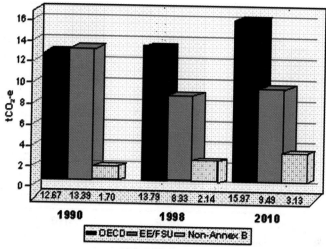

Sources: 1990, 1998 Emissions—Marland, et al., 2001; 2010 Emissions Projects—EIA, 2001; Population—World Resources Institute, 2001.

Fig. 21. 2. Projected global per capita GHG emissions under business-as-usual assumptions.

impact would be to lower and bring greater equity to emissions compared to BaU conditions. Because such a question involves forecasts of future behavior, many answers are possible. Our approach is to assume that Annex B behavior would be motivated by the core principle that animated its negotiation of the Protocol, namely, profitability. In this vein, we expect that the full volume of "hot air" would be traded – unless Russia and other EE/FSU members use their veto position regarding ratification to demand too high a price. We also anticipate that full advantage would be taken of "sink" offsets allowed under the Protocol. And we fear that CDM would deliver a reservoir of "low-lying fruit" that could adversely harm long-term Southern interests but cheaply meet the first budget period needs of the Annex B bloc (e.g., Agarwal and Narain, 1997; Lynch, 1998; Byrne et al., 2001). Under these assumptions, the prospects for improvement over BaU conditions are dim.

If Annex B members, including the U.S. and Australia,[9] take advantage of the low-cost options of the flexibility mechanisms, per capita CO_2-e emissions are likely to climb to nearly 16 tons per year by 2010 (see Fig. 21.3a). The anticipated increase in OECD emissions would be the largest single anthropocentric contributor to climate instability in 2010. At the same time, non-Annex B emissions might increase less quickly if CDM investments actually transfer the promised clean energy technology envisioned in the Kyoto Protocol. The result would be no difference between BaU and Kyoto-influenced emissions of the Annex B bloc and

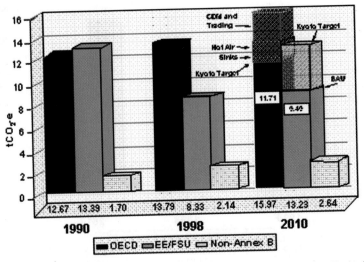

Sources : 1990,1998 Emission–Marland, et al., 2001, 2010 Emissions Projections—EIA, 2001; Population – World Resources Institute, 2001

Fig. 21.3a: Possible per capita global GHG emissions when Kyoto Flexibility Mechanisms are fully employed and the US and Australia participate.

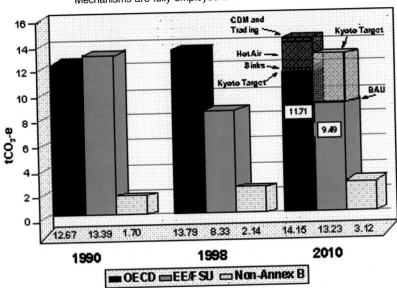

Sources: 1990,1998 Emissions – Marland, et al, 2001; 2010 Emissions Projections – EIA. 2001; Population — World Resources Institute, 2001

Fig. 21.3b: Possible per capita GHG emissions of the treaty parties when Kyoto Flexibility Mechanisms are fully employed and the US and Australia withdraw.

slower growth in per capita emissions among non-Annex B members, thereby widening the gap between the two groups.

Rather than decreasing emissions by 5.2% by 2010 (i.e., the Kyoto target for Annex B), OECD countries could increase their releases by 26% and comply with the treaty. "Hot air"would furnish over one-third of the "virtual" GHG reduction requirements needed by OECD countries for compliance (assuming that the US and Australia are participants). Sink accounting would benefit both the OECD and EE/FSU blocs, with nearly one-tenth of the OECD's "virtual" reductions possibly derived from this source. Burden-shifting via CDM projects could furnish the remaining "virtual" reductions.

Figure 21.3b depicts the case wherein the US and Australia fail to participate. In this instance, the remaining Annex B participants in the Protocol would realize smaller per capita growth, with emissions reaching 14.15 tons per capita per year. Non-Annex B per capita emissions would rise faster in this scenario because more than 80% of the Annex B reduction target (with the US and Australia not participating) is met with the purchase of "hot air". As a result, a very small amount of CDM trading is expected.

Thus, the OECD group (with or without US and Australia participation) could increase per capita emissions under Kyoto,[10] inequality could worsen as Annex B releases grow while non-Annex B emissions slow and, according to its own logic and terms, the Protocol would have successfully concluded its first phase. In effect, uncapped flexibility mechanisms are likely to nullify any substantial claim on the part of the Protocol that it advances goals of sustainability or equity. Instead a "virtual reality" of "efficient" 'emission adjustments that disguises a "real" reality of actual emission expansions would ensue (Byrne et al., 2001).

The Protocol's architects may defend the package as a "first step" and the only practical pathway politically available at this time. But even if it is supposed that an efficient allocation of resources would prevail because of the flexibility mechanisms, and emission reductions would therefore occur at considerably more cost-effective levels, this represents an untenable trade off. The scenarios depicted in Figures 21.3a and 21.3b suggest deepening social and ecological risk, especially for the least advantaged two-thirds of the world's population.

Failure to Govern: US Withdrawal and the Kyoto Protocol

COP-7's major contribution to future climate change governance was arguably its establishment of a compliance regime for the Kyoto Protocol. The basic elements of an enforcement system were delimited in Marrakech, including an effort to penalize nations failing to meet their emission reduction commitments in the first commitment period. Countries that exceed emissions quotas in the first budget period (2008-2012) will be required to compensate for the excess in the second period, 2013-2017, while assuming a penalty equal to 30% of the shortfall and being excluded from emission credits trading until compliance is realized. A basic institutional design for overseeing the compliance system was also agreed, featuring committees, expert reviewers, voting procedures, appeals, and other matters. Several commentators have commended the efforts at COP-7, some proclaiming the compliance system a breakthrough in international environmental policy (see Dessai, 2001; Ott, 2002; Wiser, 2002).

Despite COP-7's successful design of a compliance system, a basic flaw remains—there is no guidance on a key problem facing climate change governance today, namely the withdrawal of the US and Australia from the Kyoto Protocol. In this respect, the Protocol is fundamentally weakened by not having devised a penalty for the instance of refusal of major GHG emitters to participate in the emissions reduction regime.

The absence of the US and Australia from the UNFCCC process is possibly temporary. Abundant low-cost opportunities for US emission reduction through energy conservation and improved energy efficiency have been identified by leading US researchers. In fact, the best estimate of an important US research group is that 75% of the needed reductions to meet the Kyoto-assigned target to the US could be achieved by profitable domestic investments (i.e., investments with payback periods of less than five years; see IWG, 1997, 2001). It is likely that Australia has similar cost-effective, energy efficiency-based alternatives. Moreover, the market-based policy mechanisms developed under the Kyoto Protocol would spur a new and sizable global market of GHG emissions trading, creating many opportunities for the economies of both countries and their corporations, to profit from carbon commerce. Indeed, the world's first carbon trade in London was executed by the local office of the US corporate giant, DuPont (Cormier and Lowell, 2001). Far from being an aberration, US firms can be expected to participate in the profits available in the emerging carbon trading market to the extent possible under US foreign policy. Firms would lobby both governments to be allowed to participate without restraint, an activity doubtless already underway. Indeed, trading with the former Soviet bloc was anticipated by the Clinton administration to provide as much as 56% of its Kyoto commitments (Kopp and Anderson, 1998). Through such trades and other market-based policies available under the Protocol, there is the arresting possibility that the US could meet its Kyoto obligation for reducing emissions by actually increasing its carbon releases (Figure 21.3a; Flavin and Dunn, 1998; Pearce, 1998). Thus, one wonders if the renegade position of the US and Australia is a bargaining tactic to win more concessions (and profit), before returning to the Protocol.

Regardless, COP-7's failure to deal with the withdrawal of the US and Australia has several consequences that the global community needs to contemplate. Because COP-7 demurred on a domestic reduction obligation, the world has greater assurance of a burgeoning global carbon market than of real GHG emissions reduction. The US and Australia will be uniquely advantaged by their decision to withdraw from this global agreement because of the weakness of the COP-7 decisions on the compliance system. Although no longer required to incur the costs of emissions reduction that all other major economies have agreed to undertake, there are no provisions in the Kyoto Protocol to prevent the two withdrawn parties from profiting in the global carbon market. At the same time, the US and Australia can market products at higher carbon intensities, and lower prices. Clearly, the circumstance represents a failure of governance.

Not unexpectedly, the prospect of advantages accruing to the US and Australia because of their withdrawal from Kyoto has drawn sharp criticism and some efforts to prevent its occurrence. Members of the European Commission have publicly expressed their anger over the action. For example, EU Commissioner for the Environment, Margot Wallstrom, has said that President Bush's declaration is a "very, very serious statement and totally unacceptable to the outside world and I think this is what we have to make absolutely clear" (Castle, 2001).

Several civil organizations have filed a class action suit in a US district court against the US Export/Import Bank and the Overseas Private Investment Corporation, citing violation of the US National Environmental Policy Act over the global warming consequences of their loans for fossil fuel energy projects (EV World, 2001). The island nation of Tuvalu announced it would take legal action against the US and Australia regarding their responsibility for global warming and the   consequences of inundation of their homeland (Reuters News Service, 2002).

There is a pressing need to reform the compliance system so as to prevent the US, Australia, and other nations from undermining the integrity and effectiveness of policies aimed at restoring the atmosphere to commons status. Several precedents exist for cases where in nations are in contravention of international environmental agreements and offer lessons in considering penalties for US intransigence.

Under the Montreal Protocol, nations who are party to the agreement may not trade with non-Parties in substances controlled by the Protocol. Similarly, the Basel Convention on the Transboundary Movement of Hazardous Wastes prohibits the movement of waste between Parties to the Convention and non-Parties without special agreements being in place.

And the Convention on International Trade in Endangered Species of Flora and Fauna imposes strict limits on relations between participating nations and non-Parties. Researchers are investigating ways of applying such restrictions to countries electing not to participate in the Kyoto Protocol (e.g., Dannenmeier and Cohen, 2000).

Explaining Failures of the Kyoto Protocol: A Political Ecology Approach[11]

The problem of climate change has attracted the energies and ideas of a wide range of political and ecological interests. Alongside the expected involvement of nation states and the corporate tier, the COP process has itself drawn organized and sustained interventions from an incredibly diverse array of representations of civil society. From science groups to human rights movements, from "green" technologists to environmental justice activists, a myriad voices of civil concern about the evolving project to administer the sky have been raised. Yet, the COP process has hatched a plan that rhetorically acknowledges civil demands for goals of sustainability and equity, but painstakingly works on the economic problem of climate change. Why is this? We argue that the explanation lies, in part, in the informing ideology of modern liberal democracy (embodied in the politics of Europe and the US) and, in part, in economic globalization (embodied in the corporate strategy of US, European, and Japanese multinational firms). Together, these forces empower a neoliberal globalization strategy that is unable to understand the counterforces of ecological justice and sustainability (Byrne and Glover, 2002).

Obviously, there are entrenched economic interests in the US, Japan, Europe and elsewhere whose wealth and power could be threatened by a treaty to create and enforce limits on GHG emissions. But at least since the Club of Rome, there has been evidence of a significant corporate interest in the management of environmental resources (see, e.g., Golub and Townsend, 1977 and more recently, Escobar, 1996), and certainly, some policy regimes to rein in GHG emissions could be consonant with capitalist development aims. Thus, it would be oversimplifying to argue that the neglect of issues of ecological justice and sustainability in the COP process is entirely due to corporate interests.

Beyond the expected opposition from at least some sectors of industry, there is a deep-rooted problem in the democracy celebrated by liberal societies, especially with regard to relations with the natural world. The foundational principles of contemporary liberal democracy were born in an era when emancipation meant freedom from not only political tyranny, but natural constraint as well. Indeed, the salient question to 17th through 19th century architects of the new democratic society in thinking about nature was its conquest—the transformation of a stingy nature to a productive contributor to the majestic aims of a liberal democracy able to feed, clothe, and house all people. As Sheldon Wolin observed in his brilliant critique of the new democracy (1960), liberalism in the West sought to harness what it believed to be the liberatory forces of science, technology, markets, and democracy to defeat the old dynastic-feudal and natural orders. Releasing humanity from the chains of the old political, economic, intellectual, and cultural hierarchies was seen as the enterprise of science, technology and market economy. But these forces were also to be directed to the transformation of nature into a tool for use in building a cornucopic future of not only liberty but material happiness for all. The liberatory function of politics was to be situated, under liberalism, in the act of sweeping away all resistance—human and natural—to the new ideas, values, and purposes of science, technology, and market economy. From Locke and Smith, to Condorcet, Saint-Simon, and Franklin, hope was sought in a political order that would free the new productive forces to realize a wealthy modernity. Liberalism would not be satisfied with anything less: freedom and equality cannot be regarded as fully achieved until material happiness is secured (in addition to Wolin, 1960, see also Kumar, 1978, 1995).

This view led liberalism to conceive the productivity of science, technology, and economics as the alter ego of progress: one can only be gained with the other. The neoliberal revision of the last 50 years simply fine tunes this ideology to current events (Byrne and Glover, 2002). Environmental negotiations — such as those of trade (GATT) — have consistently been viewed by neoliberalism as the art of guiding the productive forces of science, technology and markets to achieve positive economic results. For an ideology that understood the natural world as unreasonable and unproductive (see especially Merchant's discussion of the mechanistic turn in Western thinking about nature in her 1980 volume), the conquest of nature meant the application of rational, productive thinking to the inspiring goal of material plenty.

Thus, it cannot be surprising that neoliberalism would be blind to the idea of sustainability when it means that the natural order should somehow be consulted to set limits on social futures. Simply put, such thinking for neoliberalism would be irrational. Ultimately, liberalism would anticipate that such a policy would halt progress.

The Kyoto negotiations follow the neoliberal script on this score. Negotiators focus on the flexibility mechanisms because they are the essential tools for organizing a global efficiency response to the prospect of climate change. At bottom, these mechanisms conceive climate change as a productivity problem: in essence, existing technology and market arrangements inefficiently use the atmosphere to dump GHG emissions. With proper market signals and with a concerted scientific effort, an efficient regime of atmospneric use will be found which allows the productive forces of modernity to resume the quest for modern progress.

While some members of the European Community have insisted that limits should be placed on the use of such mechanisms to meet emission targets, the setting of emission targets themselves has ignored sustainability questions so far. Even limiting the use of flexibility mechanisms, so that the treaty avoids becoming little more than a pact to redistribute the sources of emissions, has failed to occur.

Thus, preoccupation with efficiency mechanisms, at the risk of unsustainability, is expectable. The policies championed by the liberal democracies throughout the COP process are a continuation of liberalism's original formula—to exploit the state of nature to improve the state of commerce. This formula is traceable to liberalism's effort to effect a peculiar alliance of efficiency and equality. Although historically and currently, wealthy countries have overused the atmosphere as a CO_2 dump (see Byrne 1997), the Kyoto flexibility mechanisms permit them to avoid domestic reductions in emissions by paying others to reduce theirs so that the cornucopic ideal of government by markets, science, and technology is undisturbed. A novel ethic ensues in which those who have benefited from centuries of ecologically reckless behavior are relieved of responsibility by paying the Third World to imitate the Western progress formula, but more efficiently.

Pounded on neoliberalism's association of progress with productivity and cornucopia, the Kyoto flexibility mechanisms create climate markets controlled by polluters because they promise profitable actions to address global warming. The risk that climate markets may well accentuate, rather than reverse, the unsustainable and unequal state of nature of modernity is deemed an acceptable one in order to protect the state of commerce.

Neoliberal ideology can explain the negotiating behavior of powerful nation states such as those of North America and Europe, but there remains the question of how society-nature relations are institutionally structured to create and continue the threat of climate change. For answers to this question, we look to explanations of political economy. In our use of the term, a political economy represents a system of political and economic power that, among other things,[12] institutionalizes social access to and use of nature, propounds ideas of nature, society and their relations and broadly seeks to frame the value of nature to society. In our era, capitalist relations underpin the system of power and motivate worldwide efforts to remove obstacles that might prevent the "free flow" of goods, services, and capital. The successful spread of capitalist relations, especially over the last century, has led to an emerging global structure of capital, technology and knowledge that extends the reach and strengthens the power of this system.

Economic globalization is leading the way in rendering natural processes—not simply resources found in nature — as phenomena subject to global management. The capture of core processes of the natural order, such as the carbon cycle, in the languages of science and economics clears the way for policy regimes that intend to choose a state of nature for modem society, or at least its elites. The environment becomes an object for scientific and economic design as this capture occurs. In this way, earlier processes of the commodification of nature that transformed natural phenomena (e.g., forests) into resources that could be traded for profit are revised to valorize nature as capital.

The new valuation of nature encourages ethos to "manage environmental resources to ensure sustainable human progress and human survival" (WCED, 1987:1). This managerial   orientation perfectly matches the aims of globalization. Technical and organizational intelligence is concentrated on increasing the productivity of nature as a whole. As Sachs (1993) argued, modernity is based on the exploitation of nature (human and physical) through the collaboration of dominant actors in government, the economy, and the sciences (1964). Indeed, as Mumford (1964) noted some years ago, the agreement of these three institutional elements on the direction and purposes of society has provided the very definition of what constitutes progress for the West.

Management of nature requires global economic, bureaucratic, and technological organization. Elites who control capital, technologies, and information are the only interests, practically speaking, who could aspire to the role of global manager. Under an emergent regime of "managed nature", the global atmosphere is fast becoming, alternatively, an ecological laboratory and bank, cared for by scientific experts and financial managers. The sustainability interest in the era of globalization is to protect environmental capital for future generations, not the regenerative capacities of the natural environment. This new form of capital is to be managed in the public and scientific interest for future exploitation. In this specific sense, globalization's interest is in the "capitalization of nature" (Escobar, 1996), rather than its protection per se.

Equity and Sustainability in the Greenhouse: Beyond Kyoto

If the design of the Kyoto Protocol can be explained by powerful underlying forces such as neoliberalism and globalization, how might an alternative be constructed? Two things would seem to be needed: first, a break with the logic of the capitalization of nature; and second, the substitution of noncapitalist values for efficiency and profitability in the design of policy.

Consideration of a commons-based (rather than a capitalization-based) regime built on commitments to sustainability and equity (rather than efficiency and profitability) offers a starting point. However, this raises the broader question of defining what is meant by "sustainability" and "equity", and the treatment of the atmosphere as a "commons". Controversies regarding the definitions of these concepts abound and we do not presume that a consensus has been found. Still, we are prepared to offer operational definitions of these concepts for the purpose of investigating an alternative to the neoliberal globalization response to climate change. Hopefully, in this way a constructive debate of alternatives can be engaged.

With respect to sustainability, we suggest an operational definition that limits global GHG emissions to levels consistent with the known properties of the carbon cycle. With respect to equity, we propose a definition that determines country-specific emission targets in a manner that is broadly consistent with an assignment of per capita entitlements. Both definitions have been discussed elsewhere (see, e.g., Agarwal and Narain, 1993; Byrne, 1997; and Byrne et al., 1998).

To establish a numerical benchmark for sustainability reflecting the above definition, we turn to the work of the Intergovernmental Panel on Climate Change (IPCC). The reduction necessary to achieve long-term stabilization of atmospheric GHG concentrations has been reported by this body to be more than 60% of 1990 CO_2 (and CO_2 equivalent) emissions (IPCC, 1992, 1996). To arrest the process of climate change in the new century, we would likely need to act by mid-century to reduce emissions by the IPCC estimate.

A frequently cited equity approach to allocating the global burden of emissions reduction among nations is per capita responsibilities (see especially Agarwal and Narain, 1993). Each nation's climate action responsibility is established on the basis of a "global atmospheric commons regime" to which all peoples have equal access and share equal responsibility. Using the 1990 world population, it is possible to assign carbon "budgets" by country.

Combining these two norms – a sustainability commitment based on the IPCC's estimate of a 60% emissions reduction requirement and a democratic commitment of per capita emissions equality – an equitable and sustainable GHG emissions rate, or $ESCO_2$, can be estimated as 3.3 tons of carbon dioxide and equivalents (tCO_2^{-e}) (see Byrne et al.,1998). Using a longer term 2050 stabilization target year for per capita equity to be realized, progress can then be gauged by international action to reduce GHG emissions to eventual parity. The emissions of transitional economies (EE/FSU) and developing countries may be expected to rise above 3.3 tCO_2^{-e} in the early period, but all would eventually be asked to arrest this trend and begin a steady decline to the $ESCO_2$ rate.

In a commons-based regime with an $ESCO_2$ target, neither "hot air" trading nor sink offsets would be supportable. Such devices to virtualize GHG emission abatement only stall action on behalf of a sustainable and equitable relation between society and the global commons. Thus, the outlines of an alternative to the Kyoro Protocol can be specified: the absence of "hot air" trading, elimination of sink offsets, and setting a sustainability target for progress toward eventual ESCO$_2$ parity. Regarding a target, we propose adoption of the so-called "Toronto Target" of 20% urged in 1988 at a meeting of scientific experts and government ministers that presaged formation of the IPCC (Byrne and Inniss, 2000). Further, full Annex B participation would be obliged with specific penalties assessed for failures to comply.

Under what we have termed a "Beyond Kyoto" scenario (see below), all nations would pursue the goals of contraction and convergence (Meyer, 2000) consistent with the IPCC's findings on carbon-carrying capacity and with principles of equity and sustainability (as defined above). The purpose of collective effort in this case is to begin the process of withdrawing society from activities presumed appropriate for designing nature. Instead, humanity would embrace the goal of restoring a commons relation between society, the atmosphere, and climate.

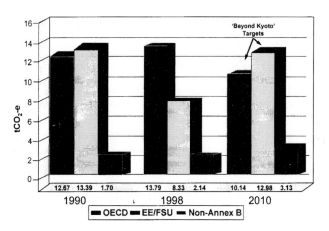

Sources: 1990,1998 Emissions – Marland, et al., 2001; 2010 Emissions Projections – EIA. 2001; Population – 'World Resources Institute. 2001

Fig. 21.4: Projected per capita global GHG emissions under a "Beyond Kyoto" scenario.

Results of the "Beyond Kyoto" approach are presented in Figure 21.4. Per capita emissions targets at 2010 for OECD nations under the "Beyond Kyoto" architecture represent a 20% reduction from 1990 levels. "For the EE/FSU group under the same target, per capita emissions meet a goal of proportional effort to that required of the OECD bloc. Just as the OECD bloc is expected to reduce emissions at roughly three times the Kyoto reduction requirement, we have set the "Beyond Kyoto" target for the EE/FSU at three times their original Kyoto obligation. Because non-Annex B countries remain below the $ESCO_2$ rate through 2010, the BaU forecast for this bloc is accepted.

It is possible to compare progress in meeting the goals of equity and sustainability between the existing Protocol and our "Beyond Kyoto" scenario by considering the ratio of per capita emissions of the wealthy and developing nations. Termed an inequality ratio, comparisons using this metric are reported in Table 21.1. While the Kyoto-Marrakech Protocol, according to our projection, may actually exacerbate inequality, our "Beyond Kyoto" proposal reduces inequality by more than one-third from the 1990 baseline, and by nearly two-thirds from our projection for the Kyoto Protocol.

Table 21.1: Allowable* per capita emissions under the existing Kyoto Protocol and a "beyond Kyoto" scenario (assuming US and Australian participation) (tCO_2^{-e})

Country blocs	BaU	Kyoto Protocol	Beyond Kyoto
OECD	15.97	15.97	10.14
EE/FSU	9.49	13.23	12.98
Non-Annex B	3.13	2.64	3.13
Inequality Ratio**	**5.10**	**6.05**	**3.27**

  * Allowable emissions are those that would be possible given a policy scenario's targets.
  ** The Inequality Ratio is formed by dividing an OECD emissions rate by a corresponding non-Annex B rate. Perfect equality would be represented by a 1:1 ratio.

Genuine, rather than virtual, GHG reductions are anticipated under our "Beyond Kyoto" proposal. While the rate of reduction in this scenario would surely be more costly to implement than the Kyoto Protocol, there is little logic in gauging such costs against BaU conditions. Doing so assumes that continued high-carbon global growth is acceptable, a notion directly challenged by norms of sustainability and equity. Further, such growth would presume commodity status of the atmosphere. In this respect, our analysis underscores the dichotomous choices before us: a Kyoto approach built on core economic values of efficiency and profit and leading to the capitalization of the atmosphere; or a socio-ecological approach built on values of sustainability and equity and conceiving society-atmosphere relations as commons-based.

Reclaiming our Atmospheric Commons

In an era of neoliberal globalism, building an equitable, sustainable, and commons-based relation with the atmosphere will not be easy. For the past 200 years, modernity has depended upon a basic formula for resolving social problems, namely, the promotion of wealth and economic opportunity. Having experienced economic growth for two centuries and now occupying the apex of the wealth pyramid, Northerners in particular often assume that the human condition, or at least their condition, has been bettered by the pursuit of cornucopic ideals. While improvements in Northerners' health and economic security have accompanied adherence to this mode of development, the world is now confronted with social and environmental problems that are uniquely the result of the North's dubious "success" — persistent social inequality and environmental unsustainability. The prospect of global warming and the inequities it augurs appears to be the inescapable threat of unchecked modem development into the 21st century.

Our best hope for a future free of the hubris of atmosphere management lies in a rejection of proposals to capitalize nature and a renewed understanding of the gifts of nature — including our atmosphere and climate — as elements of a commons of life. The sustainability of this commons depends upon societal exercise of normative constraints that enable all forms of life to prosper. Modernity has recklessly breached the constraints underpinning nature-society relations. Through efforts to shape the agenda of climate change to adhere to principles of equity and sustainability, we can take a first step in reclaiming our commons of life.

Footnotes

1   The Conference of Parties is comprised of the 161 signatories of the United Nations Framework Convention on Climate Change and is charged with negotiating revisions to the treaty and procedures for its implementation.

2   The product of COP-3 in 1997, the Kyoto Protocol set specific greenhouse gas emission reduction targets for Annex B countries (which include nations of the Organization for Economic Cooperation and Development (OECD) and those of the former Soviet Union (FSU) and Eastern Europe (EE). Membership of the OECD has expanded since 1992 when FCCC was signed. New entrants include South Korea and Mexico, neither of whom has been assigned GHG reduction targets under the Kyoto Protocol. In this chapter all references to the OECD designate the composition of the organization at the time of the signing of the UNFCCC. The Marrakech Accord was adopted at COP-7 (2001) and finalized operational language for the implementation of the Kyoto Protocol.

3   During COP-8, held in New Delhi in late October 2002, a number of nations (and researchers - see, for example, Muller, 2002) urged a shift in focus to the actions needed and costs required for developing countries, especially, to adapt to climate change. In the original draft of the so-called "Delhi Declaration," the requirement of Annex B countries to reduce emissions as a means of mitigating climate change was not mentioned. After heated negotiations, the official communication of COP-8 indicates, briefly, the existence of this obligation.

4   The Annex B nations of the Kyoto Protocol are identical to the Annex 1 nations of the UNFCCC, except for Turkey and Belarus, which are not included in the Annex B group, and Kazakhstan, which voluntarily joined Annex B.

5   Forest sink limits for Annex B nations are listed under COP-6's Appendix Z. While

most quotas are relatively small, a few nations were allocated significant sinks (notably, Canada – 12 MtC, Japan – 13 MtC, and the Russian Federation – 33 MtC).

6   Reasons for the use of per capita emissions are given below.

7   This assumes that the US and Australia participate in the Kyoto Protocol. As discussed later, the withdrawal of the US and Australia means that "hot air" can provide more than 80% of the target reduction for the remaining OECD participants. While most FSU and East European nations in Annex B have some"hot air" to sell, about 95% of "hot air" would likely be provided by Russia, Ukraine, and Romania.

8   Research (noted in IPCC, 2000) has suggested that domestic sequestration can offer low-cost emission offset options.

9   As explained below, it is possible that the US and Australia will seek to rejoin the Protocol before the first budget period is concluded. Thus, an assessment of the Protocol requires consideration of scenarios that include and exclude the participation of the US and Australia.

10   Whereas we anticipate increased GHG emissions, Nordhaus projects modest reductions of 1.5% from BaU projections in 2010; if forestry offsets are included, he expects a decrease of only 0.8% (Nordhaus, 2001).

11   This section draws heavily from Byrne and Yun (1999) and is provided with permission of the editor of Bulletin of Science, Technology and Society.

12   A political economy organizes power over many human activities. Because our focus here is on society-nature relations, we have limited our discussion to the organization of power over nature.

References

Agarwal, A.and Narain, S. l993. *Global Warming in an Unequal World: A Case of Environmental Colonialism.* Centre of Science and Environment, New Delhi.

Agarwal, A. and Narain, S. 1997 Joint implementation needs to consider needs of developing countries. Weathervane (website of Resources for the Future). Available at http://www.weathervane.rff.org/pop/popj/JT-india.html.

Byrne, J. 1997. *Equity and Sustainability in the Greenhouse:* Reclaiming our Atmospheric Commons. Parisar, Pune, India :

Byrne, J. and Yun, S.J. l999. Efficient global warming: Contradictions in liberal democratic responses to global environmental problems. Bull. Science, Technology Society, 19 (6): 493-500.

Byrne, J. and Inniss, V. 2000. Island sustainability and sustainable development in the context of climate change: Proc. Inter Conf. Sustainable Development for Island Societies, 20 - 22 April 2000, Taiwan. National Science Center and National Central University, Chungli, Taiwan

Byrne, J. and Glover, L. 2000. Climate shopping: Putting the atmosphere up for sale. Tela Environment, Economy, Society(5). from http://www.acfonline.org.au/publications/tela/intro.htm

Byrne, J., Glover, L., Inniss, V., Kulkarni, J., Mun, Y-M., Toly, N. and Y-D. Wang. 2002. Greenhouse Justice: Moving Beyond the Kyoto Protocol. Position Paper prepared for UN FCCC CoP-8, New Delhi, India. Center for Energy and Environment Policy, Newark, Delaware.

Byme, J. and Glover, L. 2002. A common future or towards a future commons: globalization and sustainable development since UNCED. Int. Rev. Environmental Strategies 3(1): 5-25.

Byme, J., Wang, Y-D., Lee, H., and Kim, J-D. 1998. An equity- and sustainability-based policy response to global climate change, Energy Policy 26(4): 335-343.

Byme, J., Glover, L., Alleng, G., Inniss, V., Mun, Y-M., and Wang, Y-D. 2001. The postmodern greenhouse: Creating virtual carbon reductions from business-as-usual politics: Bull. Science, Technology, Society, 21(6); 443 - 455.

Castle, S. 2001. European Union Sends Strong Warning to Bush Over Greenhouse Gas Emissions (March 19). Available at http://www.ecoline.m/automail/enwl-eng/200103/010320034448.txt

CEEP (Center for Energy and Environmental Policy, (2000) Ecologicat justice in the greenhouse. Position paper, Sixth Session UNFCCC Conference of Parties, The Hague, Netherlands. J. Byrne, Y.-D. Wang, G. Alleng, L. Glover, V. Inniss and Y.-M. Mun. Center for Energy and Environmental Policy, Newark, Delaware.

Cormier, L. and Lowell, R. 2001. DuPont and Marbeni Execute First UK Greenhouse Gas Emission Allowance Trade Natsource Environmental Division. New York Available at http:/www.natsource.com/print.asp?n=69&s=2

Dannenmeier, E. and Cohen, I. 2000. Promoting Meaningful Compliance with Climate Change, Commitments. Pew Center Global Climate Change, Arlington, VA.

Dessai, S. 2000. Climate regime from the Hague to Marrakech: Saving or sinking the Kyoto Protocol. Tyndall Centre for Climate Change Research Working Paper No. 12. Tyndall Centre for Climate Change Research, Norwich, UK, Available at http://www.tyndall.ac.uk/publications/working_papers/wp12.pdf

Dunn, S. 2002 Reading the Weathervane: Climate Change Policy from Rio to Johannesburg, Worldwatch Paper 160. Worldwatch Institute, Washington, DC.

EIA US. Energy Information Administration 2001. *Annual Energy Outlook 2001.* United States Department of Energy (DOE), Washington, DC. Available at http://tonto.eia.gov/ftproot/forecasting/0383(2001).pdf.

Escobar, A. 1996. Constructing Nature: Elements for a poststructural political ecology. In: Richard Peet and Michael Watts (eds.) Liberation Ecologies: Environment, Development, Social Movement. Routledge; N.Y. pp. 46-68.

EV World. 2001. Class action law suits filed against two U.S. Government Agencies. EV World (August 31). Available at http://www.evworld.com/databases/storybuilder.cfm?storyid=402.

Flavin, C. and Dunn, S. 1998. Responding to the threat of crimate change. In: State of the World 1998. W. W. Norton and Company, NY.

Golub, R. and Townsend, J. 1977. Malthus, multinationals and the Club of Rome. Social Studies of Science 7: 201-222.

IPPC. (Intergovernmental Panel on Climate Change) 1992. Climate Change 1992: The Supplementary Report to the IPCC Scientific Assessment, J.T.

IPPC. (Intergovernmental Panel on Climate Change) 1996. The IPCC Second Assessment Synthesis of Scientific-Technical Information Relevant to Interpreting Article 2 of the UN Framework Convention on Climate Change. United Nations Environment Program, New York, NY.

IPPC. (Intergovernmental Panel on Climate Change) 2000. Land Use, Land-Use Change, and Forestry: Special Report of the IPCC. R. T. Watson, I. R. Noble, B. Bolin, N. H. Ravindranath, D. J. Verardo and D. J. Dokken. Cambridge University Press, Cambridge, UK.

IWG. (Interlaboratory Working Group on Energy-Efficient and Low-Carbon Technologies). 1997. Scenarios of U.S. Carbon Reductions: Potential Impacts or EnergyTechnologies by 2010 and Beyond. United States Department of Energy, Washington, DC.

IWG (Interlaboratory Working Group on Energy-Efficient and Low-Carbon Technologies). 2001. Scenarios of U.S. Carbon Reductions: Potential Impacts or Energy Technologies by 2010 and Beyond. United States Department of Energy, Washington, DC.

Kopp, R. J. and Anderson, J. W. 1998. Estimating the Costs of Kyoto: How Plausible are the Clinton Adminstration's Figures? Resources for the Future, Washington, DC.

Kumar, K. 1978. Prophecy and Progress: The Sociology of Industrial and Post-Industrial Society. Penguin Books, New York, NY.

Kumar, K. 1995. Post-Industrial and Post-Modernism Society: New Theories of the Contemporary World. Blackwell Publi. Oxford, UK.

Kyoto Protocol to the United Nations Framework Convention on Climate Change. 1997. Available at http://unfccc.int/resource/docs/convkp/kpeng.html

Lynch, C. 998. Stormy Weather: Amicus Journal, vol. 19, (4): 15-19.

Marland, G., Boden, T., and J, A. R. Andres 2001. National CO_2 Emissions from Fossil-Fuel Burning, Cement Manufacture, and Gas Flaring: 1751-1998. Oak Ridge National Laboratory, Available at http://cdiac.esd.ornl.gov/ftp/ndp030/nation98ems.htm

Merchant, Carolyn. 1980. The Death of Nature: Women. Ecology, and the Scientific Revolution. Harper & Row, San Francisco, CA.

Meyer, A. 2000. Contraction and Convergence: The Global Solution to Climate Change. Green Books for the Schumacher Society, Foxhole, UK.

Müller, B. 2002. Equity in global climate change: The great divide. Oxford Institute for Energy Studies Paper EV 31. Available at http://www.oxfordenergy.Qrg

Mumford, L. 1964. Authoritarian and democratic techanics Technology and Culture 5: 1-8.

Nordhaus, W. D. 2001. Global Warming Economics. Science, 294 (November 2001).

Ott, H 2002. Climate Policy after the Marrakech Accords: From Legislation to Implementation. Wuppertal Institute for Climate, Environment, and Energy. Wuppertal, Germany. Available at http://www. wupperinst.org/download/ott-aftgr-marrakesh.pdf

Pearce, F. 1998. Playing dirty m Kyoto. New Scientist, 157: 48.

Reuters News Service 2002. Tuvalu seeks help in U.S. global warming suit (August 30). Available at http.//www.planetark.org/dailynewsstory.cfm/newsid/17514/story.htm.

Sachs, W. 1993. "Global ecology and the shadow of development. In: Sachs (ed.). Global Ecology: A New Arena of Political Conflict. Zed Books. Atlantic Highlands, NJ.

UNFCCC (United Nations Framework Convention on Climate Change) 2002. Report of the Conference of the Parties on its Seventh Session, Held at Marrakesh 29 October to 10 November 2001. FCCC/CP/2001/13/Add.l, Add. 2 and Add 3.

WCED (World Commission on Environment and Development). 1987. Our Common Future. Oxford University Press, New York, NY.

Winner, Langdon. 1992. Introduction. In: Langdon Winner (ed.). Democracy in a Technological Society. Kluwer Academic Publishers, Boston, MA.

Wiser, G. 2002. Kyoto Protocol Packs a Powerful Compliance Punch. The Bureau of National Affairs, Inc., Washington DC. Available at http:/www.ciel.org/publications/inter compliance.pdf

Wolin, Sheldon S. 1960. Politics and Vision: Continuity and Innovation in Western Political Thought. Little, Brown and Company, Boston, MA.

WRI (World Resources Institute). 2000. Earth Trends: The Environmental Information Portal, Demographic Indicators. Available at http://earthtrends.wri.org/pdf library/data tables/hdln 2000.pdf

Index